ANALYTICAL SOLUTIONS FOR EXTREMAL SPACE TRAJECTORIES

ANALYTICAL SOLUTIONS FOR EXTREMAL SPACE TRAJECTORIES

Dilmurat M. Azimov

ELSEVIER

Butterworth-Heinemann
An imprint of Elsevier

Butterworth-Heinemann is an imprint of Elsevier
The Boulevard, Langford Lane, Kidlington, Oxford OX5 1GB, United Kingdom
50 Hampshire Street, 5th Floor, Cambridge, MA 02139, United States

Library of Congress Cataloging-in-Publication Data
A catalog record for this book is available from the Library of Congress

British Library Cataloguing-in-Publication Data
A catalogue record for this book is available from the British Library

ISBN: 978-0-12-814058-1

Publisher: Matthew Deans
Acquisition Editor: Carrie Bolger
Editorial Project Manager: Carrie Bolger
Production Project Manager: Kiruthika Govindaraju
Designer: Matthew Limbert

Typeset by VTeX

To my parents,

Mukhammadjan Azimov and Rano Azimova

CONTENTS

PREFACE

This monograph intends to develop analytical and approximate-analytical methods for optimal control problem on determination of optimal and extremal trajectories in a Newtonian, linear central and uniform gravitational fields. This work then applies these methods to analytical trajectory synthesis for various maneuvers, including inter orbital, escape and planetary landing maneuvers. The results of this synthesis can be considered as the nominal or reference trajectory solutions for the development of onboard guidance and navigation schemes. Development of these schemes deserves separate studies and is beyond the scope of this work. This work aims to derive analytical solutions to the optimal control problem and demonstrate their application to the synthesis of optimal nominal or reference trajectories for autonomous space guidance. The main contents of this work include derivation of the solutions for thrust arcs, development of a methodology of analytical synthesis of optimal and extremal trajectories, determination of the number and sequence of thrust arcs on these trajectories and coordinates of switching points, analytical synthesis of nominal maneuver design trajectories and development of control laws and corresponding computational procedures for implementation of these laws.

ACKNOWLEDGMENTS

The author wishes to express his appreciation to all friends and colleagues who provided their feedback and suggestions to improve the quality and contents of this monograph. The author thanks Mamura D. Yuldasheva who contributed to the design of the cover page by providing many interesting ideas and optional pictures of space trajectories.

CHAPTER 1

Introduction

1.1 OPTIMAL TRAJECTORIES AND SPACE GUIDANCE

1.1.1 Synthesis of Optimal Trajectories

Investigations of the optimal control problem for a synthesis of optimal trajectories were initiated by R. Goddard, G. Oberth, H. Hamel, V. Hohmann, A. Kosmodemianski, A. Ishlinski, and others. In particular, the studies devoted to analysis of optimal trajectories in the central Newtonian field are of great importance from theoretical and practical points of view due to the sufficient accuracy of this field to simulate the gravitational field and its simplicity for analysis [1], [2], [3]. Investigation of the analytical solutions of this problem (or variation problem) for space trajectories and their synthesis has become one of the important characteristics of their applicability to solving aerospace engineering problems and to performing space operations. In particular, efficiency of space maneuvers, such as an interorbital transfer, interplanetary maneuvers, entry into a parking orbit, rendezvous and docking, and landing on a planetary body mainly depend upon characteristics of on-board guidance, navigation and control (GNC) system. One of the highly desirable features of this system is its autonomy, that is it is desired to be executed autonomously and in real-time so that the spacecraft can perform space operations.

Analytical integration of the optimal control problem equations and determination of a sequence of thrust arcs (TAs) on an optimal trajectory were investigated by D.F. Lawden, G. Leitmann, T.N. Edelbaum, H.G. Kelly, D.E. Okhotcimskii, V.A. Egorov, A.I. Lurie, V.K. Isaev, V.S. Novoselov, A.G. Azizov and N.A. Korshunova and others. Despite these studies, at present time, there is no general theory or method of analysis of TAs with constant and variable thrust, evaluation of their optimality, criteria of their applicability in practical problems. Questions on determination of new analytical solutions for TAs, their classification and optimality, optimal connection of various TAs, their sequence, design of optimal maneuvers and other questions are still remaining unsolved. The present study is devoted to some set of these questions.

Analytical Solutions for Extremal Space Trajectories
DOI: http://dx.doi.org/10.1016/B978-0-12-814058-1.00001-1

1.1.2 Autonomous Guidance Problem

The **autonomous guidance problem** can be formulated as follows: Let a spacecraft's current state be given at some time instant and characterized by some deviations of the spacecraft's parameters from their nominal values. It is required to compute and execute the guidance commands, which steer the spacecraft from its current state to a desired state at a desired time instant and keep all the deviations of the spacecraft's parameters within given boundaries without external commands. The solution of this problem includes consideration of the following sub-problems: 1) Solution of an optimal control problem and formulation of control laws; 2) Synthesis of the desired nominal or reference trajectory; 3) Design of guidance laws and determination of guidance commands and algorithms; and 4) Execution of these algorithms to achieve the desired state at the desired time. These sub-problems can be considered separately from each other and also incorporated into a sequential solution process to provide a desired flight. The solution of the autonomous guidance problem is implemented by an on-board GNC system, which represents a main part of the spacecraft's flight control system.

1.1.3 On-board Guidance Systems

An on-board guidance system, as part of the GNC system, can be characterized by the following important features:
- computation of the energetic expenses to implement a maneuver,
- capability of long-term functioning,
- continuous change of spacecraft parameters,
- simplicity and accuracy of guidance and control laws, and algorithms.

The **energetic expenses** depend on the number and duration of the TAs. Therefore, selection of an acceptable control algorithm on a TA and its simplicity are critically important requirements in the design of autonomous guidance algorithms.

The **long-term functioning** is an important capability of the modern propulsion systems that produce low thrust, and it is associated with controllability, specific impulse, principle of functionality and fuel efficiency. For example, the Deep Space 1 (DS1) flight has already demonstrated a modern electric propulsion which provides specific impulse that 10 times exceeds that of a conventional chemical propulsion. Besides that, new magneto-plasma systems of VASIMR type (Variable Specific Impulse Magnetoplasma Rocket) have been proposed for space transportation missions to Moon, Mars, and other planets. These systems produce specific impulse that may reach 100,000 seconds, which exceeds that of a conventional chemical propulsion system by 200 times. Many other projects and missions to asteroids of the Main Belt, MUSEC-C, (Jupiter Icy Moon Orbiter), ETS-V1 (Engineering Test Satellite), Dawn, ESA Earthguard 1, Pluto Orbiter Probe use low and high thrust systems. Duration of TAs with such propulsion may range from several minutes to several months. The current state of such systems necessitates 1) reconsideration of known guidance and control laws designed from short TAs, and *design of new laws for TAs with flight durations of up to several months, and 2) development of adequate methods of synthesis of space trajectories*. It should be noted that the guidance and control laws developed for systems with constant specific impulse are not applicable to the systems with variable specific impulse.

The *continuous change of spacecraft parameters* is associated with questions of convergence of iterations in real-time when using numerical methods. The specific feature of these methods is that they can count for the perturbations by multiple gravitational bodies to accurately represent the gravitational field, and provide fast solutions to complex problems of missions. Design and analysis of these missions in the context of an optimization problem for two body dynamics may reduce the accuracy of pre-mission computations and create difficulties in making reliable conclusions. At the same time,

- in numerical integration, one may face problems of convergence and continuity of the parameters when switching from one thrust regime to another (switching point), which in particular, is associated with difficulty of computing the initial values of Lagrange multipliers of the optimal control problem. For example, the values of the parameters at the end of one TA and at the beginning of another TA may not be equal because of the unknown Lagrange multipliers at the latter point. Equating the values of the parameters at this point requires an unknown number of iterations, and a rapid convergence is not guaranteed because of the arbitrariness of the initial (current) conditions for these iterations, thereby unnecessarily occupying memory, and other computational resources of a flight computer;
- the sequence of TAs and times of switches between various TAs defined by ground-based numerical integration of an optimal reference trajectory may not be the best for an on-board use due to deviations of the actual trajectory from the reference trajectory, and other factors associated with fuel, navigation, and control accuracy;
- known trajectory solutions for some maneuvers and corresponding algorithms for on-board control do not guarantee high accuracy of reaching the

desired values of the parameters within the boundaries imposed on these parameters at desired instants of time. These problems indicate that the existing numerically integrated trajectories may not always provide acceptable accuracy and continuous autonomous guidance. Consequently, to successfully solve the guidance problem, it is necessary to have reliable and simple algorithms, which do not require, in particular, iterative processes and a solution to the convergence problem at the switching points.

The *simplicity and accuracy of guidance and control laws* are important requirements for on-board guidance systems as the known laws developed for various maneuvers are based on significant simplifications in the spacecraft dynamics model. To these simplifications belong, in particular, replacement of the gravity vector by a constant vector, impulsive change of velocity without changing spacecraft's position, etc. For example, one of the Apollo guidance subroutines used an average value of the gravitational acceleration (Average G guidance) for the descent and landing maneuvers. However, development and utility of new propulsion systems, and complex goals of space missions have led to a necessity of reconsideration and enhancement of existing algorithms and software for onboard guidance computers. This imposes corresponding requirements on the accuracy of the onboard guidance and control laws. Improvement of accuracy of these laws may be achieved by removing the above mentioned simplifications in the dynamics model and establishment of explicit relationships between the spacecraft parameters.

In summary, the autonomous guidance and control algorithms must satisfy, in particular, the following requirements:
- the necessary dependences between the algorithms and propulsion system characteristics must be examined,
- the algorithms must be simple, rapid and reliable for execution,
- capability of providing the solution to the guidance problem on any TA with any spacecraft state parameters, including those deviated significantly from their nominal or reference values.

1.2 BRIEF SURVEY OF STUDIES OF OPTIMAL CONTROL PROBLEM: THRUST ARCS

1.2.1 Introductory Remarks

As is known, the basic concepts of the trajectory optimization theory were formulated by Lawden [3]. In the context of his statement of the variation problem, it was shown that when mass-flow rate is limited and exhaust velocity is constant, an optimal trajectory may contain null thrust (NT), intermediate thrust (IT), and maximum thrust (MT) arcs. This trajectory satisfies the conditions in terms of switching function and primer vector. The switching function characterizes the transfer between two arcs, and the primer vector's unit vector determines the thrust direction. The constraints, such as the mass-flow rate is limited and the exhaust velocity is constant, are usually considered in modeling the chemical propulsion systems associated with high thrust (HT) and low specific impulse [2]. It is known that there exist another type of a trajectory of motion with low magnitude of thrust referred to as low thrust (LT) arc, and this type of arc is not obtained from analysis of the constraints mentioned above. In the case of electric or thermonuclear propulsion systems, the exhaust velocity and power may be limited [2]. These types of systems produce LT with programmed values of specific impulse and power. It has been shown that an optimal trajectory of a spacecraft with a LT propulsion may consist of null power (NP) and maximum power (MP) arcs [4]. With the development of propulsion technology and its various maneuver applications, such as sample return, rendezvous, escape and capture, transfers between halo orbits, and missions to Near-Earth-Objects, more attention has been given to the studies of the fuel-optimal LT and HT trajectories [5], [6]. According to Ref. [2] and [7], note that the high and low levels of thrust are determined depending on thrust to weight ratio and specific impulse. Therefore, depending on the values of these parameters, the IT and MT arcs can be considered as LT and/or HT arcs, and vice versa.

1.2.2 Studies of Low Thrust Arcs

The LT arcs are characterized by $\frac{T}{W} \leq 10^{-3}$ and $I_{sp} \geq 1000$ s, where T, W and I_{sp} are the thrust, weight and specific impulse respectively [2], [7], [8], [9], [10], [11]. Methods for synthesizing optimal LT trajectories using power-limited exhaust-modulated propulsion systems is the topic of many papers, see for example [2–11]. One of the first works in this subject belongs to Irving, who described the motion with power-limited propulsion and the optimum thrust equations with minimizing functional [2]. Melbourne and Sauer's early investigations sought numerical solutions utilizing calculus of variations' approaches to Earth-Mars trajectories with constant and variable thrust, and with various optimization criteria [4], [12]. The variation equations included constraints for normalized power parameter and variable exhaust velocity.

Other studies can be divided into **three groups**. The first group of studies deals with various software, which use direct and indirect optimization methods to solve, in particular, the boundary value problems that involve the exhaust-modulated propulsion parameters [13], [14], [15], [16], [17], [18], [19], [20], [21], [22]. The second group of studies develops some approximations and/or simplifications in the dynamical model [23], [24]. The analytical studies of optimal LT trajectories belong to the third group, such as those provided in Ref. [25], [26].

First group of studies. These studies have developed and utilized a wide range of direct and indirect optimization methods. The advantage of the direct optimization methods is that they allow to obtain a set of necessary conditions (only), which describes a local extremum of the minimizing functional. The computation of flight trajectories to asteroids and comets using an electric propulsion was proposed by Eneev et al. [13]. Two regimes of propulsion work have been considered: the regime with no constraints on magnitude and direction of thrust vector, and the regime with constant exhaust velocity. The problems of flights to Sun by using electric systems and by implementing gravitational maneuvers near Jupiter were investigated by Malyshev and others [16], [17], [27]. Ivashkin and others have computed trajectories of transfer from Earth to asteroid Toutatis by using LT and HT [28], [29]. The direct method based on differential inclusion (DI) concepts has been developed and used to compute the Earth-Mars and Earth-Jupiter LT constant and variable specific impulse trajectories by Coverstone-Carroll [20]. These trajectories are computed using the DI software developed for obtaining the initial Lagrange multipliers, and the interactive VARITOP (VARIable thrust Trajectory Optimization Program), which optimizes a trajectory by numerical integration of the state and co-state equations [30]. Another direct optimization method, DTOM was developed by Kluever assuming that specific impulse and thrust are constants [31]. A newly released NASA software titled General Mission Analysis Tool (GMAT) is an open source software system licensed under the NASA Open Source Agreement, and designed to optimize space trajectories and perform mission design analysis [32]. The efficiency of combination of LT and HT arcs has also been demonstrated by Fedotov [33]. The flight trajectories to Phobos, the asteroid Fortuna, the planets Mercury, Jupiter, and Pluto using a solar electric propulsion have also been studied. In particular, in the problem of transfer from a satellite's orbit to the orbit of Phobos it was shown that the use of plasma propulsion systems of type SPD-100 and 140 allows us to increase the spacecraft final mass by more than 150 kg. One of the modern **indirect** methods is the method of multi-criterial double level optimization, which was used in the study of missions with direct transfer to the areas near Sun [27]. The criteria of optimization are associated with factors of cost, reliability of performance, scientific back-up and efficiency, and duration of flight. The time and controls are parameterized. COPERNICUS (A General Spacecraft Trajectory Design and Optimization System) is an example of such approach [34]. The method of collocation was proposed by Sheel and Conway, and used to solve minimum-time problems for various orbit transfers [35]. The continuous thrust orbit transfer problem and approximate analytical solutions for optimal co-states were used in the shooting method by Thorne and Hall [36]. The minimum time Mars capture problem has been considered by the collocation and nonlinear programming by Tang and Conway [37]. It is assumed that thrust acceleration is constant. Variable thrust engines with constant power were developed by Chang-Diaz and others [11]. The fuel-optimal trajectories utilizing the such engine characteristics have been considered by Vadali and others [38], [21]. As an example, the Earth-Mars trajectories including the gravitational effects of the Sun, Earth, and Mars with two phases are considered. The boundary values of variable specific impulse and engine power are given *a priori*. The problem is solved by the indirect method using Runge-Kutta-Phelberg 7-th order scheme and modified Newton's method with step size control. Optimality of the solutions is considered with respect to neighboring extremals. The studies of hybrid genetic algorithms also belong to the first group of indirect methods [39]. These algorithms are based on the selection of a coordinate system, parameters and the number of Monte-Carlo experiments, which reduce the functional of the problem to a minimum.

Second group of studies. The work of Grodzowskii and others in Ref. [23], which deals with selection of optimal relations between the weight components of the spacecraft and optimal control of the propulsion system, and with determination of the set of optimal trajectories, is a typical representative of the second group of the LT studies. Three combinations of the controls, such as the engine power, mass-flow rate, and exhaust velocity are discussed. It is shown that for an ideal power engine system, in which all the input power is transformed into a jet power, the optimization problem may be divided into two independent subproblems: (1) find an optimal ratio of the weight of the power generator to the weight of propellant, and (2) find an optimal thrust acceleration program. Some particular solutions to the problem in the cases when the gravitational acceleration is neglected or the radial component of the thrust acceleration is equal to zero are obtained. Also, the thrust acceleration is assumed parallel to the velocity, and the variational equations are linearized in the neighborhood of a certain plane Keplerian transporting trajectory. Another simplifications have been made by Markopoulos, which presents a continuous thrust program in the planar case [24]. The assumption that the thrust is tangent to the flight path allows for complete analytical solutions to the state and co-state equations.

Third group of studies. A few works represent this group. The methods of analytical mechanics can also be used to describe the LT trajectories. The method of Hamilton-Jacoby was utilized to obtain the transformed equations of LT motion [25]. The method of Poisson brackets and method of reducing the order of the canonical system using first integrals and invariant relations were applied to obtain LT spiral trajectories described analytically. It was shown that these LT spirals can be used to solve the optimal interorbital transfer problem.

In the works stated above, the only necessary (stationary) conditions have been investigated by mainly numerical methods, and the corresponding trajectories may be referred to as a class of extremals of the problem. But many other questions, such as sufficient conditions of optimality, determination of a number of switching points on the trajectory, continuity conditions at the switching points, behavior of the primer vector and its hodograph, a succession of TAs on the trajectory and etc. have been given a limited attention, or were considered in a particular manner, or were not considered at all.

Besides that, the question about convergence of integration process plays an important role, because this process may not converge for all cases or for all sets of initial and final conditions. This process and the solutions to the problem mainly depend on the proper initial values of the Lagrange multipliers and/or control variables. These values, according to the works stated above, are supposed to be guessed or found by iteration, and this often leads to difficulties in the integration process. In these cases, optimal impulsive solutions are used to provide a proper approximation which is always associated with errors of a certain order. Therefore, one may conclude that the optimization problem for spacecraft powered by power-limited exhaust-modulated propulsion system, has been considered only by using the necessary conditions of optimality and by utilizing only the methods of numerical integration. Consequently, one of the topics of the present study is the analytical investigation of spacecraft optimal and extremal trajectories and their application to the synthesis of fuel-optimal trajectories for the design of on-board guidance laws. It is expected that the analytical investigation of the problem will provide 1) a qualitative assessment of all state, co-state, and control variables; 2) analytical solutions for these variables; 3) determination of the initial values of Lagrange multipliers without using impulsive solutions; and 4) analytical determination of the number and sequence of TAs on an optimal trajectory. Moreover, such investigation also provides a possibility to test sufficient conditions of optimality.

1.2.3 Studies of High Thrust Arcs

As mentioned above, the variational problem in the Lawden's statement allows us to analyze the motion with a HT propulsion. The TAs may be represented by the IT and MT arcs [33], [2], [7]. Note that the IT arcs can also represent the LT arcs depending on $\frac{T}{W}$ and I_{sp} [40].

1.2.4 Intermediate Thrust Arcs

Existence and Optimality

In the context of the Lawden's statement of the variation problem, the IT arcs are determined by intermediate values of mass-flow rate and constant exhaust velocity, and for which $T/W > 1$, and $I_{sp} < 1000$ s. [7]. The presence of the IT arcs, sometimes called singular arcs, represents a degenerate case of the problem [22], [41], [42], [43], [44]. These arcs can not be determined by means of the first order necessary conditions of optimality, and therefore, their appearance in the problem causes significant difficulties of analytical and computational nature. Analytical solutions for IT arcs in a planar case, well known as Lawden's spirals, have been obtained by Lawden regardless the performance index [43], [44]. The mass-flow rate was considered as a control subject to an inequality constraint. The studies conducted by Kelly, where the magnitude and direction of thrust vector are considered as controls, showed the possibility of the solutions by application of Legendre-Clebsch conditions [41], [45]. However, in the references mentioned above, the main attention was given to the mechanism of obtaining the solutions, and no answer was given to the question of the IT arcs' optimality. It has been shown by Gerts and Venbeke, that if there is no constraint on angular distance or flight time, then the Lawden's spirals are not optimal [46]. Derivation of the necessary conditions of optimality of IT arcs have been considered by Kelley, Kopp, and Moyer [41], [45], Robbins [42], Johnson [47], Gabasov and Kirillova [48], [49]. Application of these conditions to the Lawden's spirals has shown that these spirals are non-optimal. The optimality conditions mentioned above were derived with one or two linear controls, on which the inequality constraints are imposed. These conditions have been generalized to a system with a finite number of controls by Goh [50], [51]. Another approach to the optimality conditions of singular controls in the case of closed sets of controls has been proposed by Gabasov and Kirillova, and Srochko [48], [49]. The main idea of these works was associated with the introduction of matrix impulses. New type of variations that generalize the special variations of Kelley, Kopp, and Moyer was proposed in Ref. [52]. Called as package of variations, this method allows for derivation of the previously obtained necessary conditions [41], [53], [54]. Also, the necessary conditions have been obtained via the magnitude of primer vector and by the method of second variations, which leads to the high-order conditions known as the conditions of Legendre-Klebsch (Kelley-Contensou test) [42], [45], [53], [54], [55]. Besides the methods of investigations of IT arcs mentioned above, the sufficient conditions of optimality have been investigated by Krotov [56], [57]. His method was then applied by Gurman to study some questions related to optimality of IT arcs [58]. It was shown that the sufficient conditions are associated with the existence of solutions to the Riccati matrix equation [55].

First Integrals

The first integrals represent an important aspect of investigations of IT arcs. Azizov and Korshunova have demonstrated that the 14-th order canonical equations admit seven first integrals, and the principal difficulty of solving the problem for IT arcs is finding only one integral [59], [60]. If an IT arc lies in a plane containing the gravitational center, the known integrals are enough to reduce the problem to quadratures [60], [61]. In particular, the first integrals played a critical role in obtaining circular [62], [63], [64], spherical [65], [66] and spiral trajectories with IT arcs [52], [61], [67], [68], [69], [70], [71], [72], [73], [74], [75], [76], [77], [78], [79], [80], [81], [82]. These trajectories have been shown to be non-optimal using Robbins' condition of optimality and simple analysis of the solutions obtained in the references above. Using the canonical equations, Letov proposed to apply the method of Poisson's brackets to derive new independent first integrals of IT arcs [83]. A more complete survey of the investigations of IT arcs have been reported by Gabasov and Kirillova [48], and by Bell and Jacobson [55].

Optimal Conjunction of Various Thrust Arcs

The optimal conjunction of various thrust arcs is another important aspect of the problem being considered. Robbins has shown that if the thrust magnitude is limited, then the IT arcs can not be connected with MT and NT arcs [42]. If the impulses are admitted, then one can show that an optimal trajectory does not include IT arcs excluding a set of particular cases, which has a zero dimension, meaning that the optimal trajectory with IT arcs occupies a limited sub-space in the space of boundary conditions. The frequency, with which the optimal trajectory oscillates between the boundary values of thrust, approaches infinity when the boundary conditions are reduced to some particular case in which the optimal trajectory may contain the IT arcs. Therefore, as Robbins concluded, although optimal trajectories with IT arcs exist, they have only theoretical value. It was shown that optimal connection of IT (singular) arcs with other (non-singular) arcs is possible only in the case of an odd order of their degeneration. It was shown in a typical example, that if a trajectory contains an IT arc, then the whole trajectory is this arc, or the IT arc degenerates into a circular NT arc. Moreover, the analyses of Johnson's and Robbins' conditions have shown that in the case of a free flight time, the IT arcs degenerate into NT arcs [84], [85], [47], [42]. A similar conclusion was reached by Makarov, who considered the mass-flow rate and exhaust velocity as controls [86].

The existing analytical solutions for IT arcs can be divided into two classes: the solutions in the case of flight with free or fixed time. In the first case, there exist the Lawden's spirals and spiral-like trajectories. In the second case, there exist only circular and spherical trajectories. The problem of spherical IT arcs and corresponding solutions for various propulsion systems has been considered by Lurie [87]. It has been shown that the spherical trajectories can be described by Levi-Civita's method of finding particular solutions [62], [66], [88].

However, despite the studies mentioned above, the issues about existence of other analytical solutions for IT arcs, their classification and optimality, optimal conjunction with other TAs and their applicability and effectiveness from an energetic point of view compared to NT and MT or finite TAs, sequence of various TAs on an optimal trajectory remain unsolved. Some of these issues have been addressed in the fourth chapter.

1.2.5 Maximum Thrust Arcs

Optimality Conditions

In the context of the variation problem under Lawden's statement, the MT arcs are determined by maximum value of mass-flow rate and constant exhaust velocity, where the ratio thrust to weight is $T/W > 1$ and specific impulse is $I_{sp} < 1000$ s. [7]. Only few works have been devoted to the studies of MT arcs. The necessary and sufficient conditions of optimality have been developed in Ref. [3], [56], [57], [59], [89], [90], [91], [92], [93], [94], [95], [22], [96]. For example, Bryson and Ho have shown that the necessary conditions of a weak minimum can be expressed in terms of Euler's equations, transversality conditions, the weak form of the Legendre-Clebsch conditions and by absence of conjugate points on an open time interval [94]. The weak form of the Weierstrass condition and the conditions mentioned above represent the necessary conditions for a strong minimum. The strengthened forms of Legendre-Clebsch and Weierstrass conditions (strong inequalities) and the requirement of the absence of the conjugate points on a closed time interval represent sufficient conditions of optimality. Note that the use of the Weierstrass condition is associated with the construction of the field of extremals [97], [94], [90], [96]. Another representation of the sufficient conditions of optimality, called "principle of optimality", has been developed by Krotov [56], [95]. He showed that the sufficient conditions of optimality can be expressed in terms of a "deriving function" and on admissible controllable process (extremal) which consists of the state and control vectors, and this allows for finding an absolute minimum of the performance index of the problem. The main task is in determining the "deriving function", which can also be used to express the connection of the "principle of optimality" with maximum principle and Bellman's optimality principle. An important advantage of this method is the possibility of determining the global minimum on the basis of the sufficient conditions [95].

Analytical Solutions and Simplifications

The analytical solutions for MT arcs in the uniform field have been obtained by Lawden, Leitmann, and Battin [3], [93], [98]. The exact solutions for the case of the Newtonian field have not yet been found. That is why it is necessary to investigate these arcs by employing approximate – analytical or numerical or purely analytical methods and assumptions. The numerical and indirect methods of investigations of MT arcs have been developed, including the methods of conjugate gradient [45], quasi-linearization [99], perturbing functions [100], [101], second variations [102], [83], and the Lagrange and maximum principle [103]. One of the important problems in using the indirect methods that imply iteration processes is strong sensitivity of the solutions to small changes in the initial conditions for the co-state vector's components. Kelly and Bryson have made progress in the application of the gradient methods because of the possibility of controlling the increments at every step according to small changes in the thrust magnitude program [45], [104]. As the results of these studies show, the numerical methods reveal the essence of the problem in a weak manner, and that is why it is obviously a necessity to address the convergence issues. An example of successful application of the numerical methods, in particular, the shooting method and obtaining qualitative results on the basis of the parametric analysis is given in Ref. [103], in which the maximum principle plays a central role. The result of applying the maximum principle is a nonlinear boundary problem. The functionals considered in these studies are "fuel expenditure – flight time" and "flight time – fuel expenditure" [105]. It was shown that the representation of the functional of the problem in this form enables to unify four types of problems: 1. Minimum time transfers with free final mass; 2. Minimum time transfers with limited final mass; 3. Limited time transfers with maximum final mass; 4. Transfers with free final time and maximum final mass [106].

There exist the studies which consider analytical simplifications. One of such methods has been proposed by Kornhauser [107]. In this work, the analytical solution for the thrust vector in the form of expansion with respect to the impulsive trajectory have been obtained. Robbins has shown that the analytical solutions for the case of limited mass-flow rate and constant exhaust velocity can be obtained in the form of expansion with respect to the quantities inverse proportional to the thrust acceleration and exhaust velocity [108]. This method requires solution of algebraic equations and has the highest accuracy for high levels of thrust, as the expansion with respect to the impulsive solution, which is assumed known. The method of transformation of impulses into the arcs of finite length and evaluation of the quality of such transformation have been proposed. The polynomial method of solving the problem represents another simplification, and the solution depends on the degree of closeness to the nominal trajectory [109]. A disadvantage of this method is the necessity of saving the coefficients of the polynomial functions of time. The questions of replacing the impulsive trajectories by MT arcs and estimation of the errors admitted have been considered by Marchall et al. [110], [111]. It was shown that such error has an order inverse proportional to the square of the ratio of the duration of MT arc to the time of flight on the impulsive trajectory. This result was demonstrated by the example of an interorbital transfer.

The analytical theory of trajectory optimization in the gravitational fields has been developed by Novoselov [91]. He proposed a methodology of approximate analytical solution of the variational problem. The analytical solutions in the neighborhood of zero order solution in the form of expansion with respect to the powers of small parameter have been obtained for several problems of interorbital transfer. The exact analytical solutions for the Newtonian and other fields were also obtained by the author and applied to some of the guidance and targeting problems [112], [113], [114].

Novoselov also reported qualitative investigation of two-impulsive trajectories between quasi-circular orbits with small inclinations and eccentricities, explicit construction of the initial approximations and general scheme of analytical approximate extremal transfers [91]. One of the integrable cases of the canonical system of equations, namely, the cases of motion with tangential thrust, were analyzed by Azizov and Korshunova [59], [115], [88]. It was shown that the TAs found can be used to form a thrust program, which provides maximum kinetic energy at final time. Another simplification is associated with the form of representation of the gravitational acceleration. Leitmann and Isaev considered a one-dimensional and planar case of motion in a uniform field, and demonstrated that the equations of the problem can be integrated completely. This leads to the complete theory of optimal trajectories in the uniform field [93], [116]. It follows from this theory that in a general case, an optimal trajectory can include one MT arc and two NT arcs. If the gravitational acceleration is represented in a form of a polynomial with respect to time, then it is possible to obtain analytical solutions to the problem of transfer with a constant thrust [117]. In this case the primer vector is a linear function of time, and the equations of the problem, in principle, can be completely integrated in quadratures. If the gravitational acceleration is represented not only in the form of a polynomial with respect to time, but also in the form of a linear function of radius vector, then the solution can be represented in the form of two thrust integrals which can be rewritten as an infinite sum of known integrals. The primer vector satisfies the linear differential equation, and the state and co-state vectors can be expressed in terms of the vectors determined at the previous cycle [118]. The control vector consisting of the primer vector, its derivative and sequence of switching points, can be determined by employing optimal impulsive solutions. It was shown that the problem of McCue can be solved by applying two MT arcs and three NT arcs [99], [118]. These solutions can be used to design trajectories with N TAs [119]. In these solutions, the gravitational acceleration is represented as a linear function of radius vector. For this case of the linear field, the solution in terms of elliptic integrals and quadratures have been obtained by Korshunova and Andrus [88], [120]. Korshunova showed that in this problem, the number of first integrals is sufficient to reduce the problem to quadratures, and the hodograph of the primer vector is a central ellipse. Andrus considered the gravitational acceleration with an additional term, represented as a function of time, and assumed the magnitude of thrust acceleration as a constant [120]. He demonstrated that his solutions can be used for a LT interorbital transfer. Besides that, he showed that the gravitational acceleration can be expressed via a polynomial with respect to a small parameter. Along with the studies on MT arcs, it has been demonstrated that the analytical solutions for the arcs with constant thrust acceleration for the uniform field can be obtained and applied to planetary landing problem [3], [121], [122], [123], [124]. The 3-dimensional trajectories have been obtained by Yang for lunar landing trajectory design [125]. In this work, the complete analytical solutions for the 3-dimensional optimal trajectories are obtained for Mars landing trajectory design in the context of the optimal control problem [126].

In summary, despite the investigations described above, many questions, such as the existence of other solutions for MT arcs, their inclusion into a specified interorbital transfer, and the questions about the behavior of the primer vector remain unsolved. It is known that the analytical difficulties of investigating the MT arcs led to the necessity of replacing them by impulses, which requires a corresponding estimation of errors under the real constraints imposed on mass-flow rate. Therefore, it is important to consider problems with analytical and approximate-analytical approach to integrate the equations of the problem for MT arcs. To date, a few studies have been devoted to address these issues [3], [59], [60], [88], [127]. Note that in the case of Newtonian field, the analysis of first integrals show that the principal difficulty in reducing the problem to quadratures is in finding additional four first integrals [59], [66]. The sixth and seventh chapters are devoted to the integration of the problem equations for MT arcs in the cases of linear and Newtonian fields by employing the analytical approach.

The studies of low (intermediate) TAs and high (maximum) TAs lead to the following conclusions. The problem of determining the low TAs is not a purely theoretical problem anymore, and there is a necessity of considering them together with high TAs under the same problem statement. In optimal control theory, this problem can be formalized as a problem of optimal control of a set of controllable dynamical systems [128]. To date, the TAs mentioned above have been investigated in the context of specific problems by using characteristics of specified or selected propulsion systems, and mainly, via the numerical methods based only on the necessary conditions of optimality.

Only a few publications have been devoted to analytical investigation of IT and MT arcs in the context of the variation problem. In particular, Azizov and Korshunova have developed a method of reducing the problem to a canonical form for NT, IT, and MT arcs, and demonstrated the advantage of applying the methods of analytical mechanics to develop analytical solutions for TAs.

Consequently, one can summarize the following aspects of solving the problem: 1. It is advantageous to analyze low and high TAs in the context of the same problem; 2. It is important to analyze the effects and influence of the sufficient conditions in further development of the problem solutions; 3. One of the main issues is the development of new analytical solutions for TAs; 4. Development of a methodology of applying the analytical solutions to the design of maneuver trajectories, including the connection of various arcs; 5. Determination of the possible structure of the trajectory is the important element of trajectory design. These aspects represent the main focus of this work. Some similar issues have been addressed in Ref. [129], [130], [131].

1.3 GENERAL STRATEGY AND MAIN CHALLENGES

The general strategy used in this study is the determination of the new analytical solutions for TAs, and their application in the design and synthesis of optimal and extremal trajectories of space maneuvers. The main challenges or key milestones associated with this strategy are

- Derivation of the solutions for TAs;
- Development of methods of applying the analytical solutions to synthesis and design of maneuver trajectories;

- Development of a methodology of analytical synthesis of optimal trajectories, determination of the sequence of TAs and coordinates of switching points;
- Synthesis of nominal interorbital trajectories, and development of control laws and corresponding algorithms for implementation of these laws in real time, and for preliminary estimation of kinematic and dynamics characteristics of spacecraft in maneuver design.

1.4 BRIEF DESCRIPTION OF CHAPTERS

Chapter 1 contains the introduction, in which the importance of the topic of this study is described, and the necessity of considering low TAs in combination with null and high thrust (intermediate and maximum) arcs. The modern problem of trajectory optimization, including investigation of TAs, the general strategy followed in this work and corresponding main issues are analyzed.

In Chapter 2, the optimal control problem as well as the variational problems, the statement of which differs from the traditional Lawden's statement in space flight dynamics, are considered. It is known, that in Lawden's statement, the constraints on controls, which characterize chemical systems of high thrust, allow one to analyze NT, IT, and MT arcs. In this chapter, the constraints that characterize power limited systems (electrical, plasma, etc.) and chemical systems are considered in the context of the same problem. The presence of such constraints allows to consider the low TAs in combination with null and high thrust (intermediate and maximum) arcs, thereby extending the group of TAs known under the traditional statement of the problem. It should be noted that this work will consider all extremals of the optimal control problem as the Pontryagin extremals. In the calculus of variations, the weak minimum is associated with weak extremals. In optimal control, the Pontryagin minimum is associated with Pontryagin extremals. It has been shown that the conditions for a Pontryagin extremal are stronger than the conditions for a weak extremal [90]. Therefore, **the present study will focus on Pontryagin extremals, which will be called simply "extremals"**. The analysis of the second differential of the extended functional is reduced to a formulation of an axillary optimization problem, which leads to a Riccati equation. The solution to this problem, in particular, the finiteness of the solution to the Riccati equation, allows for testing the presence of the conjugate points and derivation of conditions for positive definiteness of the second differential of the extended functional. The theorem on using the analytical solutions for determination of the presence of the conjugate points on TAs is formulated and proved (Theorem 2.1). The necessary and sufficient conditions of optimality, including the conditions at switching points, are described. The types of TAs, which can be optimal depending on the values of axillary control variables, are defined by testing the positive definiteness of the matrix of second order derivatives of Hamiltonian with respect to controls (Theorem 2.2). The admissible TAs are classified. The methodology of analytical determination of extremal trajectories based on the studies of optimality conditions, equations of continuity at the points of discontinuity of controls and determination of the structure of the trajectory is presented. This methodology serves as an instrument of separating the extremal trajectories, from the admissible trajectories. Note here that the extremal trajectories represent the reference trajectories used in the guidance problem.

The low TAs are considered in Chapter 3. The equations of motion in a canonical form and corresponding first integrals are derived. Classes of circular and spiral arcs are obtained (Theorem 3.1). It is shown that these arcs can find applications in the problems of an escape from and capture in a given elliptical orbit. Also, questions of the presence of the conjugate points on spiral arcs are discussed.

The next three chapters are devoted to investigation of high TAs. Chapter 4 contains the studies of IT arcs. The canonical equations, first integrals and invariant relationships are derived and analyzed. Spherical trajectories and six classes of solutions described by quadratures are obtained (Theorems 4.1 and 4.3). Lawden's spirals are studied, and their degeneration, taking place when there is no constraint on angular distance and the functional of the problem does not explicitly depend on this distance, is shown (Theorem 4.4). Besides that, two classes of extremal trajectories (or simply, extremals), the first of which represents Lawden's spirals, and the second of which represents a class, alternative to Lawden's solutions, and another two classes of trajectories with fixed final time are derived (Theorems 4.2 and 4.5). An analytical solution to the problem of a free time transfer to a given elliptical orbit by using the first class of extremals is given. The analytical solutions obtained are compared to the results of independent numerical integration of the canonical system of the problem. In the trajectory computations and in the design of optimal maneuvers, the equations of motion and costate equations for Lagrange multipliers are integrated in the rectangular Cartesian coordinate system with the origin at the center of gravitational attraction. It is concluded that the analytical solutions are similar to the numerical results of integration. The comparison mentioned above is also provided in Chapter 9. In Chapter 4, it is shown that the extremals obtained for maneuvers with free final time can also be used in the problems of escape from an elliptical orbit and transfer between elliptical orbits. The presence of conjugate points on the TAs obtained is studied. The existing approach to investigate IT arcs is analyzed, and it is proved that the testing of Robbins' necessary condition is not enough to confirm the extremality of the TAs considered. The Hamilton-Jacobi equation is considered and the derivation of its solution is presented. It is proved that the variables in this equation are not separable (Lemma 4.1). The methodology of applying the Hamilton-Jacobi equation to obtain the solutions for IT arcs is developed (Theorem 4.5). The IT arcs obtained are classified depending on the form of the functional and conditions for final time.

In Chapter 5, a class of solutions for MT arcs with four new integration constants in the case of Newtonian field is presented. The system of four equations valid for non-zero TAs is obtained (Theorem 5.1).

The studies of MT arcs in the case of linear central field are provided in Chapter 6. The Newtonian field is approximated by a liner central field. The theorem on admissibility of such an approximation and estimation of its errors is formulated and proved (Theorem 6.1). The new first integrals are derived, the trajectories for primer vector are determined, and it is proved that on MT arcs in this field the thrust direction is constant with respect to horizon (Lemma 6.1). New analytical solutions for MT arcs are obtained (Theorem 6.2). It is shown that these solutions can find applications in solving the fuel optimal transfers between circular orbits, and between circular and hyperbolic orbits in the Newtonian field.

In Chapter 7, the complete analytical solutions to the optimal control problem for the 3-dimensional minimum-fuel optimal trajectories with constant thrust acceleration in a drag-free uniform gravity field are presented for Mars descent and landing trajectory design [126]. These solutions describe the state and the Lagrange multipliers in terms of time and switching function's characteristics to determine the number and sequence of TAs on optimal landing trajectory. Obviously, these solutions can also be applied to maneuvers near other planets and celestial bodies for which a uniform gravity field can be considered.

Chapter 8 is devoted to the determination of the number of TAs on transfer trajectories between arbitrary elliptical orbits. Corresponding functional relationships are given in the form of theorems. The continuity conditions for the variables of the problem at the switching points are analyzed. These conditions together with unknowns of the problem are considered as the continuity equations. The system of equations with equal number of algebraic equations and unknowns, including integration constants, is derived. The number of continuity equations and the number of unknowns are determined on the basis of trajectories with one, two and three TAs (Theorems 8.1 and 8.2). The conclusions are generalized to arbitrary number of thrust arcs. The formulae for determining the number of TAs depending on the number of integration constants of the analytical solutions is obtained. This formulae is generalized to the case of using two and more TAs on extremal trajectory (Theorem 8.3). It is shown that, ***the possibility of applying the general solution of the canonical system for TAs with complete number of first integrals is limited, and therefore, such solution represents only theoretical or academic interest. Consequently, from a practical stand-point, the particular solutions to the problem represent the main interest in the synthesis of optimal and extremal trajectories.***

Chapter 9 contains applications of the analytical solutions obtained in the previous chapters. The methodology of applying the solutions for TAs to the problems of interorbital transfers is presented. This methodology is based on analysis of continuity and transversality conditions. It is shown that, in essence, ***the solution of the trajectory optimization problem is reduced to a solution of only a certain number of algebraic equations***. This chapter includes five problems, which contain application of the new solutions for TAs with constant and variable specific impulse obtained in the previous chapters. These solutions and their analysis are described in a format useful for practical applications. Comparisons of the analytical and numerical solutions are provided, and their analyses are given for each variable of the canonical system with illustrations and discussions of various functional relationships.

The last three chapters represent conclusions, appendix, and nomenclature used.

CHAPTER 2

Optimal and Extremal Trajectories

2.1 OPTIMAL CONTROL PROBLEM

Let a system state at any instant is determined by n quantities $x_i(i=1, ..., n)$ called as the state variables. The vector function $\mathbf{x}=(x_1, x_2, \ldots, x_n)$, $\mathbf{x}(t) \in \Re^{(n)}$ is called as the state vector and is considered to be absolutely differentiable on the interval $[t_0, t_1]$, where t_0 and t_1 are the initial and final times of the system's motion. It will be assumed that x_i, $(i=1, \ldots, n)$ are continuous, but their derivatives, in general, may have discontinuities. The behavior of the system is described by n differential equations of first order [130], [121]:

$$\dot{x}_i = f_i(x_1, x_2, \ldots, x_n, u_1, u_2, \ldots, u_k, t), \tag{2.1}$$

Eqs. (2.1) are called as the state equations valid on $[t_0, t_1]$. The vector function $\mathbf{u}$, where $\mathbf{u}=(u_1, \ldots, u_k)$, $\mathbf{u}(t) \in \Re^{(k)} \in U$, is called as the control vector, the quantities u_r $(r=1, ..., k)$ are determined on the same interval $[t_0, t_1]$ and assumed to be piece-wise continuous functions, and U is an open set of controls. The functions f_i possess continuous partial derivatives of sufficiently high order with respect to all their components.

At t_0, the components of $\mathbf{x}_0$ and t_0 satisfy q_1 constraint equations:

$$E_l(x_{01}, x_{02}, ..., x_{0n}, t_0) = 0, \qquad l=1, ..., q_1, \qquad q_1 \le n+1. \tag{2.2}$$

At t_1, the components of $\mathbf{x}_1$ and t_1 satisfy q_2 constraint equations:

$$F_m(x_{11}, x_{12}, ..., x_{1n}, t_1) = 0, \qquad m=1, ..., q_2, \quad q_2 \le n+1 \tag{2.3}$$

Here $q_1 + q_2 < 2(n+1)$. It is assumed that $\mathbf{x}$ and $\mathbf{u}$ satisfy the constraints [3], [130], [121]:

$$\begin{gathered}\Phi_s(x_1, x_2, ..., x_n, u_1, u_2, ..., u_k, \alpha_1, \alpha_2, ..., \alpha_d) = 0, \\ s=1, ..., p;\ p \le n;\ p \le k;\ d \le k,\end{gathered} \tag{2.4}$$

where $\boldsymbol{\alpha}=(\alpha_1, \alpha_2, ..., \alpha_d)$, $\boldsymbol{\alpha} \in \Re^{(d)}$ are the auxiliary control variables. The vector functions

$$\mathbf{E} \ = \ (E_1,\ E_2,\ \ldots,\ E_{q_1}), \qquad \mathbf{E} \in \Re^{(q_1)},$$

Analytical Solutions for Extremal Space Trajectories
DOI: http://dx.doi.org/10.1016/B978-0-12-814058-1.00002-3

$$\begin{aligned}\mathbf{F} &= (F_1, F_2, \ldots, F_{q_2}), \qquad \mathbf{F} \in \Re^{(q_2)}, \\ \Phi &= (\Phi_1, \Phi_2, \ldots, \Phi_p), \qquad \Phi \in \Re^{(p)}\end{aligned}$$

are continuous and possess continuous partial derivatives of sufficiently high order with respect to all their arguments. Note that Eqs. (2.2) and (2.3) can be used to express q_1 components of $\mathbf{x}_0$ and q_2 components of $\mathbf{x}_1$ in terms of remaining $n - q_1$ and $n - q_2$ components of $\mathbf{x}_0$ and $\mathbf{x}_1$ respectively. Let us determine the functional Ψ of the problem in the form:

$$\Psi = J(x_{0,q_1+1}, x_{0,q_1+2}, \ldots, x_{0,n}, x_{1,q_2+1}, x_{1,q_2+2}, \ldots, x_{1,n}, t_0, , t_1) + \int g(\mathbf{x}, \mathbf{u}, t)dt. \tag{2.5}$$

Here the scalar functions J and g are also assumed continuous and possess continuous partial derivatives of sufficiently high order with respect to all their arguments. Then it is required to determine $\mathbf{u}(t)$ and $\mathbf{x}(t)$ such that Eqs. (2.1)–(2.4) are satisfied, and the functional Ψ in Eq. (2.5) takes minimum among its all possible values. Such $\mathbf{u}(t)$ and $\mathbf{x}(t)$ are called as optimal control and optimal trajectory respectively [121].

It is known that the constraints imposed on the state and control vectors or part of these constraints can also be given in the form of inequalities, such as

$$u_{1r} \le u_r \le u_{2r}, \qquad r = 1, \ldots, q \le k.$$

In this case, the admissible region of controls may be opened by a classical method, that is by introducing auxiliary control or slack variables, and the constraint is transformed into an equality constraint (2.4), [132], [133], [92], [131]. The q_1 and q_2 components of the initial and final state vectors can be expressed in terms of $n - q_1$ and $n - q_2$ components of these vectors. In the problem being considered the equalities in Eqs. (2.4) cover, in general, both forms of equality and inequality constraints, and the auxiliary variables are denoted by $\alpha_1, \alpha_2, \ldots, \alpha_d$ [134]. The control variable, $\mathbf{u}$ is admissible if (1) $\mathbf{u}(t)$ is defined and a piece-wise continuous on $[t_0, t_1]$; (2) $\mathbf{u} = \mathbf{u}(t)$ satisfies Eqs. (2.4) [121]. The control vector, $\mathbf{u}(t)$ given on $[t_0, t_1]$ determines the system's behavior which is described by the equations

$$\dot{x}_i = z_i(x_1, x_2, \ldots, x_n, t), \tag{2.6}$$

obtained by substitution of $\mathbf{u}(t)$ into Eqs. (2.1). It is assumed that Eqs. (2.6) satisfy the conditions of the theorem of existence and uniqueness of the solutions [135]. In this case, for the given admissible control $\mathbf{u}(t)$ and given initial conditions in Eqs. (2.2), there exist a unique continuous solution, $\mathbf{x}(t)$ of Eqs. (2.1).

If it is possible to uniquely find $\mathbf{x}_0 = (x_{01}, x_{02}, ..., x_{0n})$ and t_0 from Eqs. (2.2) in the case of $q_1 = n + 1$, then the initial state and the initial time are said to be fixed. Similarly, if $q_2 = n$ and t_1 is free, one can determine the final state vector's components as functions of t_1 by employing Eqs. (2.3). If it is possible to uniquely find t_1 from Eqs. (2.3), then the final time is said to be fixed.

Below, in subsequent sections, the sufficient conditions of positive definiteness of the second differential, introduction of conditions of finiteness of the solutions to the Riccati equation, and the conditions of conjugate points associated with the statement and analysis of the auxiliary optimization problem are studied. The case of extremals with singular arcs, where $H_{uu} = 0$, is not considered. Studies of this particular case are presented in Ref. [136], [55]. It is assumed that the extremal may consist of various thrust arcs, connected at corner points. The problem being considered is generalized to the case when the extremal of the problem includes corner points. Analysis of the differentials of the extended functional allows us to obtain optimality conditions which must be satisfied for each arc including the conditions at the corner points. Furthermore, the variation problem, the formulation of which differs from the well known formulation of Lawden in flight dynamics is provided. It is shown that the presence of constraints that characterize power limited systems and chemical systems allows us to analyze the low thrust arcs together with zero and high thrust arcs thereby extending the group of thrust arcs known in the context of the conventional variation problem. Methodology of analytical determination of extremal trajectories is based on the studies of necessary conditions of optimality, continuity conditions at corner points and on the determination of structure of the trajectory. This methodology serves as a tool of determining the extremal trajectories which represent reference trajectories applicable to the guidance problem [26], [137].

2.2 SOLUTION METHODS

The existing solution methods of the problem stated in (2.1)–(2.5) can be divided into the four groups [56]:

- Lagrange's formalism;
- Hamilton-Jacobi-Bellmann's formalism;

- Direct and indirect methods;
- Unique formalism based on the sufficient conditions of absolute optimality.

By using the Lagrange's formalism, the solution of the problem is reduced to the solution of the boundary problem and to the tests of the necessary conditions of optimality. This formalism can be applied to the problem if the domain of the admissible controls is open. If this domain is given in the form of a closed set, then this set can be mapped into an open domain by consideration of axillary space of admissible controls [133], [92], [3], [94]. The Lagrange formalism enables to find the extremals of the problem, but in many practical problems these extremals are obtained by utilizing a certain number of necessary conditions. It will be shown that the analytical solutions allows one to design extremal trajectories based on this formalism.

The Hamilton-Jacobi-Bellmann's formalism is based on the sufficient conditions of optimality and reduces the solution of the problem to integration of Hamilton-Jacobi equation in partial derivatives. This is equivalent to the construction of the field of extremals. This method covers the entire domain of admissible controls and solves the problem of optimal synthesis of trajectories.

Application of the direct and indirect methods is associated with more complicated problems than in the previous two methods. The direct and indirect methods solve the problem by approximating the minimum by some given system of functions. It is important to find the right values initial Lagrange multipliers and convergence of these methods.

The unique formalism is based on the sufficient conditions of optimality, and includes the Lagrange's formalism and the Hamilton-Jacobi-Bellmann's formalism as particular cases and enables to find an absolute minimum. By utilizing this formalism, the problem is reduced to the Jacobi-Bellman equation, or to a boundary problem.

The first three methods are not applicable if a minimum in the domain of admissible controls does not exist. Also, there may exist problems when the Euler-Lagrange equations (the canonical equations) have multiple solutions. The latter method is more general and requires the existence of some function and thereby allows us to solve the problem by constructing the sequence of the pair of the state and control vectors [56], [95]. In this work the Lagrange's formalism of solving the variation problem is developed according to the methodology of determination of extremal trajectories of maneuvers which is based on the application of methods of analytical mechanics to reduce the problem to a closed form.

2.3 NEIGHBORING EXTREMALS

There may exist an unlimited set of functions $x_i(t)$ and $u_r(t)$ that satisfy (2.1)–(2.4), and for each set there exists a corresponding value of the functional (2.5). It will be assumed that among these sets there exists a pair of functions $x^0(t), u^0(t)$ which provides a minimum to the functional. In finding the minimum, it is necessary to consider the variations $\delta x_i(t)$ and $\delta u_r(t)$.

Here and below it will be assumed that the state and control variations represent weak variations, and it is possible to ignore the second and higher order terms $|\delta x|^k,\ |\delta u|^k,\ k \geq 2$.

Taking a variation of the state equation (2.1), one can have

$$\delta \dot{x}_i = \frac{\partial f_i}{\partial x_p}\delta x_p + \frac{\partial f_i}{\partial u_r}\delta u_r + \frac{\partial f_i}{\partial t} \tag{2.7}$$

Here and below the expression $\frac{\partial f_i}{\partial x_p}\eta_p$ and similar expressions will mean a summation on index p. Eqs. (2.7) are called as the equations in variations. By differentiating Eq. (2.2) and taking into account the expression

$$dx = \delta x + \frac{dx}{dt}dt, \tag{2.8}$$

one can obtain the dependent variations

$$dE_l = \frac{\partial E_l}{\partial x_{0i}}\delta x_{0i} + \left[\frac{\partial E_l}{\partial x_{0i}}\dot{x}_{0i} + \frac{\partial E_l}{\partial t_0}\right]\delta t_0 = 0.$$

Rewriting these expressions for each $l = 1, \ldots, q_1$, we have

$$\begin{aligned}
A_{11}\delta x_{01} + \ldots + A_{1,q_1}\delta x_{0,q_1} &= -(A_{1,q_1+1}\delta x_{0,q_1+1} + \ldots \\
&\quad + A_{1,n}\delta x_{0,n} + \sum_i B_{1i}\delta t_0), \\
A_{21}\delta x_{01} + \ldots + A_{1,q_1}\delta x_{0,q_1} &= -(A_{2,q_1+1}\delta x_{0,q_1+1} + \ldots \\
&\quad + A_{2,n}\delta x_{0,n} + \sum_i B_{2i}\delta t_0), \\
&\ \ldots \\
A_{q_1,1}\delta x_{01} + \ldots + A_{q_1,q_1}\delta x_{0,q_1} &= -(A_{q_1,q_1+1}\delta x_{0,q_1+1} + \ldots \\
&\quad + A_{q_1,n}\delta x_{0,n} + \sum_i B_{q_1 i}\delta t_0),
\end{aligned} \tag{2.9}$$

where $A_{li} = \partial E_l / \partial x_{0i}$. If the determinant $||A|| \neq 0$, where

$$A = \begin{bmatrix} A_{11} & A_{12} & \dots & A_{1,q_1} \\ A_{21} & A_{22} & \dots & A_{2,q_1} \\ \dots & \dots & \dots & \dots \\ A_{q_1,1} & A_{q_1,2} & \dots & A_{q_1,q_1} \end{bmatrix},$$

then the solution of Eq. (2.9) can be given in the form:

$$\begin{bmatrix} \delta x_{0,1} \\ \delta x_{0,2} \\ \dots \\ \delta x_{0,q_1} \end{bmatrix} = -A^{-1} Y \begin{bmatrix} \delta x_{0,q+1} \\ \delta x_{0,q+2} \\ \dots \\ \delta x_{0,n} \end{bmatrix} + -A^{-1} \begin{bmatrix} \sum_i B_{1i} \\ \sum_i B_{2i} \\ \dots \\ \sum_i B_{q_1 i} \end{bmatrix} \delta t_0, \qquad (2.10)$$

where

$$Y = \begin{bmatrix} A_{1,q+1} & A_{1,q+2} & \dots & A_{1n} \\ A_{2,q+1} & A_{2,q+2} & \dots & A_{2n} \\ \dots & \dots & \dots & \dots \\ A_{q,q+1} & A_{q,q+2} & \dots & A_{qn} \end{bmatrix}.$$

This means that only $n - q_1$ variations δx_0 and δt_0 are independent. The $n - q_2$ independent variations δx_1 can be obtained analogically by varying F_m in Eq. (2.3). Besides that, the variations of Eq. (2.4) with respect to controls are given in the form:

$$\frac{\partial \Phi_s}{\partial x_i} \delta x_i + \left[\frac{\partial \Phi_s}{\partial x_i} \dot{x}_i + \frac{\partial \Phi_s}{\partial t} \right] \delta t + \frac{\partial \Phi_s}{\partial u_r} \delta u_r + \frac{\partial \Phi_s}{\partial \alpha_a} \delta \alpha_a = 0, \qquad (2.11)$$

$$r = 1, \dots, k, \quad a = 1, \dots, d, \quad s = 1, \dots, p.$$

The variations δx and δu, defined in Eq. (2.7), and which satisfy Eq. (2.10) and Eq. (2.11), are said to be admissible variations for the minimum $x^0(t), u^0(t)$ [3], [130].

2.4 FIRST DIFFERENTIAL OF EXTENDED FUNCTIONAL

Consider the trajectories which do not include corner points. Then the results will be generalized for the case of the trajectories with corner points. In order to obtain the necessary conditions of optimality, consider the extended functional of the problem stated above:

$$K(\mathbf{x},\mathbf{u},\boldsymbol{\gamma},\boldsymbol{\alpha},\boldsymbol{\mu},\boldsymbol{\nu})=\varepsilon J(\mathbf{x}_0,\mathbf{x}_1,t_0,t_1)+\int_{t_0}^{t_1}g(\mathbf{x},\mathbf{u},t)dt+$$
$$+\boldsymbol{\mu}^T\mathbf{E}(\mathbf{x}_0,t_0)+\boldsymbol{\nu}^T\mathbf{F}(\mathbf{x}_1,t_1)+\int_{t_0}^{t_1}\boldsymbol{\gamma}^T\boldsymbol{\Phi}(\mathbf{x},\mathbf{u},\boldsymbol{\alpha})dt, \tag{2.12}$$

where

$$\begin{aligned}
\mathbf{x}_0 &= (x_{01},x_{02},...,x_{0n}),\\
\mathbf{x}_1 &= (x_{11},x_{12},...,x_{1n}),\\
\mathbf{u} &= (u_1,u_2,...,u_k),\\
J(\mathbf{x}_0,\mathbf{x}_1,t_0,t_1) &= J(x_{0,q_1+1},x_{0,q_1+2},...,x_{0,n},x_{1,q_2+1},x_{1,q_2+2},...,x_{1,n},t_0,t_1),\\
\mathbf{E}(\mathbf{x}_0,t_0) &= \mathbf{E}(x_{01},x_{02},...,x_{0n},t_0),\\
\mathbf{F}(\mathbf{x}_1,t_1) &= \mathbf{F}(x_{11},x_{12},...,x_{1n},t_1),\\
\boldsymbol{\Phi}(\mathbf{x},\mathbf{u},\boldsymbol{\alpha}) &= \boldsymbol{\Phi}(x_1,x_2,...,x_n,u_1,u_2,...,u_k,\alpha_1,\alpha_2,...,\alpha_d),\\
\boldsymbol{\mu} &= (\mu_1,\mu_2,...,\mu_{q_1}),\\
\boldsymbol{\nu} &= (\nu_1,\nu_2,...,\nu_{q_2}),\\
\boldsymbol{\gamma} &= (\gamma_1,\gamma_2,...,\gamma_p),
\end{aligned}$$

the scalar quantity ε is assumed positive, $\varepsilon>0$, and the vectors $\boldsymbol{\mu}$, $\boldsymbol{\nu}$, and $\boldsymbol{\gamma}$ are considered as multipliers to be determined. The vectors $\boldsymbol{\mu}$ and $\boldsymbol{\nu}$ are assumed constants. Here and below the indexes "0" and "1" will denote the initial and final values of the variables. If the functions

$$H(\mathbf{x},\mathbf{u},t,\boldsymbol{\lambda},\boldsymbol{\alpha},\boldsymbol{\gamma})=\boldsymbol{\lambda}^T\mathbf{f}(\mathbf{x},\mathbf{u},t)+\boldsymbol{\gamma}^T\Phi(\mathbf{x},\mathbf{u},\boldsymbol{\alpha})+g(\mathbf{x},\mathbf{u},t), \tag{2.13}$$
$$G(\mathbf{x}_0,\mathbf{x}_1,t_0,t_1,\varepsilon,\boldsymbol{\mu},\boldsymbol{\nu})=\varepsilon J(\mathbf{x}_0,t_0,\mathbf{x}_1,t_1)+\boldsymbol{\mu}^T\mathbf{E}(\mathbf{x}_0,t_0)+\boldsymbol{\nu}^T\mathbf{F}(\mathbf{x}_1,t_1), \tag{2.14}$$

are introduced, then Eq. (2.12) is simplified as

$$K(\mathbf{x}_0,t_0,\mathbf{x}_1,t_1,\boldsymbol{\mu},\boldsymbol{\nu},\boldsymbol{\gamma},\boldsymbol{\alpha})=G(\mathbf{x}_0,t_0,\mathbf{x}_1,t_1,\varepsilon,\boldsymbol{\mu},\boldsymbol{\nu})$$
$$+\int_{t_0}^{t_1}[H(\mathbf{x},\mathbf{u},t,\boldsymbol{\lambda},\boldsymbol{\gamma},\boldsymbol{\alpha})-\boldsymbol{\lambda}^T\dot{\mathbf{x}}]dt. \tag{2.15}$$

Following [90], the function $H(\mathbf{x},\mathbf{u},t,\boldsymbol{\lambda},\boldsymbol{\alpha},\boldsymbol{\gamma})$ will be called as Pontryagin function, although this function in its original form does not include the term $\boldsymbol{\gamma}^T\Phi(\mathbf{x},\mathbf{u},\boldsymbol{\alpha})$. Accepting the symbolics $\partial()/\partial x=()_x$ to denote a partial derivative and using the equality for the differential of integral $I = \int_{t_0}^{t_1} S dt$ in the form [131]

$$dI = [Sdt]_{t_0}^{t_1} + \int_{t_0}^{t_1} \delta S dt, \tag{2.16}$$

one can write the first differential of the extended functional Eq. (2.15) as

$$\begin{gathered} dK = G_{\mathbf{x}_0}^T d\mathbf{x}_0 + G_{\mathbf{x}_1}^T d\mathbf{x}_1 + G_{t_0} dt_0 + G_{t_1} dt_1 + G_{\boldsymbol{\mu}}^T d\boldsymbol{\mu} + G_{\boldsymbol{\nu}}^T d\boldsymbol{\nu} + \\ +[(H(\mathbf{x}, \mathbf{u}, t, \boldsymbol{\lambda}, \boldsymbol{\gamma}, \boldsymbol{\alpha}) - \boldsymbol{\lambda}^T \dot{\mathbf{x}}) dt]_{t_0}^{t_1} + \\ \int_{t_0}^{t_1} (H_{\mathbf{x}} \delta \mathbf{x} + H_{\boldsymbol{\lambda}} \delta \boldsymbol{\lambda} + H_{\boldsymbol{\gamma}} \delta \boldsymbol{\gamma} + H_{\mathbf{u}} \delta \mathbf{u} + H_{\boldsymbol{\alpha}} \delta \boldsymbol{\alpha} - \delta \boldsymbol{\lambda}^T \dot{\mathbf{x}} - \boldsymbol{\lambda}^T \delta \dot{\mathbf{x}}) dt \end{gathered}$$

where

$$\begin{aligned} G_{t_0} &= \varepsilon \frac{\partial J}{\partial t_0} + \boldsymbol{\mu}^T \frac{\partial \mathbf{E}}{\partial t_0}, & G_{t_1} &= \varepsilon \frac{\partial J}{\partial t_1} + \boldsymbol{\nu}^T \frac{\partial \mathbf{F}}{\partial t_1}, \\ G_{x_0} &= \varepsilon \frac{\partial J}{\partial \mathbf{x}_0} + \boldsymbol{\mu}^T \frac{\partial \mathbf{E}}{\partial \mathbf{x}_0}, & G_{x_1} &= \varepsilon \frac{\partial J}{\partial \mathbf{x}_1} + \boldsymbol{\nu}^T \frac{\partial \mathbf{F}}{\partial \mathbf{x}_1}. \end{aligned}$$

Using the equalities

$$\begin{gathered} d\mathbf{x}_0 = \delta \mathbf{x}_0 + \dot{\mathbf{x}}_0 dt_0, \\ d\mathbf{x}_1 = \delta \mathbf{x}_1 + \dot{\mathbf{x}}_1 dt_1, \\ \int_{t_0}^{t_1} \boldsymbol{\lambda}^T \delta \dot{\mathbf{x}} dt = (\boldsymbol{\lambda}^T \delta \mathbf{x})_{t_0}^{t_1} - \int_{t_0}^{t_1} \dot{\boldsymbol{\lambda}}^T \delta \mathbf{x} dt, \\ H_\lambda = \mathbf{f}^T, \quad G_\mu = \mathbf{E}, \quad G_\nu = \mathbf{F}, \quad H_\gamma = \boldsymbol{\Phi}, \end{gathered}$$

the expression for dK is reduced to the form:

$$\begin{gathered} dK = G_{\mathbf{x}_0}^T \delta \mathbf{x}_0 + G_{\mathbf{x}_1}^T \delta \mathbf{x}_1 + G_{t_0} dt_0 + G_{t_1} dt_1 + G_{\boldsymbol{\mu}}^T d\boldsymbol{\mu} + G_{\boldsymbol{\nu}}^T d\boldsymbol{\nu} + G_{\mathbf{x}_0}^T \dot{\mathbf{x}}_0 dt_0 + \\ G_{\mathbf{x}_1}^T \dot{\mathbf{x}}_1 dt_1 + (H_1 - \boldsymbol{\lambda}_1^T \dot{\mathbf{x}}_1) dt_1 - \boldsymbol{\lambda}_1^T \delta \mathbf{x}_1 - (H_0 - \boldsymbol{\lambda}_0^T \dot{\mathbf{x}}_0) dt_0 + \boldsymbol{\lambda}_0^T \delta \mathbf{x}_0 + \\ \int_{t_0}^{t_1} (H_{\mathbf{x}} \delta \mathbf{x} + \dot{\boldsymbol{\lambda}}^T \delta \mathbf{x} + (H_\lambda - \dot{\mathbf{x}}^T) \delta \boldsymbol{\lambda} + H_{\mathbf{u}} \delta \mathbf{u} + H_{\boldsymbol{\alpha}} \delta \boldsymbol{\alpha}) dt \end{gathered}$$

or

$$\begin{gathered} dK = (G_{x_0} + \boldsymbol{\lambda}_0^T)^T \delta \mathbf{x}_0 + (G_{x_1} - \boldsymbol{\lambda}_1^T)^T \delta \mathbf{x}_1 + (G_{t_0} + H_0) dt_0 + (G_{t_1} + H_1) dt_1 \\ +[(G_{x_0} - \boldsymbol{\lambda}_0^T)^T \dot{\mathbf{x}}_0] dt_0 + [(G_{x_1} - \boldsymbol{\lambda}_1^T)^T \dot{\mathbf{x}}_1] dt_1 \\ + \int_{t_0}^{t_1} [(H_{\mathbf{x}} + \dot{\boldsymbol{\lambda}}^T)^T \delta \mathbf{x} + (H_\lambda - \dot{\mathbf{x}}^T)^T \delta \boldsymbol{\lambda} + H_{\mathbf{u}}^T \delta \mathbf{u} + H_{\boldsymbol{\alpha}}^T \delta \boldsymbol{\alpha}] dt. \end{gathered}$$

It is easy to see that the coefficients of $d\boldsymbol{\mu}$, $d\boldsymbol{\nu}$ and $d\boldsymbol{\gamma}$ are zero due to the constraints (2.2)–(2.4). Using the equalities $d\mathbf{x}_0 = \delta\mathbf{x}_0 + \dot{\mathbf{x}}_0 dt_0$ and $d\mathbf{x}_1 = \delta\mathbf{x}_1 + \dot{\mathbf{x}}_1 dt_1$ in this expression, one can obtain:

$$dK = (G_{\mathbf{x}_0} + \boldsymbol{\lambda}_0^T)^T d\mathbf{x}_0 + (G_{\mathbf{x}_1} - \boldsymbol{\lambda}_1^T)^T d\mathbf{x}_1 + (G_{t_0} - H_0)dt_0 + (G_{t_1} + H_1)dt_1 + \int_{t_0}^{t_1} [(H_{\mathbf{x}} + \dot{\boldsymbol{\lambda}}^T)^T \delta\mathbf{x} + (H_{\boldsymbol{\lambda}} - \dot{\mathbf{x}}^T)^T \delta\boldsymbol{\lambda} + H_{\mathbf{u}}^T \delta\mathbf{u} + H_{\boldsymbol{\alpha}}^T \delta\boldsymbol{\alpha}]dt. \tag{2.17}$$

The necessary conditions of optimality can be obtained from the condition $dK = 0$. The coefficient of $\delta\boldsymbol{\lambda}$ is zero due to Eq. (2.1). The coefficients of $d\mathbf{x}_0$, $d\mathbf{x}_1$, dt_0 and dt_1 can be equated to zero by choosing $\boldsymbol{\lambda}_0$, $\boldsymbol{\lambda}_1$, $\boldsymbol{\mu}$ and $\boldsymbol{\nu}$. As H is defined with respect to $\mathbf{u}$ and $\boldsymbol{\alpha}$ in the open set, then for arbitrary $\delta\mathbf{u}$ and $\delta\boldsymbol{\alpha}$, one can obtain

$$\int_{t_0}^{t_1} (H_{\mathbf{u}}\delta\mathbf{u} + H_{\boldsymbol{\alpha}}\delta\boldsymbol{\alpha})dt = 0. \tag{2.18}$$

The condition $dK = 0$ can be satisfied only and only if Eqs. (2.18) are satisfied along with all conditions obtained from equating to zero all terms of Eq. (2.17) except for the last two terms of its last integral expression.

The first order necessary conditions of optimality can be written in the form:

$$\dot{\mathbf{x}}^T = H_{\lambda}, \quad \dot{\boldsymbol{\lambda}}^T = -H_x, \quad \int_{t_0}^{t_1} (H_{\mathbf{u}}^T \delta\mathbf{u} + H_{\boldsymbol{\alpha}}^T \delta\boldsymbol{\alpha})dt = 0, \quad \Phi = 0, \tag{2.19}$$

$$\mathbf{E} = 0, \quad \mathbf{F} = 0, \quad \boldsymbol{\lambda}_0 = -G_{\mathbf{x}_0}^T, \quad \boldsymbol{\lambda}_1 = G_{\mathbf{x}_1}^T, \quad H_0 = G_{t_0}, \quad H_1 = -G_{t_1}, \tag{2.20}$$

where $\mathbf{x} = (x_1, ..., x_n)$ and $\boldsymbol{\lambda} = (\lambda_1, ..., \lambda_n)$. The first two equations in Eq. (2.19) are said to be canonical equations, where the Pontryagin function is determined from Eq. (2.13). The last three equations in Eqs. (2.20) are said to be the transversality conditions. Note that when t_1 is given, the last equation in Eq. (2.20) is not applicable. Consider the condition

$$\int_{t_0}^{t_1} (H_{\mathbf{u}}^T \delta\mathbf{u} + H_{\boldsymbol{\alpha}}^T \delta\boldsymbol{\alpha})dt = 0$$

in Eq. (2.19). Here it should be noted that due to Eq. (2.11), the total number of the independent controls u and α is equal to $k + d - p$. Noting that $p < k$ and $d \leq k$, the controls u_r, $(r = 1, ..., k)$ and auxiliary controls

α_a, $(a=1, ..., d)$ can be transformed into the new variables υ_j, $(j=1, ..., k+d)$:

$$\upsilon_1 = u_1, \qquad \upsilon_2 = u_2, \; ..., \upsilon_k = u_k,$$
$$\upsilon_{k+1} = \alpha_1, \qquad \upsilon_{k+2} = \alpha_2, \; ..., \upsilon_{k+d} = \alpha_d.$$

If $\upsilon = (\upsilon_1, \upsilon_2, ..., \upsilon_{k+d})$, then the functions in Eq. (2.1) and Eq. (2.4), and also H and g take the form:

$$\begin{aligned} \bar{\mathbf{f}} &= \mathbf{f}(x, \upsilon), \\ \bar{\Phi} &= \Phi(\upsilon), \\ \bar{H} &= H(x, \upsilon, \lambda, t), \\ \bar{g} &= g(x, \upsilon, t). \end{aligned}$$

It is easy to show that the integral condition in Eq. (2.19) is transformed into the form:

$$\int_{t_0}^{t_1} (H_u \delta \mathbf{u} + H_\alpha \delta \boldsymbol{\alpha}) dt = \int_{t_0}^{t_1} (\bar{H}_\upsilon \delta \boldsymbol{\upsilon}) dt = 0. \tag{2.21}$$

Now γ_s can be found such that the equalities $\frac{\partial \bar{H}}{\partial \upsilon_s} = 0$, $s = 1, ..., p < k < n$ or

$$\begin{aligned} \frac{\partial \bar{H}}{\partial \upsilon_1} &= \frac{\partial \bar{g}}{\partial \upsilon_1} + \sum \lambda_i \frac{\partial \bar{f}_i}{\partial \upsilon_1} + \sum \gamma_s \frac{\partial \bar{\Phi}_s}{\partial \upsilon_1} = 0, \\ \frac{\partial \bar{H}}{\partial \upsilon_2} &= \frac{\partial \bar{g}}{\partial \upsilon_2} + \sum \lambda_i \frac{\partial \bar{f}_i}{\partial \upsilon_2} + \sum \gamma_s \frac{\partial \bar{\Phi}_s}{\partial \upsilon_2} = 0, \\ &\cdots\cdots\cdots \\ \frac{\partial \bar{H}}{\partial \upsilon_p} &= \frac{\partial \bar{g}}{\partial \upsilon_p} + \sum \lambda_i \frac{\partial \bar{f}_i}{\partial \upsilon_p} + \sum \gamma_s \frac{\partial \bar{\Phi}_s}{\partial \upsilon_p} = 0, \end{aligned} \tag{2.22}$$

are satisfied. One can accept $||X|| \neq 0$, where

$$X = \begin{bmatrix} \frac{\partial \bar{\Phi}_1}{\partial \upsilon_1} & \cdots & \frac{\partial \bar{\Phi}_p}{\partial \upsilon_1} \\ \cdots & \cdots & \cdots \\ \frac{\partial \bar{\Phi}_1}{\partial \upsilon_p} & \cdots & \frac{\partial \bar{\Phi}_p}{\partial \upsilon_p}. \end{bmatrix}$$

These equations can be used to find γ_s, $s = 1, ..., p$. The remaining γ_e, $(e = p+1, ..., k+d)$ can be chosen arbitrarily. Based on the above described, the condition (2.21) is reduced to the form:

$$\int_{t_0}^{t_1} \frac{\partial \bar{H}}{\partial v_e} \delta v_e dt = 0.$$

In this equation the functions $\delta v_{p+1}, ..., \delta v_{k+d}$ can be chosen arbitrarily. Then from this equation it follows that

$$\frac{\partial \bar{H}}{\partial v_e} = 0, \quad e = p+1, ..., k.$$

But taking into account the condition

$$\frac{\partial \bar{H}}{\partial v_s} = 0, \quad s = 1, ..., p,$$

and the previous conditions, one can have:

$$\frac{\partial \bar{H}}{\partial v_j} = 0, \quad j = 1, ..., k+d. \tag{2.23}$$

Transforming the variables v_j, $(j = 1, ..., k+d)$ back to the controls u_r, $(r = 1, ..., k)$ and α_a, $(a = 1, ..., d)$, it can be obtained that

$$\begin{aligned} \frac{\partial H}{\partial u_r} &= 0, \quad r = 1, ..., k, \\ \frac{\partial H}{\partial \alpha_a} &= 0, \quad a = 1, ..., d. \end{aligned} \tag{2.24}$$

Consequently, the integral condition in Eq. (2.19) can be replaced by the conditions (2.24). The conditions (2.24) are said to be the conditions of local maximality [90]. The first two equations (2.19) and Eqs. (2.24), after the corresponding integration, define $2n + k + d$ variables x_i, λ_i, u_r, and α_a together with $2n$ integration constants. The $2n$ constants, the q_1 variables μ_l $(l = 1, ..., q_1)$, the q_2 variables $\nu_m (m = 1, ..., q_2)$ and the final time t_1 (or $2n + q_1 + q_2 + 1$ variables) can be defined using q_1 conditions (2.2), the q_2 conditions (2.3) and the $2n+1$ conditions of transversality Eq. (2.20). As it was mentioned above, γ_s, $(s = 1, ..., p)$ can be determined from Eq. (2.22).

2.5 MAXIMALITY OR MINIMALITY OF HAMILTONIAN

Consider an optimal trajectory with **x** and **u**, and the equations

$$\delta \dot{\mathbf{x}} = \mathbf{f}_x \delta \mathbf{x} + \mathbf{f}_u \delta \mathbf{u}, \quad \dot{\boldsymbol{\lambda}}^T = -H_x,$$

where $\mathbf{f} = \mathbf{f}(\mathbf{x}, \mathbf{u}, t)$ and the variations $\delta\mathbf{x}$ and $\delta\mathbf{u}$ are considered at fixed t, and so $\delta t = 0$. By multiplying the first equation by $\boldsymbol{\lambda}$ and the second equation – by $\delta\mathbf{x}$, and subtracting the second equation from the first equation, one can obtain $d(\boldsymbol{\lambda}\delta\mathbf{x})/dt = H_u\delta\mathbf{u}$. As $H_u = 0$ on optimal trajectory, from this equation it follows that

$$\boldsymbol{\lambda}\delta\mathbf{x} = const, \tag{2.25}$$

where *const* means an integration constant. Consider now a minimum time problem with $J = t_1 - t_0 \to min$, where t_0 and t_1 are the initial and final times. As previously mentioned, the state vector is given by $\mathbf{x} = (x_1, x_2, ..., x_n)$. Let us introduce new variables, x_{n+1} and λ_{n+1} defined by $x_{n+1} = J$ and $\dot{\lambda}_{n+1} = -\partial H/\partial x_{n+1}$. According to the transversality conditions, using

$$G = J + \boldsymbol{\mu}^T\mathbf{E} + \boldsymbol{\nu}^T\mathbf{F},$$

at t_1 one can have

$$\lambda_{n+1}(t_1) = -\frac{\partial G}{\partial x_{n+1}(t_1)} = -1. \tag{2.26}$$

Let us assume that $\delta x_i(t_1) = 0,\ i = 1, ..., n$. In this case Eq. (2.25) can be rewritten in the form:

$$\boldsymbol{\lambda}(t_1)\delta\mathbf{x}(t_1) = \sum_{j=1}^{n+1}\lambda_j(t_1)\delta x_j(t_1) = \lambda_{n+1}(t_1)\delta x_{n+1}(t_1) = -\delta x_{n+1} = -\delta J \le 0, \tag{2.27}$$

as $\delta J = \delta t_1 \ge 0$. Consequently, using Eq. (2.25), one can obtain

$$\boldsymbol{\lambda}^T\delta\mathbf{x} \le 0. \tag{2.28}$$

Now, consider a small interval of time, $\tau - \epsilon < t < \tau$, where $\epsilon > 0$ and $\tau \in (t_0, t_1)$. Also consider arbitrary variations of $\mathbf{x}$ and $\mathbf{u}$, defined as $\delta\mathbf{x} = \tilde{\mathbf{x}} - \mathbf{x}^0$ and $\delta\mathbf{u} = \tilde{\mathbf{u}} - \mathbf{u}^0$ on $\tau - \epsilon < t < \tau$. Here $\delta\mathbf{u}$ is arbitrary and not necessarily small. The vectors $\mathbf{x}^0$ and $\mathbf{u}^0$ describe an optimal trajectory, and the sign " " means any admissible variable. Deviation of the trajectory due to this variation can be approximated as

$$\tilde{x}_i(\tau) - x_i^0(\tau) \approx (\dot{\tilde{x}}_i(\tau) - \dot{x}_i^0(\tau))\epsilon = [f_i(\tilde{\mathbf{x}}(\tau), \tilde{\mathbf{u}}(\tau), \tau) - f_i(\mathbf{x}^0(\tau), \mathbf{u}^0(\tau), \tau)]\epsilon. \tag{2.29}$$

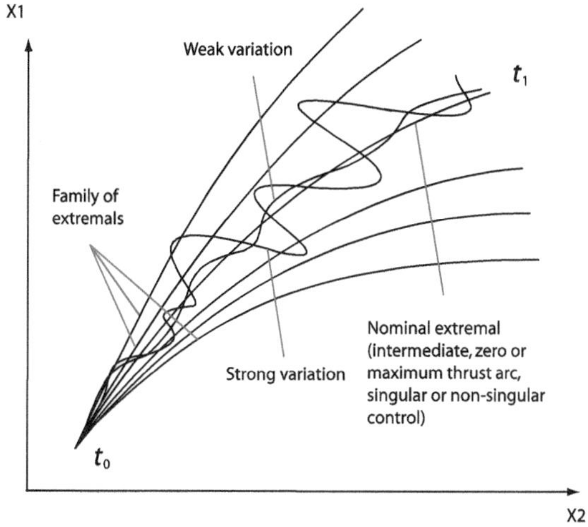

Figure 2.1 Illustration of nominal extremals, weak and strong variations with fixed initial conditions and free final conditions.

Noting that $\delta\mathbf{x} = \tilde{\mathbf{x}} - \mathbf{x}^0$ and using Eq. (2.28), one can have

$$\begin{aligned}\boldsymbol{\lambda}(\tau)^T\delta\mathbf{x}(\tau) &= \sum \boldsymbol{\lambda}^T(\tilde{\mathbf{x}} - \mathbf{x}^0)\\ &= \sum \boldsymbol{\lambda}(\tau)[f_i(\tilde{\mathbf{x}}(\tau), \tilde{\mathbf{u}}(\tau), \tau) - f_i(\mathbf{x}^0(\tau), \mathbf{u}^0(\tau), \tau)]\epsilon \leq 0.\end{aligned}$$

As $\tau \in (t_0, t_1)$, this inequality means that

$$\boldsymbol{\lambda}^T\mathbf{f}(\tilde{\mathbf{x}}, \tilde{\mathbf{u}}, t) \leq \boldsymbol{\lambda}^T\mathbf{f}(\tilde{\mathbf{x}}^0, \mathbf{u}^0, t) \qquad \forall t \in [t_0, t_1]$$

Due to the continuity of $\mathbf{f}(\tilde{\mathbf{x}}, \tilde{\mathbf{u}}, t)$, the difference

$$\mathbf{f}(\mathbf{x}^0, \tilde{\mathbf{u}}, t) - \mathbf{f}(\mathbf{x}^0, \mathbf{u}^0, t)$$

will have the same order of ϵ, and this inequality can also be written as

$$\boldsymbol{\lambda}^T\mathbf{f}(\mathbf{x}^0, \mathbf{u}, t) \leq \boldsymbol{\lambda}^T\mathbf{f}(\mathbf{x}^0, \mathbf{u}^0, t),$$

or

$$H(\mathbf{x}^0, \mathbf{u}, \boldsymbol{\lambda}, t) \leq H(\mathbf{x}^0, \mathbf{u}^0, \boldsymbol{\lambda}, t). \tag{2.30}$$

As $\delta\mathbf{u}$ is an arbitrary variation, and therefore, Eq. (2.30) is valid for both weak or strong variations of the control (see Fig. 2.1). Also, as τ is arbitrary quantity, Eq. (2.30) is valid at any time on the interval $[t_0, t_1]$. Note that

$x_{n+1} = J$ can be used for any functional, J. Consequently, the condition in Eq. (2.30) is applicable to all problems with minimizing functionals in an arbitrary form. It can be easily shown that if these functionals are to be maximized, then the inequality sign in Eq. (2.30) is reversed.

2.6 WEIERSTRASS NECESSARY CONDITION

As it was shown above, by introducing a new variable, x_{n+1} as

$$x_{n+1} = t, \qquad \dot{x}_{n+1} = 1,$$

Eqs. (2.1) can be rewritten in the form

$$\dot{x}_i = f_i(x_1, x_2, \ldots, x_n, x_{n+1}, u_1, u_2, \ldots, u_k, t), \quad i = 1, \ldots, n+1. \tag{2.31}$$

In this case it can be shown that

$$\dot{x}_{n+1} = H_{\lambda_t}, \quad \dot{\lambda}_t = \dot{\lambda}_{n+1} = -H_{x_{n+1}} = -H_t.$$

Also, using Eq. (2.19), one can also show that

$$\frac{dH}{dt} = H_{\mathbf{x}}^T \dot{\mathbf{x}} + H_{\boldsymbol{\lambda}}^T \dot{\boldsymbol{\lambda}} + H_t = H_t,$$

and therefore,

$$\dot{\lambda}_t = -\frac{dH}{dt} = -H_t. \tag{2.32}$$

Integration of this equation yields

$$H = -\lambda_t + \lambda_c, \tag{2.33}$$

where λ_c is the arbitrary integration constant. As Eq. (2.33) directly follows from Eqs. (2.19), it can also be considered as a necessary condition. Let in the problem being considered $\mathbf{x}^0(t)$, $\mathbf{u}^0(t)$, $\boldsymbol{\lambda}(t)$ represent an optimal trajectory. Consequently, if

$$\max_{\mathbf{u} \in U} H(\mathbf{x}^0, \mathbf{u}, \boldsymbol{\lambda}, t) = H(\mathbf{x}^0, \mathbf{u}^0, \boldsymbol{\lambda}, t), \tag{2.34}$$

then on optimal trajectory the following condition is satisfied:

$$H(\mathbf{x}^0, \mathbf{u}, \boldsymbol{\lambda}, t) \leq H(\mathbf{x}^0, \mathbf{u}^0, \boldsymbol{\lambda}, t) = -\lambda_t + \lambda_c. \tag{2.35}$$

Note that this condition is the same as the Weierstrass necessary condition [90], [130]. Utilizing Eq. (2.33), one can have

$$H(\mathbf{x}^0, \mathbf{u}, t) + \lambda_t \leq \lambda_c,$$
$$H(\mathbf{x}^0, \mathbf{u}^0, t) + \lambda_t = \lambda_c.$$

If $H = H(\mathbf{x}^0, \mathbf{u}, \boldsymbol{\lambda}, t)\ \forall \mathbf{u} \in U$, where $\mathbf{u}$ is the admissible control vector, and $H^0 = H(\mathbf{x}^0, \mathbf{u}^0, \boldsymbol{\lambda}, t)$, then assuming that $\lambda_c = 0$ due to its arbitrariness, from Eqs. (2.35) it follows that

$$\begin{aligned} H(\mathbf{x}^0, \mathbf{u}, \boldsymbol{\lambda}, t) + \lambda_t &\leq 0 \quad \forall \mathbf{u} \in U, \quad t \in [t_0, t_1], \\ H(\mathbf{x}^0, \mathbf{u}^0, \boldsymbol{\lambda}, t) + \lambda_t &= 0, \quad t \in [t_0, t_1]. \end{aligned} \tag{2.36}$$

Also note that the Weierstrass necessary condition must be satisfied on all points of the optimal trajectory, including the corner points. The sign of the inequality is changed if the functional is maximized. Note that the strict sign of the inequality in Eq. (2.36) transforms this condition into the Weierstrass condition for a strong minimum [97].

In summary, Eqs. (2.19), (2.20), (2.32) and (2.36) are called the necessary conditions for the optimal control and optimal trajectory. These conditions are very similar to those of the maximum principle for the problem [90]. A trajectory which satisfies these conditions is called a Pontryagin minimum, and consequently, the member of the family of Pontryagin extremals.

2.7 WEAK EXTREMALS AND PONTRYAGIN EXTREMALS

Definition of extremals in the sense of calculus of variations. A triple of vector-functions $\mathbf{x}, \boldsymbol{\lambda}, \mathbf{u}$ satisfying the equations

$$\dot{\mathbf{x}}^T = H_\lambda, \quad \dot{\boldsymbol{\lambda}}^T = -H_x, \quad H_\mathbf{u} = 0 \tag{2.37}$$

is called a weak extremal of the problem Eqs. (2.1)–(2.5). This type of extremals is defined in the sense of calculus of variations [90], [131].

Assume that it is possible to find controls $\mathbf{u}$ and $\boldsymbol{\alpha}$ from Eq. (2.24). Then, by substituting them into Eq. (2.13), one can have the equality

$$\tilde{H}(\mathbf{x}, \boldsymbol{\lambda}, t) = H(\mathbf{x}, \mathbf{u}, \boldsymbol{\lambda}, t, \boldsymbol{\alpha}, \boldsymbol{\gamma}).$$

Consequently, on the basis of Eq. (2.19) one can conclude that the extremals of the problem can be determined by the canonical equations of the form:

$$\dot{\mathbf{x}}^T = \tilde{H}_\lambda, \quad \dot{\boldsymbol{\lambda}}^T = -\tilde{H}_x. \tag{2.38}$$

Definition of weak variations. If the variations $\delta\mathbf{x}$ and $\delta\mathbf{u}$ are so small that it is possible to ignore the higher order terms, $|\delta x|^k, \ |\delta u|^k, \ k \geq 2$, then these variations are said to be weak [3], [94].

Definition of strong variations. If the variations $\delta\mathbf{x}$ are small, but the variations $\delta\mathbf{u}$ are arbitrary, then these variations are said to be strong [94].

It is known that there exist two types of minima: a weak minimum and a Pontryagin minimum. The corresponding variations of the state and control vectors are called weak and strong variations respectively. Note that the strong variations, known in the calculus of variations, have been shown to be a particular case of the Ponryagin variations [90]. The weak minimum is a member of a family of weak extremals of the problem, that is the trajectories that satisfy the state and costate equations, and the conditions of local optimality. So, in the calculus of variations, the weak minimum is associated with weak extremals. Similarly, in optimal control, the Pontryagin minimum is associated with Pontryagin extremals. It has been shown that the conditions for a Pontryagin extremal are stronger than the conditions for a weak extremal [90].

Definition of a Pontryagin extremal. The vector-functions $\mathbf{x}, \boldsymbol{\lambda}, \lambda_t, \mathbf{u}$ satisfying the equations

$$\begin{gathered}\dot{\mathbf{x}}^T = H_\lambda, \quad \dot{\boldsymbol{\lambda}}^T = -H_x, \quad \dot{\lambda}_t = -H_t, \\ H(\mathbf{x}^0, \boldsymbol{\lambda}, \mathbf{u}^0, t) + \lambda_t(t) = 0, \qquad H(\mathbf{x}^0, \boldsymbol{\lambda}, \mathbf{u}, t) + \lambda_t(t) \leq 0, \quad \forall u \in U\end{gathered} \tag{2.39}$$

where U is an open set of controls, is called a Pontryagin extremal of the problem Eqs. (2.1)–(2.5). This type of extremals is defined in the sense of optimal control [90]. The conditions (2.19) and (2.36) define the Pontryagin extremal, which is stronger type of an extremal than an extremal satisfying Eqs. (2.24).

2.8 ADVANTAGES OF PONTRYAGIN EXTREMALS OVER TYPICAL EXTREMALS

Eq. (2.34) shows that the Pontryagin extremal is valid for all $\mathbf{u} \in U$, where U is an open set of controls. In comparison, as it follows from Eqs. (2.24), the typical extremal is valid for all $\mathbf{u} \in N$, where N is a bounded set of controls yielding $[\partial H/\partial \mathbf{u}]^T = 0$. Consequently, the Pontryagin extremal is valid for a wider set of controls, which include the bounded set of controls, and allows for consideration of a wider families of trajectories. In a theoretical sense, this improves the possibility of finding a better extremal compared to a typical extremal. This means that this type of extremal corresponds to strong variations defined in the open set, U which obviously contains the bounded set, N. This allows us to conclude that it is more advantageous to consider Eqs. (2.19) and (2.36), which determine a Pontryagin extremal, rather than considering Eqs. (2.19) and (2.24). Therefore, in this work, all extremals will be considered as Pontryagin extremals and one of these extremals is called as a Pontryagin minimum. Obviously, the Pontryagin minimum is a stronger type of a minimum compared to a minimum defined in the sense of calculus of variations. As was shown in Ref. [90], any trajectory that represents a Pontryagin minimum is also a weak extremal. The opposite statement is not necessarily true [90].

One more advantage of the Pontryagin extremals over the typical extremals can be shown by the following contradiction. As previously mentioned, let us also consider a minimum time problem with $J = t_1 - t_0 \to min$, where t_0 and t_1 are the initial and final times. So, the final time is not fixed. The transversality condition yields (see Eq. (2.20))

$$\left(\sum \lambda_i \dot{x}_i\right)_{t_1} = -\frac{\partial J}{\partial t_1} = -1.$$

Note that this condition is applicable when the final time is not fixed or free. If the Hamiltonian does not explicitly depend on time, then it can be shown that $H = \sum \lambda_i \dot{x}_i = C$ and consequently, $C = -1$ for the case when the final time is not fixed and free. Consider now a minimum fuel problem with $J = m_0 - m_1$ and free final time. By using the transversality condition and $H = \sum \lambda_i \dot{x}_i = C$, one can obtain that

$$C = \left(\sum \lambda_i \dot{x}_i\right)_{t_1} = -\frac{\partial J}{\partial t_1} = 0.$$

The contradiction here is that for both problems with free final time, the same transversality condition provides different values for the constant of Hamiltonian. The use of Pontryagin extremals would allow us to consider the condition $C = H(t_1) = \lambda_t(t_1)$ to compute the constant C and avoid this contradiction.

In the majority cases of optimization problems, it is more beneficial to analyze a family of trajectories rather than one numerically obtained minimum trajectory. This would provide an opportunity to better analyze the qualitative picture of all possible ways to minimize the performance index. In particular, it is more advantageous to consider Pontryagin extremals over Pontryagin minimum of the problem. The latter usually includes a bang-bang solution which includes only boundary values of controls and does not include intermediate values of controls. But a family of Pontryagin extremals allow us to consider both boundary values and the intermediate values of the controls. For example, the studies have shown that the numerical methods never reveal the existence of intermediate thrust (singular controls) arcs as part of an optimal trajectory. At the same time, as was shown above, the family of extremals would necessarily include such thrust arcs as part of extremal trajectories. This family can provide a best extremal which may not be noticeably different from the minimum, including impulsive trajectories. The problem of turning an elliptical orbit's plane via IT arc may serve as example of this advantage of the family of extremals over a minimum trajectory. The studies have shown that the Pontryagin extremals considered in this work can perform nearly as well as optimal trajectories computed using numerical methods. It is also helpful to note that if a trajectory represents a Pontryagin minimum, then the additional condition of Weierstrass-Erdmann is satisfied at the corner points. In all sections below, the Pontryagin extremals will be considered and called simply as the extremals of the optimal control problem.

2.9 LEGENDRE-CLEBSCH NECESSARY CONDITION

As it has been shown above, the following condition is satisfied on an optimal trajectory [94], [131]:

$$\frac{\partial H}{\partial u_r} = 0.$$

The analogical conditions for α can also be considered, but they will be skipped here. Let $\delta u_r = \tilde{u}_r - u_r$ and the set of values $\tilde{\mathbf{u}} = (\tilde{u}_1, \tilde{u}_2, ..., \tilde{u}_k)$ form a neighborhood of an optimal control $\mathbf{u} = (u_1, u_2, ..., u_k)$ with a radius $\delta_0 > 0$, if

$$\sqrt{\sum_{r=1}^{k}(\tilde{u}_r - u_r)^2} = \sqrt{\sum_{r}\delta u_r^2} < \delta_0 \tag{2.40}$$

for all small δ_0 *a priori* chosen. Consequently, the admissible controls $\tilde{u}_r$ are considered as a neighborhood u_r. If δu_r satisfy Eq. (2.40) and the Weierstrass condition is satisfied, then by expanding

$$H(x, u, t, \lambda, \alpha, \gamma) - H(x, \tilde{u}, t, \lambda, \alpha, \gamma) \geq 0$$

into a Taylor series about u, one can obtain

$$H(x, u, t, \lambda, \alpha, \gamma) - H(x, \tilde{u}, t, \lambda, \alpha, \gamma) = H_u \delta u + \delta u^T H_{uu} \delta u + \ldots \geq 0 \quad (2.41)$$

On an optimal trajectory $H_u = 0$ and the form $\delta \mathbf{u}^T H_{uu} \delta \mathbf{u}$ is a positive semi-definite if

$$H_{uu} \geq 0. \quad (2.42)$$

If the functional is maximized, then the elements H_{uu} must be negative semi-definite. The condition (2.42) is said to be the necessary condition of Legendre-Clebsch and represents a necessary condition for a weak minimum [131]. By leaving the only one sign of inequality in Eq. (2.42), this condition is transformed into the necessary condition of Legendre-Clebsch for a strong minimum [94], [90]. As is known, this condition is a consequence of the maximum principle, and unlike the Weierstrass condition, is not equivalent to this principle [90].

2.10 SECOND DIFFERENTIAL OF EXTENDED FUNCTIONAL

Rewrite the first differential of the extended functional, dK in the form:

$$dK = (G_{x_0} + \boldsymbol{\lambda}_0^T)^T d\mathbf{x}_0 + \mathbf{E}^T d\boldsymbol{\mu} + (G_{x_1} - \boldsymbol{\lambda}_1^T) dx_1 + \mathbf{F}^T d\boldsymbol{\nu} + \Omega_0 dt_0 + \Omega_1 dt_1 +$$
$$\int_{t_0}^{t_1} [(H_x + \dot{\boldsymbol{\lambda}}^T)^T \delta \mathbf{x} + H_u^T \delta \mathbf{u} + H_\alpha^T \delta \boldsymbol{\alpha} + \boldsymbol{\Phi}^T \delta \boldsymbol{\gamma} + (\mathbf{f}^T - \dot{\mathbf{x}}^T)^T \delta \boldsymbol{\lambda}] dt, \quad (2.43)$$

where

$$\Omega_0 = [\Omega(x, u, \mu, t_0, \gamma)]_0 = [G_{t_0} - (g - \boldsymbol{\gamma}^T \Phi + G_{x_0}^T \mathbf{f}]_0,$$
$$\Omega_1 = [\Omega(x, u, \nu, t_1, \gamma)]_1 = [G_{t_1} + g + \boldsymbol{\gamma}^T \Phi + G_{x_1}^T \mathbf{f}]_1.$$

Note that the third term in the expression for Ω_1 can be skipped due to Eq. (2.4). Using the definition of $Y(x, u, \lambda, t, \alpha)$ and the equalities

$$dE = E' dt_0 + E_{x_0}^T \delta \mathbf{x}_0,$$
$$E' = E_{t_0} + E_{x_0}^T \dot{\mathbf{x}}_0$$
$$dF = F' dt_1 + F_{x_1}^T \delta \mathbf{x}_1,$$

$$F' = \frac{dF}{dt_1} = F_{t_1} + F_{x_1}^T \dot{\mathbf{x}}_1, \quad F_{t_1} = \frac{\partial F}{\partial t_1},$$
$$\delta \dot{\mathbf{x}} = f_x^T \delta \mathbf{x} + f_u^T \delta \mathbf{u},$$
$$\Omega_\nu = F',$$

the expression for the second differential, d^2K, ignoring the variations of the second and higher order terms can be given in the form:

$$\begin{aligned}
d^2K = &(\Omega_{x_1}^T dx_1 + \Omega_{u_1}^T d\mathbf{u}_1 + \Omega_\nu^T \delta \boldsymbol{\nu} + \Omega_\gamma^T \delta \boldsymbol{\gamma} + \Omega_{t_1} dt_1) dt_1 + \\
&(\Omega_{x_0}^T dx_0 + \Omega_{u_0}^T d\mathbf{u}_0 + \Omega_\mu^T \delta \boldsymbol{\mu} + \Omega_\gamma^T \delta \boldsymbol{\gamma} + \Omega_{t_0} dt_0) dt_0 + \\
&(G_{x_0 x_0}^T d\mathbf{x}_0 + G_{x_0 \mu}^T d\boldsymbol{\mu} - d\boldsymbol{\lambda}_0^T + E_{x_0}^T d\boldsymbol{\mu}) \delta \mathbf{x}_0^T + \\
&+(G_{x_1 x_1}^T d\mathbf{x}_1 + G_{x_1} dt_1 + G_{x_1 \nu}^T d\boldsymbol{\nu} - d\boldsymbol{\lambda}_1^T)^T \delta \mathbf{x}_1 + \\
&+(F_{x_1}^T d\mathbf{x}_1 + F_{t_1} dt_1) d\boldsymbol{\nu} + \\
&+[(H_x + \dot{\boldsymbol{\lambda}}^T)^T \delta \mathbf{x} + H_u^T \delta \mathbf{u} + (\mathbf{f}^T - \dot{\mathbf{x}}^T)^T \delta \boldsymbol{\lambda}]_1 dt_1 + \\
&+ \int_{t_0}^{t_1} [(H_{xx}^T \delta \mathbf{x} + H_{xu}^T \delta \mathbf{u} + H_{xt}^T \delta t + H_{x\lambda}^T \delta \boldsymbol{\lambda} + \delta \dot{\boldsymbol{\lambda}}^T)^T \delta \mathbf{x} + \\
&+(H_{ux}^T \delta \mathbf{x} + H_{uu}^T \delta \mathbf{u} + H_{ut}^T \delta t + H_{u\lambda}^T \delta \boldsymbol{\lambda} + H_{u\alpha}^T \delta \boldsymbol{\alpha})^T \delta \mathbf{u} + \\
&+(H_{\alpha x}^T \delta \mathbf{x} + H_{\alpha u}^T \delta \mathbf{u} + H_{\alpha t} \delta t + H_{\alpha \lambda}^T \delta \boldsymbol{\lambda} + H_{\alpha\alpha}^T \delta \boldsymbol{\alpha}) \delta \boldsymbol{\alpha}] dt,
\end{aligned}$$

or

$$\begin{aligned}
d^2K = &[\delta \mathbf{x}_0^T \quad dt_0] \begin{bmatrix} G_{x_0 x_0} & \Omega_{x_0}^T \\ \Omega_{x_0} & \Omega' \end{bmatrix} \begin{bmatrix} \delta \mathbf{x}_0 \\ dt_0 \end{bmatrix} \\
&+[\delta \mathbf{x}_1^T \quad dt_1] \begin{bmatrix} G_{x_1 x_1} & \Omega_{x_1}^T \\ \Omega_{x_1} & \Omega' \end{bmatrix} \begin{bmatrix} \delta \mathbf{x}_1 \\ dt_1 \end{bmatrix} + \\
&+ \int_{t_0}^{t_1} [\delta \mathbf{x}^T \quad \delta \mathbf{u}^T] \begin{bmatrix} H_{xx} & H_{xu} \\ H_{ux} & H_{uu} \end{bmatrix} \begin{bmatrix} \delta \mathbf{x} \\ \delta \mathbf{u} \end{bmatrix},
\end{aligned} \tag{2.44}$$

where

$$\Omega_0' = \Omega_{t_0} + E_{x_0} \dot{\mathbf{x}}_0, \quad \Omega_1' = \Omega_{t_1} + E_{x_1} \dot{\mathbf{x}}_1,$$

and the matrix with ij elements $\frac{\partial f_i}{\partial x_j} = \frac{\partial f}{\partial x} = f_x$ is denoted as an $[n \times n]$ matrix, and the matrix with ir elements $\frac{\partial f_i}{\partial u_r} = \frac{\partial f}{\partial u} = f_u$ is denoted as an $[n \times r]$ matrix, and

$$\frac{\partial^2 H}{\partial x_i \partial x_j} = \frac{\partial^2 H}{\partial x \partial x} = H_{xx}$$

$$\frac{\partial^2 H}{\partial x_i \partial u_r} = \frac{\partial^2 H}{\partial x \partial u} = H_{xu}$$
$$\frac{\partial^2 H}{\partial u_r \partial u_k} = \frac{\partial^2 H}{\partial u \partial u} = H_{uu}.$$

2.11 AUXILIARY OPTIMIZATION PROBLEM AND RICCATI EQUATION

For the statement of the auxiliary optimization problem, first obtain the equations that describe the neighboring arcs. The equations of the extremal have the following form:

$$\begin{aligned} \dot{\mathbf{x}}^T &= \mathbf{f}, \\ \dot{\boldsymbol{\lambda}}^T &= -H_x, \\ H_u^T &= 0. \end{aligned} \tag{2.45}$$

After varying these equations, one can obtain

$$\delta\dot{\mathbf{x}} = f_x^T \delta\mathbf{x} + f_u^T \delta\mathbf{u}, \tag{2.46}$$
$$\delta\dot{\boldsymbol{\lambda}} = -H_{xx}^T \delta\mathbf{x} - H_{xu}^T \delta\mathbf{u} - f_x^T \delta\boldsymbol{\lambda}, \tag{2.47}$$
$$H_{ux}^T \delta\mathbf{x} + H_{uu}^T \delta\mathbf{u} + f_u^T \delta\boldsymbol{\lambda} = 0. \tag{2.48}$$

In these equations and below it will be assumed that the matrix H_{uu} is positive definite, $H_{uu} > 0$. From this it follows that $det|H_{uu}| \neq 0$ and the inverse matrix H_{uu}^{-1} exists. Finding δu from Eq. (2.48) in the form

$$\delta\mathbf{u} = -H_{uu}^{-1}(H_{xu}^T \delta\mathbf{x} + f_u^T \delta\boldsymbol{\lambda}) \tag{2.49}$$

and substituting it into Eq. (2.46) and Eq. (2.47), it can be shown that

$$\begin{aligned} \delta\dot{\mathbf{x}} &= A\delta\mathbf{x} - B\delta\boldsymbol{\lambda}, \\ \delta\dot{\boldsymbol{\lambda}} &= C\delta\mathbf{x} - A^T\delta\boldsymbol{\lambda}, \end{aligned} \tag{2.50}$$

where

$$A = f_x - f_u H_{uu}^{-1} H_{xu}^T, \quad B = f_u H_{uu}^{-1} f_u^T, \quad C = H_{xu} H_{uu}^{-1} H_{xu}^T - H_{xx}.$$

The initial and final equations of Eqs. (2.50) can be obtained from the boundary conditions:

$$G_{x0} = -\boldsymbol{\lambda}_0^T, \qquad E = 0,$$

$$G_{x_1} = \boldsymbol{\lambda}_1^T, \qquad F = 0, \qquad \Omega = G_{t_1} + g_1 + \boldsymbol{\gamma}^T \Phi_1 + G_{x_1} f_1 = 0. \tag{2.51}$$

Here the index "1" means that the corresponding functions are computed at final time t_1. Using the variation and differentiation operations in Eq. (2.51), one can obtain the following equalities:

$$\begin{aligned}
\delta\boldsymbol{\lambda}_0 &= -G_{x_0x_0}\delta\mathbf{x}_0 - G_{x_0\mu}\delta\boldsymbol{\mu} - G_{x_0t_0}\delta t_0, \\
\delta E &= E_{x_0}\delta\mathbf{x}_0 + E_{t_0}\delta t_0 = 0, \\
\delta\boldsymbol{\lambda}_1 &= G_{x_1x_1}\delta\mathbf{x}_1 + G_{x_1\nu}\delta\boldsymbol{\nu} + G_{x_1t_1}\delta t_1, \\
\delta F &= F_{x_1}\delta\mathbf{x}_1 + F_{t_1}\delta t_1 = 0, \\
\delta\Omega &= \Omega_{x_1}\delta\mathbf{x}_1 + \Omega_{x_1}\delta\boldsymbol{\nu} + \Omega_{t_1}\delta t_1 + [\Omega_\alpha\delta\boldsymbol{\alpha}]_1 + [\Omega_u\delta\mathbf{u}]_1 = 0.
\end{aligned} \tag{2.52}$$

Note that the necessary condition of a minimum is given by $d^2K \geq 0$. The auxiliary optimization problem states that it is required to find $\delta\mathbf{x}$, $\delta\mathbf{u}$, $\delta\boldsymbol{\lambda}$ and $\delta\boldsymbol{\alpha}$ such that Eqs. (2.50), and (2.52) are satisfied, and the functional d^2K takes a minimum among all possible values.

Suppose we have sets of n solutions of Eqs. (2.50):

$$\delta\mathbf{x} = [\delta x_1, \delta x_2, \ldots, \delta x_n], \qquad \delta\boldsymbol{\lambda} = [\delta\lambda_1, \delta\lambda_2, \ldots, \delta\lambda_n]. \tag{2.53}$$

For $|\delta\mathbf{x}| \neq 0$, we define a matrix R as

$$R = \delta\boldsymbol{\lambda}[\delta\mathbf{x}]^{-1}. \tag{2.54}$$

By eliminating $\delta\dot{\mathbf{x}}$ from Eq. (2.46) and Eq. (2.47) and using $\delta\boldsymbol{\lambda} = R\delta\mathbf{x}$ obtained from Eq. (2.54), one can have

$$[\dot{R} - (Rf_u + H_{xu})H_{uu}^{-1}(f_u^T R + H_{xu}^T) + Rf_x + f_x^T R + H_{xx}]\delta\mathbf{x} = 0,$$

or since $|\delta\mathbf{x}| \neq 0$, the latter equation yields

$$\dot{R} - (Rf_u + H_{xu})H_{uu}^{-1}(f_u^T R + H_{xu}^T) + Rf_x + f_x^T R + H_{xx} = 0 \tag{2.55}$$

This equation is called as the accessory Riccati equation. If $\delta x_i(t)$ are the solutions to the accessory equations with initial conditions $\delta x_i(t_0)$, then the conjugate point on the optimal trajectory exists at $t = t'$ $(t_0 < t < t_1)$ if and only if $|\delta x_i(t')| = 0$. If the solution R matrix exists and if it is finite, then the corresponding accessory problem has a solution which satisfies the necessary conditions (Jacobi condition).

If there is no conjugate point to the initial point t_0 on the trajectory, then $\delta x_j(t) \neq 0$, $\delta \mathbf{x}$ and its inverse $\delta \mathbf{x}^{-1}$ exist, and the Riccati equation, Eq. (2.55) must have a finite solution $R = \delta \boldsymbol{\lambda}[\delta \mathbf{x}]^{-1}$, where $\delta \mathbf{x} \neq 0$ $(\delta x_j(t) \neq 0)$.

2.12 SUFFICIENT CONDITIONS

It is known that if d^2K is strictly positive for all admissible variations $\delta \mathbf{x}(t)$ and $\delta \mathbf{u}(t)$, that satisfy Eqs. (2.46)–(2.48) (or Eq. (2.49)), then the corresponding solutions $x(t)$ and $u(t)$ provide at least a local minimum of K with respect to variations in $x(t)$ and $u(t)$. If the initial and final conditions are fixed, then these variations do not affect the end states. From Eq. (2.44) it follows that

$$d^2K = \Omega_0 + \Omega_1 + \int_{t_0}^{t_1} (\mathbf{x}^T H_{xx}\mathbf{x} + 2\mathbf{x}^T H_{xu}\mathbf{u} + \mathbf{u}^T H_{uu}\mathbf{u})\, dt, \tag{2.56}$$

where

$$\Omega_0 = \Omega(\mathbf{x}_0, t_0) = [\delta \mathbf{x}_0^T \quad dt_0] \begin{bmatrix} G_{x_0x_0} & \Omega_{x_0}^T \\ \Omega_{x_0} & \Omega' \end{bmatrix} \begin{bmatrix} \delta \mathbf{x}_0 \\ dt_0 \end{bmatrix}, \tag{2.57}$$

$$\Omega_1 = \Omega(\mathbf{x}_1, t_1) = [\delta \mathbf{x}_1^T \quad dt_1] \begin{bmatrix} G_{x_1x_1} & \Omega_{x_1}^T \\ \Omega_{x_1} & \Omega' \end{bmatrix} \begin{bmatrix} \delta \mathbf{x}_1 \\ dt_1 \end{bmatrix} \tag{2.58}$$

and

$$\Omega_0' = \Omega_{t_0} + E_{x_0}\dot{\mathbf{x}}_0, \quad \Omega_1' = \Omega_{t_1} + F_{x_1}\dot{\mathbf{x}}_1. \tag{2.59}$$

First, consider the case when the initial and final states are fixed and the end conditions are $\delta \mathbf{x}_0 = \delta \mathbf{x}_1 = 0$. In this chapter we have shown that if there is no point conjugate to the initial point on the extremal, then the Riccati equation (2.55) must possess the finite solution, R. Following [130], consider the derivative

$$\frac{d(\delta \mathbf{x}^T R \delta \mathbf{x})}{dt} = \delta \mathbf{x}^T \dot{R} \delta \mathbf{x} + 2\delta \mathbf{x}^T R \delta \dot{\mathbf{x}}.$$

Substituting Eq. (2.46), it is reduced to

$$\frac{d(\delta \mathbf{x}^T R \delta \mathbf{x})}{dt} = \delta \mathbf{x}^T \dot{R} \delta \mathbf{x} + 2\delta \mathbf{x}^T R f_x \delta \mathbf{x} + 2\delta \mathbf{x}^T R f_u \delta \mathbf{u}. \tag{2.60}$$

As $\delta\mathbf{x}(t_0) = \delta\mathbf{x}(t_1) = 0$, then integration of this equation over (t_0, t_1) results in

$$(\delta\mathbf{x}^T R\delta\mathbf{x})_{t=t_1} - (\delta\mathbf{x}^T R\delta\mathbf{x})_{t=t_0} = 0$$
$$= \int (\delta\mathbf{x}^T \dot{R}\delta\mathbf{x} + 2\delta\mathbf{x}^T Rf_x\delta\mathbf{x} + 2\delta\mathbf{x}^T Rf_u\delta\mathbf{u})dt. \tag{2.61}$$

By adding the second derivative

$$d^2K = \int_{t_0}^{t_1} (\delta\mathbf{x}^T H_{xx}\delta\mathbf{x} + 2\delta\mathbf{x}^T H_{xu}\delta\mathbf{u} + \delta\mathbf{u}^T H_{uu}\delta\mathbf{u})dt \tag{2.62}$$

Eq. (2.61) yields

$$\begin{aligned} d^2K &= \int_{t_0}^{t_1} (\delta\mathbf{x}^T H_{xx}\delta\mathbf{x} + 2\delta\mathbf{x}^T H_{xu}\delta\mathbf{u} \\ &+ \delta\mathbf{u}^T H_{uu}\delta\mathbf{u} + \delta\mathbf{x}^T \dot{R}\delta\mathbf{x} + 2\delta\mathbf{x}^T Rf_x\delta\mathbf{x} + 2\delta\mathbf{x}^T Rf_u\delta\mathbf{u})dt = \\ &= \int (\delta\mathbf{u}^T H_{uu}\delta\mathbf{u} + 2\delta\mathbf{x}^T (H_{xu} + Rf_u)\delta\mathbf{u} + \delta\mathbf{x}^T (H_{xx} + \dot{R} + 2Rf_x)\delta\mathbf{x})dt. \end{aligned} \tag{2.63}$$

It can be shown that

$$\delta\mathbf{x}^T (H_{xu} + Rf_u)\delta\mathbf{u} = \delta\mathbf{u}^T (H_{xu}^T + f_u^T R)\delta\mathbf{x}, \quad (Rf_x)^T = f_x^T R^T. \tag{2.64}$$

Substituting Eq. (2.64) into Eq. (2.63) results in

$$\begin{aligned} d^2K &= \int (\delta\mathbf{u}^T H_{uu}\delta\mathbf{u} + 2\delta\mathbf{x}^T (H_{xu} + Rf_u)\delta\mathbf{u} \\ &+ \delta\mathbf{u}^T (H_{xu}^T + f_u^T R)\delta\mathbf{x} + \delta\mathbf{x}^T (H_{xx} + \dot{R} + 2Rf_x + f_x^T R))dt \end{aligned}$$

Taking into account Eq. (2.55), this integral can be reduced to

$$\begin{aligned} d^2K &= \int (\delta\mathbf{u}^T H_{uu}\delta\mathbf{u} + 2\delta\mathbf{x}^T (H_{xu} + Rf_u)\delta\mathbf{u} + \delta\mathbf{u}^T (H_{xu}^T + f_u^T R)\delta\mathbf{x} + \\ &\delta\mathbf{x}^T (\dot{R}f_u + H_{xu})H_{uu}^{-1}(H_{xu}^T + f_u^T R))dt = \int_{t_0}^{t_1} \Theta^T H_{uu}\Theta dt, \end{aligned} \tag{2.65}$$

where

$$\Theta = \delta\mathbf{u} + H_{uu}^{-1}(H_{xu}^T + f_u^T R)\delta\mathbf{x}. \tag{2.66}$$

In the previous section we assumed that $H_{uu} > 0$ which means that $d^2K = 0$ only if $\Theta = 0$, and otherwise $d^2K > 0$. It will be shown below that $\Theta \neq 0$. Here it is assumed that Eqs. (2.46)–(2.48) have finite solutions, that is there is no conjugate point on the extremal. Consider the following cases:

1. If the initial and final states are fixed, and the initial and final times are also fixed, that is $\Omega_0 = \Omega(t_0) = 0$ and $\Omega_1 = \Omega(t_1) = 0$, then as it follows from Eq. (2.66)

$$\delta\mathbf{u} = -H_{uu}^{-1}(H_{xu}^T + f_u^T R)\delta\mathbf{x}. \tag{2.67}$$

Multiplying Eq. (2.55) on the right by $\delta\mathbf{x}$ and using Eq. (2.67), we obtain

$$\dot{R}\delta\mathbf{x} + (Rf_u + H_{xu})\delta\mathbf{u} + (Rf_x + f_x^T R + H_{xx})\delta\mathbf{x} = 0. \tag{2.68}$$

Use of the expression $\delta\boldsymbol{\lambda} = R\delta\mathbf{x}$ and its derivative,

$$\frac{d\delta\boldsymbol{\lambda}}{dt} = \dot{R}\delta\mathbf{x} + R\delta\dot{\mathbf{x}} = \dot{R}\delta\mathbf{x} + R(f_x^T\delta\mathbf{x} + f_u^T\delta\mathbf{u})$$

and substitution of these expressions into Eq. (2.68) result in

$$\frac{d\delta\boldsymbol{\lambda}}{dt} = -H_{xu}\delta\mathbf{u} + H_{xx}\delta\mathbf{x} + f_x^T\delta\boldsymbol{\lambda}.$$

In the same manner, by substituting Eq. (2.67) into (2.48) we obtain an identity. This is Eq. (2.47), and therefore we conclude that Eq. (2.67) satisfies Eqs. (2.46) and (2.47). But the solution in Eq. (2.67) satisfies the end conditions $\delta\mathbf{x}(t_0) = \delta\mathbf{x}(t_1) = 0$, which lead to $\delta\mathbf{u}(t_0) = \delta\mathbf{u}(t_1) = 0$, $\delta\boldsymbol{\lambda}(t_0) = \delta\boldsymbol{\lambda}(t_1) = 0$. This implies that the end points were conjugate, which is contrary to the assumption that there is no conjugate point on the extremal. Based on this, Eq. (2.67) is not valid over (t_0, t_1) and consequently $\Theta \neq 0$ and $d^2K > 0$.

We conclude that if the initial and final states are fixed, and the initial and final times are also fixed, then $\Omega_0 = 0$ and $\Omega_1 = 0$, and then the sufficiency condition $d^2K > 0$ is satisfied if $H_{uu} > 0$ and there is no conjugate point on the extremal.

On the contrary, if the initial and/or final states are not fixed, and the initial and/or final times are not fixed, thats is $\Omega_0 \neq 0$ and/or $\Omega_1 \neq 0$, then assuming

$$\int_{t_0}^{t_1} \Theta^T H_{uu}\Theta dt > 0,$$

the condition $d^2K > 0$ can be satisfied under the following circumstances (see Eq. (2.56)):

2. If the initial states and/or time are not fixed, and the final states and/or time are fixed, that is $\Omega_0 \neq 0$ and $\Omega_1 = 0$, then from Eq. (2.56) it follows that

$$d^2K = \Omega_0 + \int_{t_0}^{t_1} \Theta^T H_{uu} \Theta dt.$$

From Eq. (2.57) one can find that the condition $d^2K > 0$ is satisfied if and only if all the main minors of the following matrix are non-negative:

$$\begin{bmatrix} G_{x_0x_0} & \Omega_{x_0}^T \\ \Omega_{x_0} & \Omega' \end{bmatrix} \geq 0.$$

3. If the initial states and/or time are fixed, and the final states and/or time are not fixed, that is $\Omega_0 = 0$ and $\Omega_1 \neq 0$, then from Eq. (2.56) one can obtain that

$$d^2K = \Omega_1 + \int_{t_0}^{t_1} \Theta^T H_{uu} \Theta dt$$

From Eq. (2.58) it follows that the condition $d^2K > 0$ is satisfied if and only if all the main minors of the following matrix are non-negative:

$$\begin{bmatrix} G_{x_1x_1} & \Omega_{x_1}^T \\ \Omega_{x_1} & \Omega' \end{bmatrix} \geq 0.$$

4. If the initial and final states are not fixed, and the initial and final times are not fixed, that is $\Omega_0 \neq 0$ and $\Omega_1 \neq 0$, then from Eq. (2.56) one can obtain that

$$d^2K = \Omega_0 + \Omega_1 + \int_{t_0}^{t_1} \Theta^T H_{uu} \Theta dt.$$

From Eq. (2.58) it follows that the condition $d^2K > 0$ is satisfied if and only if all the main minors of the following matrices are non-negative:

$$\begin{bmatrix} G_{x_0x_0} & \Omega_{x_0}^T \\ \Omega_{x_0} & \Omega' \end{bmatrix} \geq 0, \qquad \begin{bmatrix} G_{x_1x_1} & \Omega_{x_1}^T \\ \Omega_{x_1} & \Omega' \end{bmatrix} \geq 0.$$

2.13 JACOBI NECESSARY CONDITION: CONJUGATE POINTS AND THEIR EXISTENCE

In this subsection, the known results on conjugate points of extremals are briefly described, and the expression determining the presence of conjugate points on thrust arcs by employing corresponding analytical solutions is presented. Let $\delta \mathbf{x}_0 = \delta \mathbf{x}(t_0) = 0$ and $H_{uu} > 0$. Then if at some instant $t = \tau$, $(\tau \in (t_0, t_1])$ the matrix $\bar{R}$ is infinite, then from the expression $\delta \mathbf{x} = \bar{R}^{-1} \delta \boldsymbol{\lambda}$ it follows that $\delta \mathbf{x}(\tau) = 0$. The corresponding instant of time, τ is said to be conjugate to t_0 [90], [131]. Detailed research of conjugate points is presented in Milyutin's and Osmolovskii's work [90]. It has been shown that if $\tau \in (t_0, t_1)$, then it is always possible to find such a neighboring arc on which the condition $d^2K < 0$ is satisfied [131]. If $\bar{R}$ is infinite at $\tau = t_0$, then it can be shown that by appropriate selection of $\delta \mathbf{u}$ one can have $d^2K = 0$. Here the conjugate point $\tau = t_1$ is not considered as $\bar{R}(t_1)$ is finite. Consequently, the finiteness of $\bar{R}$ on $[t_0, t_1]$ means the absence of the conjugate points on $(t_0, t_1]$ with respect to the initial point t_0. Note that when $H_{uu} > 0$, the condition of absence of the conjugate points on $(t_0, t_1]$ on an extremal is said to be the strengthened Jacobi condition. The absence of the conjugate points on (t_0, t_1) is said to be the classical Jacobi condition [90], [131].

For the arbitrary variations $\delta \boldsymbol{\lambda}$, from

$$\delta \mathbf{x} = \bar{R}^{-1} \delta \boldsymbol{\lambda} \tag{2.69}$$

it follows that if $\bar{R}$ is finite, then $\delta \mathbf{x} \neq 0$ (that is not all δx_{0i}, $(i = 1, \ldots, n)$) and vice versa.

As it follows from the above described, the condition $\delta \mathbf{x} \neq 0$ (or $\delta \boldsymbol{\lambda} \neq 0$) allows us to determine the presence of the conjugate points on extremal thrust arcs by employing the corresponding analytical solutions of Eqs. (2.45), if such solutions exist. It should be noted that one of the advantages of having the analytical solutions is that these solutions allow us to determine the presence of the conjugate points without requiring the condition $H_{uu} > 0$. Let us prove now the theorem on application of the analytical solutions for thrust arcs in determining the existence of conjugate points on corresponding thrust arcs:

Theorem 2.1. *If $\mathbf{x} = \mathbf{x}(c_1, c_2, \ldots, c_m, t)$ and $\boldsymbol{\lambda} = \boldsymbol{\lambda}(c_1, c_2, \ldots, c_m, t)$, $(m \leq 2n)$, $n = 7$, where c_j, $(j = 1, \ldots, m)$ are the integration constants, represent the analytical solutions for the extremal in Eqs. (2.1)–(2.5) (or Eqs. (2.45)), then the presence of conjugate points on this extremal can be determined by the following* *equalities:*

$$
\begin{aligned}
\delta\mathbf{x} &= \mathbf{N}\left[\frac{\partial\mathbf{x}}{\partial\mathbf{c}}\right]^T, \\
\delta\boldsymbol{\lambda} &= \mathbf{L}\left[\frac{\partial\boldsymbol{\lambda}}{\partial\mathbf{c}}\right]^T,
\end{aligned}
\tag{2.70}
$$

where **N** *and* **L** *are the matrices of constants.*

Proof of Theorem 2.1. Let the analytical solutions of the equations of an extremal given by

$$\dot{x}_i = \frac{\partial H}{\partial \lambda_i}, \tag{2.71}$$

$$\dot{\lambda}_i = -\frac{\partial H}{\partial x_i}, \qquad i = 1, \dots, n. \tag{2.72}$$

$$\frac{\partial H}{\partial u_r} = 0, \qquad r = 1, \dots, k, \tag{2.73}$$

with

$$H = \lambda_i f_i + \gamma_s \Phi_s + g \tag{2.74}$$

contain m ($\leq 2n$) constants of integration:

$$x_i = x_i(t, c_1, \dots, c_m), \quad \lambda_i = \lambda_i(t, c_1, \dots, c_m), \quad u_r = u_r(t, c_1, \dots, c_m), \tag{2.75}$$

with $i = 1, \dots, n$ and $r = 1, \dots, k$. If c is one of these constants, then its variation gives the family of solutions $x = x(t, c)$, $\lambda = \lambda(t, c)$, $u = u(t, c)$. From $\dot{x}_i = f(x, u, t)$ it follows that

$$\frac{\partial \dot{x}_i}{\partial c} = \frac{d}{dt}\left(\frac{\partial x_i}{\partial c}\right) = \frac{\partial f_i}{\partial x_j}\frac{\partial x_j}{\partial c} + \frac{\partial f_i}{\partial u_r}\frac{\partial u_r}{\partial c}, \tag{2.76}$$

$$\frac{\partial \dot{\lambda}_i}{\partial c} = \frac{d}{dt}\left(\frac{\partial \lambda_i}{\partial c}\right) = -\frac{\partial^2 H}{\partial x_i \partial x_j}\frac{\partial x_j}{\partial c} - \frac{\partial^2 H}{\partial x_i \partial x_j}\frac{\partial x_j}{\partial c} - \frac{\partial^2 H}{\partial x_i \partial \lambda_j}\frac{\partial \lambda_j}{\partial c}, \tag{2.77}$$

$$\frac{\partial^2 H}{\partial x_i \partial u_r}\frac{\partial x_i}{\partial c} + \frac{\partial^2 H}{\partial u_r \partial u_k}\frac{\partial u_k}{\partial c} + \frac{\partial^2 H}{\partial u_r \partial \lambda_i}\frac{\partial \lambda_i}{\partial c} = 0, \tag{2.78}$$

where

$$\frac{\partial f_j}{\partial x_i} = \frac{\partial^2 H}{\partial x_i \partial \lambda_j}, \quad \frac{\partial^2 H}{\partial u_r \partial \lambda_i} = \frac{\partial f_i}{\partial u_r}. \tag{2.79}$$

It can be shown that Eqs. (2.76)–(2.78) have the same form as Eqs. (2.46), (2.47), and (2.48). Consequently, the expressions [130]:

$$\begin{aligned} \delta x_i &= \frac{\partial x_i}{\partial c}, \\ \delta \lambda_i &= \frac{\partial \lambda_i}{\partial c}, \\ \delta u_r &= \frac{\partial u_r}{\partial c}, \end{aligned} \tag{2.80}$$

represent the solutions of Eqs. (2.46), (2.47), and (2.48) of the auxiliary optimization problem, which satisfy the necessary conditions of optimality. As these equations are linear and of order $2n$, their solutions on optimal trajectory can be written by using the principle of superposition in the form:

$$\begin{aligned} \delta x_i &= \sum_{j=1}^{m} N_j \frac{\partial x_i}{\partial c_j}, \\ \delta \lambda_i &= \sum_{j=1}^{m} N_j \frac{\partial \lambda_i}{\partial c_j}, \\ \delta u_r &= \sum_{j=1}^{m} U_j \frac{\partial u_r}{\partial c_j}, \end{aligned} \tag{2.81}$$

where c_j, $(j = 1, \ldots, m)$ are the constants of integration in the solutions of the original problem given in Eqs. (2.1)–(2.5), and N_j, L_j, U_j are the constants. The first two equalities in Eq. (2.81) may be used to determine the presence of conjugate points on the extremal of the problem without requiring the condition $H_{uu} > 0$. End of the proof. □

This theorem will be used in the chapters below to determine the existence of the conjugate points on thrust arcs to be studied.

2.14 NECESSARY CONDITION FOR THE CASE WHEN $H_{uu} = 0$

Note that the Riccati equation (2.55) take place under the assumption that $H_{uu} \neq 0$. The H_{uu}, coefficient of the control variable, $\delta \mathbf{u}$, is very important from point of view of practical problems, because in most aerospace problems H_{uu} identically vanishes and Eqs. (2.55) is no longer applicable. Here we consider the case when $H_{uu} = 0$ over a nonzero time interval.

As it has been shown in the previous sections, if the Pontryagin function does not depend on time explicitly, then it represents the first integral of the system. For the accessory problem being considered, this first integral is

$$H = \delta\mathbf{x}^T H_{xx}\delta\mathbf{x} + 2\delta\mathbf{x}^T H_{xu}\delta\mathbf{u} + (f_x\delta\mathbf{x})^T\delta\boldsymbol{\lambda} + (f_u\delta\mathbf{u})^T\delta\boldsymbol{\lambda} = B. \tag{2.82}$$

From Eq. (2.82) we find

$$\delta\mathbf{u} = (2H_{xu}\delta\mathbf{x} + \delta\boldsymbol{\lambda} f_u)^T(B - \delta\mathbf{x}^T H_{xx}\delta\mathbf{x} - \delta\boldsymbol{\lambda} f_x\delta\mathbf{x}), \tag{2.83}$$

where it is assumed that $2H_{xu}\delta\mathbf{x} + \delta\boldsymbol{\lambda} f_u \neq 0$, and from Eq. (2.48) it follows that

$$\delta\boldsymbol{\lambda} = -(f_u^T)^{-1}(H_{xu}^T\delta\mathbf{x}), \qquad \delta\boldsymbol{\lambda} f_u = -H_{xu}^T\delta\mathbf{x}. \tag{2.84}$$

Substituting Eq. (2.84) into Eq. (2.83) leads to

$$\delta\mathbf{u} = (2H_{xu}\delta\mathbf{x} - H_{xu}^T\delta\mathbf{x})^{-1}P = \delta\mathbf{x}^{-1}(2H_{xu} - H_{xu}^T)^{-1}P \tag{2.85}$$

where

$$P = (B - \delta\mathbf{x}^T H_{xx}\delta\mathbf{x} + (H_{xu}^T\delta\mathbf{x})f_u^{-1}f_x\delta\mathbf{x})$$

and where it is assumed that $\delta\mathbf{x} \neq 0$ and $H_{xu} \neq 0$. Then elimination of $\delta\mathbf{u}$ from Eq. (2.46) and (2.47) results in

$$\delta\dot{\mathbf{x}} = f_x\delta\mathbf{x} + f_u(2H_{xu}\delta\mathbf{x} - H_{xu}^T\delta\mathbf{x})^{-1}P, \tag{2.86}$$

$$\delta\dot{\boldsymbol{\lambda}} = -H_{xx}\delta\mathbf{x} - H_{xu}(2H_{xu}\delta\mathbf{x} - H_{xu}^T\delta\mathbf{x})^{-1}P - (H_{xu}^T\delta\mathbf{x})f_u^{-1}f_x. \tag{2.87}$$

Note that Eq. (2.86) and (2.87) are $2n$ first order differential equations. Eq. (2.82) can replace one of these two equations. Eq. (2.87) is associated with two variables, $\delta\boldsymbol{\lambda}$, $\delta\mathbf{x}$, while Eq. (2.86) is first order differential equation with respect to $\delta\mathbf{x}$.

As it has been shown in the previous section, the conjugate point on the extremal trajectory exists at $t = t'$, $(t_0 < t' < t_1)$ if and only if $|\delta x_i(t')| = 0$. If $|\delta x_i(t')| \neq 0$, $H_{xu} \neq 0$, then there is no conjugate point on the trajectory and the vector matrix $\delta\mathbf{x}(t')$ and its inverse $\delta\mathbf{x}^{-1}$ does exist. Consequently Eqs. (2.86) and (2.87) must have finite solutions $\delta\mathbf{x}$ and $\delta\boldsymbol{\lambda}$. The inverse statement is also true, that is if Eq. (2.86) (and/or Eq. (2.87)) has no finite solution over a nonzero time interval, then a conjugate point must exist in the interval and therefore the extremal trajectory is not optimal.

It should be noted that if $H_{xu} = 0$, then in the case when $H_{uu} = 0$, from Eq. (2.48) it follows that $f_u^T\delta\boldsymbol{\lambda} = 0$ or $f_u = 0$. But this contradicts with

$\mathbf{f} = \mathbf{f}(\mathbf{x}, \mathbf{u}, t)$. Therefore, the condition $H_{xu} \neq 0$ has to be satisfied to be consistent with the statement of the original problem and may not be used together with $|\delta\mathbf{x}(t')| \neq 0$.

As mentioned above, for the case $H_{uu} = 0$ one can have

$$H = \delta\mathbf{x}^T H_{xx}\delta\mathbf{x} + 2\delta\mathbf{x}^T H_{xu}\delta\mathbf{u} + \delta\boldsymbol{\lambda}(f_x\delta\mathbf{x} + f_u\delta\mathbf{u}) = B = constant. \tag{2.88}$$

If the initial and final states are fixed, and the initial and final times are also fixed, that is $\Omega_0 = \Omega(t_0) = 0$ and $\Omega_1 = \Omega(t_1) = 0$, then

$$d^2K = \int A dt = \int_{t_0}^{t_1} (\delta\mathbf{x}^T H_{xx}\delta\mathbf{x} + 2\delta\mathbf{x}^T H_{xu}\delta\mathbf{u}) dt \tag{2.89}$$

where

$$A = \delta\mathbf{x}^T H_{xx}\delta\mathbf{x} + 2\delta\mathbf{x}^T H_{xu}\delta\mathbf{u}. \tag{2.90}$$

Substitution of Eq. (2.84) into Eq. (2.88) leads to

$$\begin{aligned}
H &= A + \boldsymbol{\lambda}\delta\dot{\mathbf{x}} = \\
&= \delta\mathbf{x}^T H_{xx}\delta\mathbf{x} + (2\delta\mathbf{x}^T H_{xu} - (f_u^T)^{-1} H_{xu}^T \delta\mathbf{x} f_u)\delta\mathbf{u} - (f_u^T)^{-1} H_{xu}^T \delta\mathbf{x} f_x \delta\mathbf{x} \\
&= \delta\mathbf{x}^T H_{xx}\delta\mathbf{x} + (2\delta\mathbf{x}^T H_{xu} - f_x^T f_u^T)^{-1} (H_{xu}^T \delta\mathbf{x})^T)\delta\mathbf{u} - \delta\mathbf{x}^T f_x^T (f_u^T)^{-1} H_{xu}^T \delta\mathbf{x} \\
&= \delta\mathbf{x}^T H_{xx}\delta\mathbf{x} + (2\delta\mathbf{x}^T H_{xu} - \delta\mathbf{x}^T H_{xu})\delta\mathbf{u} - \delta\mathbf{x}^T f_x^T (f_u^T)^{-1} H_{xu}^T \delta\mathbf{x} \\
&= \delta\mathbf{x}^T (H_{xx} - f_x^T (f_u^T)^{-1} H_{xu}^T)\delta\mathbf{x} + \delta\mathbf{x}^T H_{xu}\delta\mathbf{u} = B = constant.
\end{aligned} \tag{2.91}$$

As it was assumed, at t_0 we have $\eta(t_0) = 0$ and consequently we can find that

$$B = H(t_0) = 0. \tag{2.92}$$

From Eq. (2.91) and (2.92) it follows that

$$\delta\mathbf{u}^T H_{xu}\delta\mathbf{u} = -\delta\mathbf{x}^T (H_{xx} - f_x^T (f_u^T)^{-1} H_{xu}^T)\delta\mathbf{x}. \tag{2.93}$$

Substituting Eq. (2.93) into (2.89), one can show that

$$\begin{aligned}
d^2K &= \int A dt = \int_{t_0}^{t_1} (\delta\mathbf{x}^T H_{xx}\delta\mathbf{x} + 2\delta\mathbf{x}^T H_{xu}\delta\mathbf{u}) dt \\
&= \int [\delta\mathbf{x}^T H_{xx}\delta\mathbf{x} - 2\delta\mathbf{x}^T (H_{xx} - f_x^T (f_u^T)^{-1} H_{xu}^T)\delta\mathbf{x}] dt \\
&= \int [\delta\mathbf{x}^T \left(2 f_x^T (f_u^T)^{-1} H_{xu}^T - H_{xx}\right) \delta\mathbf{x}] dt.
\end{aligned}$$

This implies that if $d^2K > 0$ and then using the previous assumption $\delta\mathbf{x} \neq 0$, we obtain

$$2f_x^T(f_u^T)^{-1}H_{xu}^T - H_{xx} > 0. \tag{2.94}$$

Condition Eq. (2.94) can be rewritten as

$$2(f_u^T)^{-1}H_{xu}^T > (f_x^{-1})^T H_{xx}$$

or

$$2(H_{xu}f_u^{-1}f_x)^T > H_{xx}. \tag{2.95}$$

This inequality can be considered as a new necessary condition of optimality for the case when $H_{uu} = 0$.

We can conclude that if (1) there is no point conjugate to the initial point on the extremal trajectory (Jacobi condition), that is if $|\delta x_i| \neq 0$, $H_{xu} \neq 0$ on (t_0, t_1); and (2) Eq. (2.95) is satisfied, then $d^2K > 0$. Eqs. (2.46)–(2.48) are the necessary conditions and the conditions given in (1) and (2) are sufficient conditions for $d^2K > 0$. These are the results from the accessory optimization problem.

Based on these conclusions, we now can formulate the sufficient conditions for the original problem (where $H_{uu} = 0$) in the following form.

If the conditions (1) and (2) are satisfied, then the extremal is local optimal with respect to small variations that do not change the end conditions.

The condition (1) can be treated independently from the condition (2). The condition (1) can be tested using (see next subsection)

$$\delta x_i(t) = \frac{\partial x_i}{\partial c_j}, \qquad \delta \lambda_i(t) = \frac{\partial \lambda_i}{\partial c_j}, \qquad \delta u_r = \frac{\partial u_r}{\partial c_j},$$

where x_i, λ_i and u_r are the solutions obtained using the necessary conditions of the original problem and termed as the extremals. Then the condition (2.95) can be checked using the equations of the original problem. Note that the sufficiency conditions obtained above are valid in the case $H_{uu} = 0$. If $H_{uu} \neq 0$, then the results of the previous section are used to test the extremals of the problem.

2.15 EXTREMALS WITH CORNER POINTS

2.15.1 Statement of the Problem With Corner Points

Assume that an extremal contains m_1 points of control discontinuities of the first kind at τ_j, $(j = 1, \ldots, m_1)$ and m_2 different thrust arcs that take place on $[t_0, \tau_1]$, $[\tau_1, \tau_2]$, ..., $[\tau_{m_1}, t_f]$. We use index "i" to denote boundary values of time of flight on the thrust arcs, and we use "j" for the corner points. Consider the following problem: let the system state be given by equalities:

$$\dot{x}_i = f_i(\mathbf{x}, u_j, t), \qquad i = 1, \ldots, m_1 + 1; \qquad j = 1, \ldots, m_1. \tag{2.96}$$

The boundary conditions are given in Eq. (2.2) and (2.3), and the control vector is constrained by Eq. (2.4). It is necessary to find $\mathbf{x} = \mathbf{x}(t)$ and $\mathbf{u} = \mathbf{u}(t)$ that minimize the functional of the form:

$$K = \varepsilon J(\mathbf{x}_f, t_f) + \int_{t_0}^{\tau_1} g(\mathbf{x}, \mathbf{u}, t)dt + \int_{\tau_1}^{\tau_2} g(\mathbf{x}, \mathbf{u}, t)dt + +\ldots + \int_{\tau_{m_1}}^{t_f} g(\mathbf{x}, \mathbf{u}, t)dt. \tag{2.97}$$

As it has been shown in the previous subsections, the functional can be extended in the form:

$$K = G(x_0, x_f, t_f, \mu, \nu) + \sum_{i=1}^{m_1+1} X_i, \tag{2.98}$$

where

$$G = G(\mathbf{x}_0, \mathbf{x}_1, t_1, \mu, \nu) = \varepsilon J(\mathbf{x}_1, t_1) + \boldsymbol{\mu}^T\mathbf{E} + \boldsymbol{\nu}^T\mathbf{F},$$
$$H = H(\mathbf{x}, \boldsymbol{\lambda}, \mathbf{u}, t, \boldsymbol{\alpha}, \boldsymbol{\gamma}) = g(\mathbf{x}, \mathbf{u}, t) + \boldsymbol{\lambda}^T\mathbf{f} + \boldsymbol{\gamma}^T\boldsymbol{\Phi},$$
$$X_i = \int_{t_{i-1}}^{t_i} [H(\mathbf{x}, \boldsymbol{\lambda}, \mathbf{u}, t, \boldsymbol{\alpha}, \boldsymbol{\gamma}) - \boldsymbol{\lambda}^T\dot{\mathbf{x}}]dt, \qquad i = 1, \ldots, m_1 + 1,$$
$$\tau_j = t_j, \qquad j = 1, \ldots, m_1, \qquad t_0 < \tau_1 = t_1, \qquad \tau_{m_1} < t_{m_1+1} = t_f.$$

2.15.2 First Differential of the Extended Functional

On the basis of the previous subsections one can show that the first differential of the extended functional, Eq. (2.98) can be reduced to the form:

$$dK = (G_{t_f} + H_f)dt_f + (G_{x_0} + \boldsymbol{\lambda}_0^T)d\mathbf{x}_0 + (G_{x_f} + \boldsymbol{\lambda}_f^T)d\mathbf{x}_f +$$
$$+ (H_{j-} - H_{j+})dt_j + (\boldsymbol{\lambda}_{j-} - \boldsymbol{\lambda}_{j+})dx_j + \sum_{i=1}^{m_1+1} Z_i, \tag{2.99}$$

where

$$Z_i = \int_{i-1}^{i} [(H_x + \dot{\boldsymbol{\lambda}}^T)\delta\mathbf{x} + H_u\delta\mathbf{u} + (\mathbf{f}^T - \dot{\mathbf{x}}^T)\delta\boldsymbol{\lambda}]dt.$$

It can be shown that the computation of the first differential of the extended functional leads to the following necessary conditions for each extremal thrust arc [131]:

$$\dot{x}_i = f_i(\mathbf{x}, u_j, t), \qquad i = 1, ..., m_1 + 1, \quad j = 1, ...m_1, \forall t, t \in [t_{i-1}, t_i].$$
$$\dot{\lambda}_i = -\frac{\partial H}{\partial x_i}, \quad \frac{\partial H}{\partial u_j} = 0, \quad \frac{\partial H}{\partial \alpha_a} = 0, \qquad a = 1, ..., d. \tag{2.100}$$

The following boundary conditions are also satisfied:

$$\boldsymbol{\lambda}_0^T = -G_{x_0}, \quad \boldsymbol{\lambda}_f^T = G_{x_f}, \quad H_0 = G_{t_0}, \quad H_f = -G_{t_f}. \tag{2.101}$$

Besides that, the Weierstrass-Erdmann's condition is satisfied at each corner point [90], [131]:

$$H(\mathbf{x}_{j-}, \mathbf{u}_{j-}, \boldsymbol{\lambda}_{j-}, t_j) = H(\mathbf{x}_{j+}, \mathbf{u}_{j+}, \boldsymbol{\lambda}_{j+}, t_j),$$
$$\boldsymbol{\lambda}_{j-} = \boldsymbol{\lambda}_{j+}, \tag{2.102}$$

where the indexes $j-$ and $j+$ mean that the values of the variables are computed respectively right before and after the corner point with index j.

2.15.3 Additional Condition of Weierstrass-Erdmann

It is helpful to note that if a trajectory represents a Pontryagin minimum, then the additional condition of Weierstrass-Erdmann is satisfied at the corner points, besides Eq. (2.102) [90]:

$$D^j(\hat{H}) \geq 0, \qquad \forall t_j, t_j \in \Theta, \tag{2.103}$$

where Θ represents a set of all corner points (see Fig. 2.2),

$$\hat{H} = \boldsymbol{\lambda}^T\mathbf{f},$$

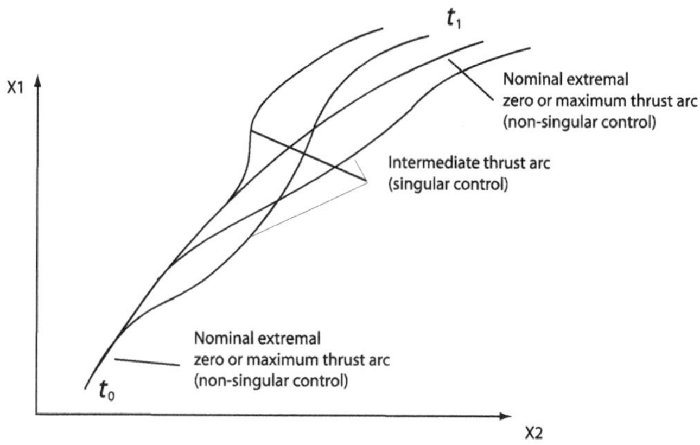

Figure 2.2 Illustration of nominal extremals with corner points, fixed initial conditions and free final conditions.

$$D^j(\hat{H}) = \hat{H}_x^{j+}\hat{H}_\lambda^{j-} - \hat{H}_x^{j-}\hat{H}_\lambda^{j+} + [\hat{H}_t]^j. \qquad (2.104)$$

It can be shown that due to the canonical equations of the extremals and due to the satisfaction of the condition of Weierstrass-Erdmann, the following equality takes place at the corner point:

$$\hat{H}_x^{j+}\hat{H}_\lambda^{j-} - \hat{H}_x^{j-}\hat{H}_\lambda^{j+} = 0,$$

and consequently, Eq. (2.104) is reduced to the form:

$$D^j(\hat{H}) = [\hat{H}_t]^j.$$

If it is assumed that $\hat{H}$ does not explicitly depend on time (as it takes place in most problems of space flight dynamics), then $\hat{H}_t = 0$. Then one can have

$$D^j(\hat{H}) = 0, \qquad \forall t_j,\, t_j \in \Theta. \qquad (2.105)$$

This, in particular, means that for the problems considered in this monograph, the additional condition of Weierstrass-Erdmann is satisfied at a corner point.

As was shown in the Ref. [90], any trajectory that represents a Pontryagin minimum is also a weak extremal. The opposite statement is not necessarily true [90].

2.15.4 Legendre Condition

As it was mentioned above, if a trajectory defined by functions $\mathbf{x}=\mathbf{x}^0$ and $\mathbf{u}=\mathbf{u}^0$ is optimal, then the Legendre condition is satisfied on this trajectory, that is $\hat{H}_{uu}(\mathbf{x}_0, \mathbf{u}_0, \boldsymbol{\lambda}, t) \geq 0$, $\forall t \in [t_0, t_f]$ not including the corner points of the set Θ. Using Eq. (2.102) it can be shown that the following conditions are satisfied at each corner point of an optimal trajectory [90]:

$$\begin{aligned} \hat{H}_{uu}(\mathbf{x}^0_{j-}, \mathbf{u}^0_{j-}, \boldsymbol{\lambda}^0_{j-}, t^j) = \hat{H}^{j-}_{uu} \geq 0, \qquad \forall t_j \in \Theta, \\ \hat{H}_{uu}(\mathbf{x}^0_{j+}, \mathbf{u}^0_{j+}, \boldsymbol{\lambda}^0_{j+}, t^j) = \hat{H}^{j+}_{uu} \geq 0, \qquad \forall t_j \in \Theta. \end{aligned} \tag{2.106}$$

These conditions are said to be the classical Legendre conditions. The sign of strict inequality transforms them into the strengthened Legendre condition. In terms of notations introduced above, the condition $\forall t_j \in \Theta$ can be considered as $\forall t_j,\ t_j \in [t_1, t_{m_1}]$.

First Differential of Extended Functional

By performing the same operations that were presented in the previous subsections, from Eq. (2.99) one can obtain that

$$\begin{aligned} d^2K &= \Omega_0 + \Omega_1 + [\delta\mathbf{x}^T_{j-} \quad dt_j] \begin{bmatrix} G_{x_jx_j} & \Gamma^T_{x_j} \\ \Gamma_{x_j} & \Gamma' + \boldsymbol{\lambda}^T_{j+}(f_{j-} - f_{j+}) \end{bmatrix} \begin{bmatrix} \delta\mathbf{x}_{j-} \\ dt_j \end{bmatrix} \\ &+ \sum_{i=1}^{m_1+1} W_i, \end{aligned} \tag{2.107}$$

where Ω_0 and Ω_1 are determined using Eq. (2.57) and (2.58), and

$$\Gamma = G_{t_j} + g_{j-} + G_{x_j}\mathbf{f}_{j-} - g_{j+} + \boldsymbol{\lambda}^T_{j+}(\mathbf{f}_{j-} - \mathbf{f}_{j+}),$$

$$\Gamma' = \Gamma_{t_j} + \Gamma_{x_{j-}}\dot{\mathbf{x}}_{j-},$$

$$\Omega = G_{t_f} + g_f + G_{x_f}\mathbf{f}_f,$$

$$\Omega' = \Omega_{t_j} + \Omega_{x_j}\dot{\mathbf{x}}_{j-},$$

$$W_i = \int_{t_{i-1}}^{t_1} [\delta\mathbf{x}^T \quad \delta\mathbf{u}^T] \begin{bmatrix} H_{xx} & H_{xu} \\ H_{ux} & H_{uu} \end{bmatrix} \begin{bmatrix} \delta\mathbf{x} \\ \delta\mathbf{u} \end{bmatrix} dt, \qquad i = 1, ..., m_1 + 1,$$

or after some transformations [131],

$$d^2K = \Omega_0 + \Omega_1 + \sum_{i=1}^{m_1+1} \int_{t_{i-1}}^{t_i} \Theta_i^T H_{uu} \Theta_i dt, \tag{2.108}$$

where R satisfies the Riccati equation, Eq. (2.55), Θ satisfies Eq. (2.66), H_{uu}, H_{ux} and $\mathbf{f}_u$ are estimated on the current interval $[t_{i-1}, t_i]$. The studies, conducted in the previous subsections for the analysis of $dK = 0$ and $d^2K > 0$ on extremals which do not contain the conjugate points may also take place in the problem being considered, where the extremals have the conjugate points. This means that in the given problem the necessary and sufficient conditions are considered for each arc. So, for each extremal arc (that is $\forall t,\ t \in [t_{i-1}, t_i],\ i = 1, ..., m_1 + 1$) there take place the following necessary and sufficient conditions for the Pontryagin extremals [90], [131].

Necessary Conditions of Optimality

1. Equations of extremals (the canonical equations and the local optimality conditions, see Eq. (2.45));
2. Weierstrass condition (see Eq. (2.36));
3. Classical Weierstrass-Erdmann conditions for conjugate points (see Eq. (2.102));
4. Classical Legendre-Clebsch conditions ($H_{uu} \geq 0$), excluding the conditions at the corner points (see Eq. (2.42));
5. Legendre conditions for the corner points (see Eq. (2.106));
6. Additional Legendre-Clebsch conditions for the corner points $D^j(\hat{H}) \geq 0$ (see Eq. (2.103));
7. R is finite on $[t_0, t_1)$ (see Eq. (2.69) and (2.81));
8. Boundary conditions (see Eq. (2.101)).

Sufficient Conditions of Optimality

1. Equations of extremals (the canonical equations and the local optimality conditions, see Eq. (2.45));
2. Classical Weierstrass-Erdmann conditions for conjugate points (see Eq. (2.102));
3. Strengthened Legendre-Clebsch conditions ($H_{uu} > 0$), excluding the conditions at the corner points (see Eq. (2.42));
4. Legendre conditions with strict sign of inequality unlike the necessary conditions listed above (see Eq. (2.106));
5. Additional Legendre-Clebsch conditions for the corner points with strict sign of inequality, $D^j(\hat{H}) > 0$ (see Eq. (2.103));
6. R is finite on $[t_0, t_1)$ (see Eq. (2.69));
7. Boundary conditions (see Eq. (2.101)).

As mentioned above, the finiteness of R on all intervals $(t_{i-1}, t_i]$ means the absence of conjugate points on the same interval of time, and this can be verified utilizing the analytical solutions for each arc on extremal.

Consequently, if the above listed conditions are satisfied for each arc, then it is possible to consider the task of constructing the extremals of the problem stated, and it is done in the Chapter 8 of this monograph. The comments on satisfaction of the optimality conditions are presented in the next subsection. In general, the theory developed in this chapter allows us to analyze the extremals constructed on the basis of connecting various thrust arcs in the presence of the corresponding analytical solutions. This shows the importance of having the analytical solutions in constructing the extremals.

2.15.5 How to Satisfy the Necessary and Sufficient Conditions of Optimality?

Necessary Conditions of Optimality

1. The extremals of the problem are obtained in the form of analytical solutions (not excluding the possibility of numerical integration), in particular, by using the canonical equations and the conditions of local optimality (see Eq. (2.45), and subsection "Stationary conditions and admissible arcs");
2. The extremals of the problem must be selected among all possible solutions to the equations of extremals on the basis of the Weierstrass condition, Eq. (2.36) and the classical Legendre-Clebsch conditions, ($H_{uu} \geq 0$), excluding the conditions at the corner points (see Eq. (2.45), and subsection "Stationary conditions and admissible arcs");
3. The classical Legendre-Clebsch conditions at corners, Eq. (2.102) are one of the main aspects of construction of extremals with corner points. The integration constants and unknown parameters (for instance, the thrust angle, true anomaly at corner points) are found from Eq. (2.102), that is the parameters of two connecting arcs are equated to each other at the corner points (see subsection "Method of application of analytical solutions for thrust arcs", Chapter 7);
4. The additional Weierstrass-Erdmann condition at corner points, $D^j(\hat{H}) \geq 0$ (see Eq. (2.103)) can be easily tested by the Weierstrass-Erdmann condition at corner points. As was shown above, for the problems considered in this monograph, $D^j(\hat{H}) = 0$ (see Eq. (2.105));
5. The Legendre condition at corner points, Eq. (2.106) means a positive definiteness of the matrix of the second order partial derivatives from

Hamiltonian for the points located before and after the corner point, that is the satisfaction of this condition on both arcs connected at this corner point. As both of these arcs are the extremals of the problem, and satisfy the classical Legendre-Clebsch conditions, ($H_{uu} \geq 0$), then one can conclude that the Legendre conditions at corner points are satisfied (see subsection "Stationary conditions and admissible arcs");

6. The condition that R is finite on $[t_0, t_1)$ can be tested by using Eq. (2.69) and on the basis of Theorem 2.1 (see subsection "Corner points");
7. The boundary conditions (2.101) must be satisfied in the construction of the maneuver trajectory in the context of the problem considered as some unknowns and Lagrange multipliers are found from these conditions (see the problems analyzed in Chapter 8).

As it follows from the above described, all necessary conditions listed above can be tested in the presence of extremals of the problem. For the problems considered in this monograph, these conditions are satisfied as demonstrated in the previous sections and chapters.

Sufficient Conditions of Optimality

Note that when needed, we will refer to the necessary conditions of optimality in order to test the conditions described below.

1. The equations of extremals allow us to obtain the extremals of the problem (see Eq. (2.45));
2. The classical Weierstrass-Erdmann condition at corner points, Eq. (2.102) need to be satisfied as one of the main parts of the solution process;
3. The strengthened Legendre-Clebsch conditions, ($H_{uu} > 0$), excluding the corner points (see Eq. (2.42)), may be satisfied only for a certain group of extremals ($I_{sp} = const,\ P = const$);
4. As mentioned above, for the problems considered in this monograph, the additional Legendre-Clebsch conditions with strict sign of inequality, $D^j(\hat{H}) > 0$ (see Eq. (2.103)) are not satisfied;
5. The Legendre-Clebsch conditions with strict sign of inequality, Eq. (2.106) may be satisfied only for a certain group of extremals ($I_{sp} = const,\ P = const$);
6. The condition which states that $\bar{R}$ is finite on $[t_0, t_1)$ (see Eq. (2.69)) can be tested by using Eq. (2.69) and the statement of the Theorem 2.1;
7. Satisfaction of the boundary conditions, Eq. (2.101) is one of the final steps of the trajectory construction.

It follows from the above described that for the problems considered in this monograph, the sufficient conditions listed above can be satisfied for only a certain group of extremals, except for the Legendre-Clebsch conditions with strict sign of inequality at the corners.

However, this does not mean that there are no optimal trajectories in the given problem. For the problems considered in this monograph, as it was shown above in the previous subsections, the satisfaction of the sufficient conditions is followed from the satisfaction of the necessary conditions. Below we will focus on only extremals of the variation problem with an alternative statement.

2.16 LAWDEN'S STATEMENT OF THE MAYER'S VARIATIONAL PROBLEM FOR A NEWTONIAN GRAVITY FIELD

The Mayer's variational problem on determination of rocket's optimal trajectories with constant exhaust velocity and limited mass-flow rate in a central Newtonian field was formulated by Lawden in the following form [3], [59]. Let the equations of motion are given in a vector form:

$$\dot{\mathbf{v}} = \frac{c\beta}{m}\mathbf{e} - \frac{\mu}{r^3}\mathbf{r}, \qquad \dot{\mathbf{r}} = \mathbf{v}, \qquad \dot{m} = -\beta, \tag{2.109}$$
$$\dot{\boldsymbol{\lambda}} = -\boldsymbol{\lambda}_r, \qquad \dot{\boldsymbol{\lambda}}_r = \frac{\mu}{r^3}\boldsymbol{\lambda} - 3\frac{\mu}{r^5}(\boldsymbol{\lambda}^T\mathbf{r})\mathbf{r}, \qquad \dot{\lambda}_7 = \frac{c\beta}{m^2}\boldsymbol{\lambda}^T\mathbf{e},$$

where $\mathbf{r}$ is radius vector, $\mathbf{v}$ is velocity vector, $\mathbf{e}$ is unit thrust vector, m is mass, β $(0 \le \beta \le \bar{\beta})$ is mass-flow rate, c is exhaust velocity, $\boldsymbol{\lambda}$ is Lagrange multiplier vector (the primer vector) for velocity vector, $\boldsymbol{\lambda}_r$ is Lagrange multiplier vector for the radius vector, and λ_7 is Lagrange multiplier associated with mass, and β and the components of $\mathbf{e}$ satisfy the constraints:

$$\beta(\bar{\beta} - \beta) - q^2 = 0, \qquad e_1^2 + e_2^2 + e_3^2 = 1.$$

Consider β, q, e_i $(i = 1, 2, 3)$ as piece-wise continuous control functions. For simplicity, all variables are denoted as x_i, $i = 1, \cdots, 7$, that is the components of $\mathbf{v}$ are denoted as x_1, x_2, x_3, the components of $\mathbf{r}$ are denoted as x_4, x_5, x_6, and the mass is denoted as x_7. It is assumed that the initial and final conditions are given in the form:

$$x_j = x_{j0}, \qquad x_n = x_{n1}, \; j = 1, \cdots, 7; \; n = 1, \cdots, l, \quad l < 7.$$

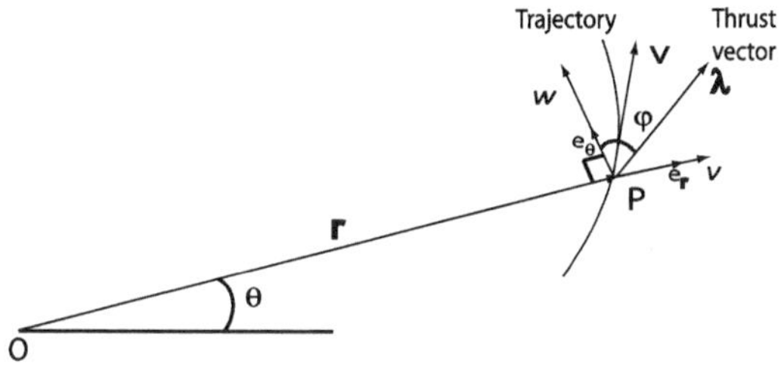

Figure 2.3 The coordinate system introduced by Lawden [3].

It is required to determine β, q, and e_i, so that x_i satisfy the equations of motion, the constraints, the boundary conditions by minimizing the given functional

$$J(x_{l+1,1}, x_{l+2,1}, \cdots, x_{7,1}).$$

It follows from the stationary conditions and as mentioned in the introduction, an optimal trajectory may consist of zero thrust (ZT) arcs, intermediate thrust (IT) are and maximum thrust (MT) arcs. As it follows from the Weierstrass condition (see Fig. 2.3),

$$\mathbf{e} = \frac{\boldsymbol{\lambda}}{\lambda},$$

with the help of which the system Eqs. (2.109) can be rewritten in a canonical form [59]:

$$\dot{x}_i = \frac{\partial H}{\partial \lambda_i}, \qquad \dot{\lambda}_i = -\frac{\partial H}{\partial x_i}, \qquad (i = 1, ..., 7) \tag{2.110}$$

with Hamiltonian

$$H = -\frac{\mu}{r^3}\boldsymbol{\lambda}^T\mathbf{r} + \boldsymbol{\lambda}_r^T\mathbf{v} + \chi\beta,$$

where

$$\chi = \frac{c}{m}\boldsymbol{\lambda}^T e - \lambda_7$$

is the switching function determined by the following conditions [3]:

1. $\chi \le 0$ on ZT arcs, where $\beta = 0$
2. $\chi = 0$ on IT arcs, where $0 < \beta < \bar{\beta}$
3. $\chi \ge 0$ on MT arcs, where $\beta = \bar{\beta}$.

If $q \neq 0$, the intermediate values of β are possible. It has been shown that if the thrust is limited, then the optimal trajectory may not contain IT arcs, excluding some particular cases [42]. An IT arc can be used if an impulsive change of velocity is allowed. At the same time, the sequence of ZT and MT arcs can be realized without impulses at the switching points (the corners). Some examples of the use of IT arcs, not including impulses, have been discussed in the Ref. [67], [68].

2.17 AN ALTERNATIVE STATEMENT OF THE MAYER'S VARIATIONAL PROBLEM

The problem of trajectory optimization for a spacecraft center of mass can be formulated in the context of the variational problem stated in the beginning of this chapter [26]. Among the piece-wise continuous functions $\mathbf{e}$, P, I_{sp} and continuous vector-functions $\mathbf{r}, \mathbf{v}$ and scalar function, m, satisfying the differential equations

$$\begin{aligned} \dot{\mathbf{v}} &= -\frac{\mu}{r^3}\mathbf{r} + \frac{2P}{I_{sp}gm}\mathbf{e}, \\ \dot{\mathbf{r}} &= \mathbf{v}, \\ \dot{m} &= -\frac{2P}{I_{sp}^2 g^2} \end{aligned} \tag{2.111}$$

and the constraints

$$\begin{aligned} \Phi_1 = e_1^2 + e_2^2 + e_3^2 - 1 &= 0, \\ \Phi_2 = P(P_{max} - P) - \gamma^2 &= 0, \\ \Phi_3 = (I_{sp,max} - I_{sp})(I_{sp} - I_{sp,min}) - \eta^2 &= 0, \end{aligned} \tag{2.112}$$

it is necessary to find those which transfer the spacecraft center of mass from the initial configuration

$$E_l(x_{01}, x_{02}, ..., x_{07}) = 0, \qquad l = 1, ..., q_1, \qquad q_1 \leq 7 + 1 \tag{2.113}$$

at initial time t_0 to the final configuration

$$F_m(x_{11}, x_{12}, ..., x_{17}, t_1) = 0, \quad m = 1, ..., q_2, \quad q_2 \leq 7 + 1 \tag{2.114}$$

at final time t_1 while minimizing the functional

$$J = J(x_{0,q_1+1}, x_{0,q_1+2}, ..., x_{0,7}, x_{1,q_2+1}, x_{1,q_2+2}, ..., x_{1,7}, t_0, t_1), \tag{2.115}$$

where $q_1 + q_2 < 2(n+1)$, and η and γ are additional controls, I_{sp} is specific impulse, P is spacecraft power, which can be computed as $P = \frac{1}{2}\beta I_{sp}^2 g^2$, and β is the mass-flow rate. Corresponding $\mathbf{u}(t)$ and $\mathbf{x}(t)$ are called the optimal control and optimal trajectory respectively. The controls can be considered as the components of the control vector: $\mathbf{u} = \left[P, I_{sp}, e_1, e_2, e_3, \gamma, \eta\right]^T$.

The variation problem stated above is given in the form which is alternative to Lawden's statement [3] (see subsection 2.9). The constraints considered in this statement characterize the systems with low and high thrust. This generalizes the variational problem and allows to extend the thrust arcs to the case of low thrust arcs.

Note that this problem statement differs from the Lawden's statement by the equations of the constraints and forms of the initial and final conditions. All results presented in this monograph are described in the context of the alternative problem in Eqs. (2.111)–(2.115) stated above.

2.17.1 Stationary Conditions and Admissible Arcs

Define the Hamiltonian in the form [130], [137]

$$\begin{aligned} H = \boldsymbol{\lambda}(-\frac{\mu}{r^3}\mathbf{r} + \frac{2P}{I_{sp} g m}\mathbf{e}) + \boldsymbol{\lambda}_r \mathbf{v} - \lambda_7 \frac{2P}{I_{sp}^2 g^2} + \mu_1[1 - e_1^2 - e_2^2 - e_3^2] + \\ \mu_2[P(P_{max} - P) - \gamma^2] + \mu_3[(I_{sp,max} - I_{sp})(I_{sp} - I_{sp,min}) - \eta^2], \end{aligned} \quad (2.116)$$

and noting that $u = \left[P, I_{sp}, e_1, e_2, e_3, \gamma, \eta\right]^T$, write the costate equations and the conditions of local maximality in the form [26]:

$$\begin{aligned} \dot{\boldsymbol{\lambda}} &= -\boldsymbol{\lambda}_r, \\ \dot{\boldsymbol{\lambda}}_r &= \frac{\mu}{r^3}\boldsymbol{\lambda} - 3\frac{\mu}{r^5}\mathbf{r}(\boldsymbol{\lambda}\mathbf{r}), \\ \dot{\lambda}_7 &= \frac{2P}{cm^2}\boldsymbol{\lambda}\mathbf{e}, \end{aligned} \quad (2.117)$$

$$\frac{\partial H}{\partial P} = \frac{2}{cm}\boldsymbol{\lambda}\mathbf{e} - \frac{2}{c^2}\lambda_7 + \mu_2(P_{max} - 2P) = 0, \quad (2.118)$$

$$\frac{\partial H}{\partial I_{sp}} = -\frac{2Pg}{c^2 m}\boldsymbol{\lambda}\mathbf{e} + \frac{4Pg}{c^3}\lambda_7 + \mu_3(I_{sp,max} - 2I_{sp} + I_{sp,min}) = 0, \quad (2.119)$$

$$\frac{\partial H}{\partial e_1} = \frac{2P}{cm}\lambda_1 - 2\,\mu_1 e_1 = 0, \quad (2.120)$$

$$\frac{\partial H}{\partial e_2} = \frac{2P}{cm}\lambda_2 - 2\,\mu_1 e_2 = 0, \quad (2.121)$$

$$\frac{\partial H}{\partial e_3} = \frac{2P}{cm}\lambda_3 - 2\,\mu_1 e_3 = 0, \quad (2.122)$$

$$\frac{\partial H}{\partial \gamma} = -2\gamma\ \mu_2 = 0, \tag{2.123}$$

$$\frac{\partial H}{\partial \eta} = -2\eta\ \mu_3 = 0, \tag{2.124}$$

where

$$c = I_{sp}g. \tag{2.125}$$

Any trajectory described by the functions $x = x(t), \lambda = \lambda(t), u = u(t)$, that satisfy Eqs. (2.111), (2.112), (2.117), (2.118)–(2.124), is said to be an *extremal trajectory* or an *extremal* [121]. As is known, an extremal is not necessarily an optimal trajectory, but an optimal trajectory must belong to a family of extremals and represents the best extremal. It should be assumed that a corresponding solution satisfying all conditions of the problem exists [121].

It can easily be seen from Eqs. (2.120)–(2.122) that on thrust arcs $\boldsymbol{\lambda}$ is always collinear to the thrust direction, **e**:

$$\boldsymbol{\lambda} = \mu_1 \frac{cm}{P}\mathbf{e}, \tag{2.126}$$

except for the case when thrust is off. From Eqs. (2.112) and (2.123) it follows that depending on $\gamma,\ \mu_2$, an optimal trajectory may contain the following arcs [137]:
1) Arcs of zero power, $P = 0$ or maximum power $P = P_{max}$, when $\gamma = 0,\ \mu_2 \neq 0$.
2) Arcs of variable power, $0 < P < P_{max}$, when $\gamma \neq 0,\ \mu_2 = 0$. In this case from (2.118) it follows that $\frac{\lambda}{m} I_{sp}g - \lambda_7 = 0$ on a non-zero time interval [3].

Accepting Eqs. (2.112), (2.124) and (2.125), depending on η and μ_3 there may exist the following conditions for specific impulse, I_{sp}:
1) Constant specific impulse, $I_{sp} = I_{sp_{max}}$ or $I_{sp} = I_{sp_{min}}$, when $\eta = 0,\ \mu_3 \neq 0$.
2) Variable specific impulse, $I_{sp_{min}} < I_{sp} < I_{sp_{max}}$, when $\eta \neq 0,\ \mu_3 = 0$. In this case from Eq. (2.119) it follows that $\frac{\lambda}{m} I_{sp}g - 2\lambda_7 = 0$.

If $\gamma \neq 0,\ \mu_2 = 0$, in which P satisfies that condition $0 < P < P_{max}$, then in the case $\eta \neq 0,\ \mu_3 = 0$ from Eqs. (2.118) and (2.119) the equalities $\lambda = 0,\ \lambda_7 = 0$ that take place on a non-zero time interval will follow. However, these equalities are inadmissible as they contradict the condition of collinearity of $\boldsymbol{\lambda}$ to the thrust direction, **e** on thrust arcs, that is they contradict to Eq. (2.126). Therefore, if $\gamma \neq 0,\ \mu_2 = 0$ then the conditions $\eta = 0,\ \mu_3 \neq 0$, which mean constant vales of I_{sp} will take place: $I_{sp} = I_{sp_{max}}$ or $I_{sp} = I_{sp_{min}}$.

Similarly, if $\eta \neq 0,\ \mu_3 = 0$, in which I_{sp} satisfies the condition $I_{sp_{min}} < I_{sp} < I_{sp_{max}}$, then when $\gamma \neq 0,\ \mu_2 = 0$ from Eqs. (2.118) and (2.119) one can obtain the equalities $\lambda = 0,\ \lambda_7 = 0$, which, as it has been shown above, is inadmissible. Consequently, if $\eta \neq 0,\ \mu_3 = 0$ one can accept $\gamma = 0,\ \mu_2 \neq 0$, that is $P = 0$ or $P = P_{max}$.

From the above described one can conclude that the following conditions may take place:

1) $\gamma = 0,\ \mu_2 \neq 0,\ \eta = 0,\ \mu_3 \neq 0$. These conditions correspond to the case $P = 0$ or $P = P_{max}$ with $I_{sp} = I_{sp_{max}}$ or $I_{sp} = I_{sp_{min}}$.

2) $\gamma = 0,\ \mu_2 \neq 0,\ \eta \neq 0,\ \mu_3 = 0$. These conditions correspond to $P = 0$ or $P = P_{max}$ with I_{sp}, which satisfies the inequality $I_{sp_{min}} < I_{sp} < I_{sp_{max}}$.

3) $\gamma \neq 0,\ \mu_2 = 0,\ \eta = 0,\ \mu_3 \neq 0$. In this case the power satisfies the condition $0 < P < P_{max}$, and one can also have $I_{sp} = I_{sp_{max}}$ or $I_{sp} = I_{sp_{min}}$.

Note that in the case when $\gamma \neq 0,\ \mu_2 = 0,\ \eta \neq 0,\ \mu_3 = 0$, that is when P and I_{sp} take intermediate values, corresponding to $0 < P < P_{max}$ and $I_{sp_{min}} < I_{sp} < I_{sp_{max}}$, the conditions (2.118) and (2.119) yield again to the equalities $\lambda = 0,\ \lambda_7 = 0$. Therefore, this case is not considered further.

Now, based on the results obtained above, consider the Weierstrass condition. This condition can be written in the form:

$$\sum \lambda_i f_i(x_i, u_s, t) \geq \sum \lambda_i f_i(x_i, \tilde{u}_s, t)$$

or

$$\frac{2P}{I_{sp}^2}\left(\frac{\boldsymbol{\lambda}\mathbf{e}}{m} I_{sp} g - \lambda_7\right) \geq \frac{2\tilde{P}}{\tilde{I}_{sp}^{\,2}}\left(\frac{\boldsymbol{\lambda}\tilde{\mathbf{e}}}{m} \tilde{I}_{sp} g - \lambda_7\right). \tag{2.127}$$

This yields to the following:

- If $P = 0$, then $\frac{\boldsymbol{\lambda}\tilde{\mathbf{e}}}{m}\tilde{I}_{sp} g - \lambda_7 \leq 0$ for any $\tilde{I}_{sp}$ and $\tilde{\mathbf{e}}$. As the maximum of $\boldsymbol{\lambda}\tilde{\mathbf{e}}$ can be obtained when $\lambda = \boldsymbol{\lambda}\tilde{\mathbf{e}}$, then $\chi = \frac{\lambda}{m} I_{sp} g - \lambda_7 \leq 0$ along the arcs of zero power.
- If $0 < P < P_{max}$, then $\tilde{P}$ can be less or greater than P. If $\tilde{I}_{sp} = I_{sp} = const$ then the condition (2.127) may be satisfied when $\frac{\boldsymbol{\lambda}\mathbf{e} I_{sp} g}{m} - \lambda_7 = 0$ and $\frac{\boldsymbol{\lambda}\tilde{\mathbf{e}} \tilde{I}_{sp} g}{m} - \lambda_7 = 0$ for all admissible $\tilde{\mathbf{e}}$. Then accepting that $\tilde{\mathbf{e}} = \mathbf{e}$, one can have $\frac{\lambda}{m} I_{sp} g - \lambda_7 = 0$ along the arcs of variable power with constant I_{sp}. Note that in this case, as it has been shown above, the variable values of I_{sp} do not satisfy the stationary conditions and that is they are not considered here.

- If $P = P_{max}$, then accepting that $\tilde{P} = P$ and $\tilde{I_{sp}} = I_{sp}$, from Eq. (2.127) one can have

$$\boldsymbol{\lambda e} \geq \boldsymbol{\lambda \tilde{e}}. \tag{2.128}$$

If $\tilde{\mathbf{e}} = \mathbf{e}$, then noting that $P_{max} \geq \tilde{P}$, with $\tilde{I_{sp}} = I_{sp}$ we obtain

$$\frac{\boldsymbol{\lambda e}}{m} I_{sp} g - \lambda_7 \geq 0. \tag{2.129}$$

Eq. (2.128) is satisfied for all admissible $\tilde{\mathbf{e}}$, if $\boldsymbol{\lambda e}$ takes its maximum value, that is if $\mathbf{e}$ is directed along $\boldsymbol{\lambda}$, or if $\lambda = \boldsymbol{\lambda \tilde{e}}$. Then the condition (2.129) is rewritten as

$$\frac{\lambda}{m} I_{sp} g - \lambda_7 \geq 0. \tag{2.130}$$

Eqs. (2.128) and (2.130) are enough to satisfy Eq. (2.127).
If $\tilde{I_{sp}} \neq I_{sp}$, then taking into account that $\tilde{P} = P$ and $\tilde{\mathbf{e}} = \mathbf{e}$, from Eq. (2.127) one can have

$$\frac{\boldsymbol{\lambda e}}{m}\left(\frac{1}{I_{sp} g} - \frac{1}{\tilde{I_{sp}} g}\right) - \lambda_7\left(\frac{1}{I_{sp}^2 g^2} - \frac{1}{\tilde{I_{sp}^2} g^2}\right) \geq 0$$

or

$$\frac{(\tilde{I_{sp}} - I_{sp})}{\tilde{I_{sp}^2} I_{sp}^2}\left[\frac{\boldsymbol{\lambda e}}{m} \tilde{I_{sp}} I_{sp} g + \lambda_7 (I_{sp} + \tilde{I_{sp}})\right] \geq 0. \tag{2.131}$$

It is seen from the equation for λ_7 of Eqs. (2.117) that if $\mathbf{e}$ is along $\boldsymbol{\lambda}$, then λ_7 is a monotonically increasing function on zero power arcs, and $\lambda_7 = const$. Assuming that $\lambda_7 \geq 0$, one can show that Eq. (2.131) is satisfied if

$$\frac{\boldsymbol{\lambda e} I_{sp} g}{m} - \lambda_7 \geq \frac{\boldsymbol{\lambda \tilde{e}} \tilde{I_{sp}} g}{m} - \lambda_7$$

or

$$I_{sp} \boldsymbol{\lambda e} \geq \tilde{I_{sp}} \boldsymbol{\lambda \tilde{e}}.$$

This condition is satisfied if and only if $\lambda = \boldsymbol{\lambda e} \geq \boldsymbol{\lambda \tilde{e}}$. Consequently, in the case when $P > \tilde{P}$, $\tilde{\mathbf{e}} \neq \mathbf{e}$, $\tilde{I_{sp}} > 0$, Eq. (2.127) is satisfied if $\frac{P}{I_{sp}^2} > \frac{\tilde{P}}{\tilde{I_{sp}^2}}$, $\lambda = \boldsymbol{\lambda e} \geq \boldsymbol{\lambda \tilde{e}}$ and $\frac{\lambda I_{sp} g}{m} - \lambda_7 \geq 0$.
Summarizing he above described, one can conclude that if $P = P_{max}$ and $I_{sp\,min} \leq I_{sp} \leq I_{sp\,max}$, then Eq. (2.127) can be satisfied when $\frac{\lambda I_{sp} g}{m} - \lambda_7 \geq 0$. It can be easily seen that this case covers two thrust arcs: 1) $P = P_{max}$ and

$I_{sp} = I_{sp_{max}}$ if $I_{sp} = I_{sp_{min}}$; 2) $P = P_{max}$ and $I_{sp_{min}} < I_{sp} < I_{sp_{max}}$. In particular, the case $P = P_{max}$ and $I_{sp} = const$, where $I_{sp} \neq I_{sp_{max}}$, $I_{sp} \neq I_{sp_{min}}$ is related to the second case, but can be analyzed similar to the first case and therefore, will be skipped from further consideration. Besides that, the case $P = P_{max}$ and $I_{sp} = I_{sp_{min}}$ or $I_{sp} = I_{sp_{max}}$ can describe the maximum thrust arcs that can be obtained from the problem with Lawden's statement [3].

So, based on the stationary conditions and Weierstrass condition, one may conclude that

1) on thrust arcs the thrust vector is along the primer vector $\boldsymbol{\lambda}$;
2) If $\chi = \frac{\lambda}{m} I_{sp} g - \lambda_7$, then it is necessary that

- $\chi \leq 0$ on arcs with zero power, where $P = 0$.
- $\chi = 0$ on thrust arcs where $0 < P < P_{max}$ and $I_{sp} = I_{sp_{max}}$ or $I_{sp} = I_{sp_{min}}$.
- $\chi \geq 0$ on thrust arcs where $P = P_{max}$ and $I_{sp_{min}} < I_{sp} < I_{sp_{max}}$.
- $\chi \geq 0$ on thrust arcs where $P = P_{max}$ and $I_{sp} = I_{sp_{max}}$ or $I_{sp} = I_{sp_{min}}$.

Here χ is known as the switching function [3]. As it was mentioned in the introduction, the thrust arcs include the low and high thrust arcs. The definitions of these arcs are given below [2], [11], [7], [5].

Definition of a low thrust arc. A segment of a trajectory on which the ratio of thrust to weight is $\frac{T}{W} \ll 1$ and the specific impulse is $I_{sp} > 1000$ in the presence of constraints $P \leq P_{max}$ and $I_{sp,min} \leq I_{sp} \leq I_{sp,max}$ is said to be a low thrust arc.

Definition of a high thrust arc. A segment of a trajectory on which the ratio of thrust to weight is $\frac{T}{W} \geq 1$ and the specific impulse is $I_{sp} \leq 1000$ in the presence of constraints $P \leq P_{max}$ and $I_{sp,min} \leq I_{sp} \leq I_{sp,max}$ is said to be a high thrust arc.

It can be easily seen that the constrains on power and specific impulse mentioned above, in principle, include the constraints of the form $0 \leq \beta \leq \beta_{max}$, $I_{sp} = const$, that is, as known, characterize the chemical propulsion systems.

2.17.2 Classification of Extremal Thrust Arcs

The analysis of the results obtained in the previous subsections allows us to classify the thrust arcs depending on the values of P and I_{sp}. This classification can be given based on the definitions of thrust and power, that is $T = \beta I_{sp} g = \frac{2P}{I_{sp} g}$ and $P = \frac{1}{2} \beta I_{sp}^2 g^2$, definitions of low and high thrust. Note that in the general analytical formulation it is not necessary to divide the systems of low and high thrust as they can be described by the same parameters [7]. Besides that, as will be shown below, both types of thrust arcs can be divided into a minimum, intermediate and maximum thrust arcs depending on the thrust magnitude [138]. The classification of possible thrust arcs can be given in the following form:

- Arcs of zero thrust (power): $T=0,\ (P=0)$
- Thrust arcs:
  - arcs of minimum thrust $P=P_{max}$ and $I_{sp}=I_{sp_{max}}$:

$$T_{min}=\frac{2P_{max}}{I_{sp_{max}}g}$$

  - arcs of intermediate thrust:
    * The case when $P=P_{max}$ and $I_{sp}\neq const,\ I_{sp}=I_{sp_{min}}<I_{sp}<I_{sp_{max}}$:

$$T=\frac{2P_{max}}{I_{sp}g}$$

    * The case when $P<P_{max}$ and $I_{sp}=const,\ I_{sp}\geq I_{sp_{min}},\ I_{sp}\leq I_{sp_{max}}$:

$$T=\frac{2P}{I_{sp}g}$$

  - arcs of maximum thrust $P=P_{max}$ and $I_{sp}=I_{sp_{min}}$:

$$T_{max}=\frac{2P_{max}}{I_{sp_{min}}g}.$$

All thrust arcs described above can represent low or high thrust arcs depending on the ratio of thrust to weight and specific impulse. However, it can be defined *a priori* that a thrust arc characterized by the first case of the intermediate thrust arcs represents low-thrust arc [7], [5]. As it can be seen from this classification that the arcs of minimum thrust and maximum thrust are described in the same way, that is by constant values of power and specific impulse. Therefore, if a maximum thrust arc is considered, the qualitative results will also be applicable to a minimum thrust arc excluding the numerical values of the parameters. Both thrust arcs can be similar to the thrust arcs in the context of the problem with Lawden's statement [3]. The second case of intermediate thrust arcs in the classification may be similar to the arcs of intermediate thrust obtained in the Lawden's theory [3].

2.17.3 Functionals for Low and High Thrust Arcs

Consider the question about the functionals. In the case of high thrust systems we considered the minimization of the integral $\int a_t(t)dt$, which is equivalent to the velocity in the vacuum or the expression $m_0 - m_f$. Here a_t is the thrust acceleration. In the case of low-thrust systems, the functional to be minimized is the integral $\int a_t^2(t)dt$ [2]. Excluding c from $P = \frac{1}{2}\beta c^2$ and $a_t = \frac{\beta c}{m}$, one can have

$$\frac{a_t^2}{2P} = \frac{\beta}{m^2} = \frac{d}{dt}\left(\frac{1}{m}\right) \quad \rightarrow \quad \frac{m_0}{m_1} = 1 + m_0 \int_{t_0}^{t_1} \frac{a_t^2(t)}{2P(t)} dt.$$

As it follows from this, in particular, regardless of the type of the propulsion systems, the minimization of the integral $\int a_t^2(t)dt$ while holding the power at its maximum value yields the minimization of the ratio $\frac{m_0}{m_1}$. Consequently, the selection of $P = P_{max}$ is effective with respect to the intermediate values of power for any thrust arcs when $\int a_t^2(t)dt$ is minimized.

From the other hand, it can be shown that for $c = I_{sp}g = const$ one can have

$$\mathcal{J} = \int_{t_0}^{t_1} a_t(t)dt = \int \frac{c\beta}{m} dt = c \ln \frac{m_0}{m_1},$$

which is equivalent to the minimization of $m_0 - m_f$ [91]. From what has been described above, it follows that, in particular, when $P = P_{max}$, the minimization of the integral $\int a_t(t)dt$ with $c = const$ is equivalent to the minimization of $\int a_t^2(t)dt$. Consequently, one can conclude that the low and high thrust arcs can be considered under the same problem statement.

2.17.4 Legendre-Clebsch Necessary Condition

As it has been shown in the previous subsections, one of the sufficient conditions is expressed in terms of main minors of the matrix of second-order partial derivatives of the Hamiltonian with respect to controls, $H_{uu} > 0$. The elements of this matrix are defined as

$$[H_{uu}] = \frac{\partial^2 H}{\partial u_i \partial u_j}, \quad (i, j = 1, \ldots, 7)$$

where H is determined by Eq. (2.116) and $\mathbf{u} = \left[P, I_{sp}, e_1, e_2, e_3, \gamma, \eta\right]^T$. Compute the elements of H_{uu} in the form:

$$\frac{\partial^2 H}{\partial P^2} = -2\mu_2, \quad \frac{\partial^2 H}{\partial P \partial I_{sp}} = -2\frac{2g\vec{e}\vec{\lambda}}{mc^2} + 4\frac{g\lambda_7}{c^3},$$

$$\frac{\partial^2 H}{\partial P \partial e_1} = 2\frac{\lambda_1}{mc}, \quad \frac{\partial^2 H}{\partial P \partial e_2} = 2\frac{\lambda_2}{mc}, \quad \frac{\partial^2 H}{\partial P \partial e_3} = 2\frac{\lambda_3}{mc},$$

$$\frac{\partial^2 H}{\partial P \partial \gamma} = \frac{\partial^2 H}{\partial P \partial \eta} = 0,$$

$$\frac{\partial^2 H}{\partial I_{sp} \partial P} = \frac{\partial^2 H}{\partial P \partial I_{sp}}, \quad \frac{\partial^2 H}{\partial I_{sp}^2} = 4\frac{Pg\vec{e}\vec{\lambda}}{mc^3} - 12\frac{Pg^2\lambda_7}{c^4} - 2\mu_3,$$

$$\frac{\partial^2 H}{\partial I_{sp} \partial e_1} = -2\frac{gP\lambda_1}{mc^2}, \quad \frac{\partial^2 H}{\partial I_{sp} \partial e_2} = -2\frac{gP\lambda_2}{mc^2}, \quad \frac{\partial^2 H}{\partial I_{sp} \partial e_3} = -2\frac{gP\lambda_3}{mc^2},$$

$$\frac{\partial^2 H}{\partial I_{sp} \partial \gamma} = \frac{\partial^2 H}{\partial I_{sp} \partial \eta} = 0,$$

$$\frac{\partial^2 H}{\partial e_1 \partial P} = \frac{\partial^2 H}{\partial P \partial e_1}, \quad \frac{\partial^2 H}{\partial e_1 \partial I_{sp}} = \frac{\partial^2 H}{\partial I_{sp} \partial e_1},$$

$$\frac{\partial^2 H}{\partial e_1^2} = 2\mu_1, \quad \frac{\partial^2 H}{\partial e_1 \partial e_2} = \frac{\partial^2 H}{\partial e_1 \partial e_3} = \frac{\partial^2 H}{\partial e_1 \partial \gamma} = \frac{\partial^2 H}{\partial e_1 \partial \eta} = 0,$$

$$\frac{\partial^2 H}{\partial e_2 \partial P} = \frac{\partial^2 H}{\partial P \partial e_2}, \quad \frac{\partial^2 H}{\partial e_2 \partial I_{sp}} = \frac{\partial^2 H}{\partial I_{sp} \partial e_2},$$

$$\frac{\partial^2 H}{\partial e_2 \partial e_1} = \frac{\partial^2 H}{\partial e_1 \partial e_2}, \quad \frac{\partial^2 H}{\partial e_2^2} = 2\mu_1, \quad \frac{\partial^2 H}{\partial e_2 \partial e_3} = \frac{\partial^2 H}{\partial e_2 \partial \gamma} = \frac{\partial^2 H}{\partial e_2 \partial \eta} = 0,$$

$$\frac{\partial^2 H}{\partial e_3 \partial P} = \frac{\partial^2 H}{\partial P \partial e_3}, \quad \frac{\partial^2 H}{\partial e_3 \partial I_{sp}} = \frac{\partial^2 H}{\partial I_{sp} \partial e_3},$$

$$\frac{\partial^2 H}{\partial e_3 \partial e_1} = \frac{\partial^2 H}{\partial e_1 \partial e_3}, \quad \frac{\partial^2 H}{\partial e_3 \partial e_2} = 0, \quad \frac{\partial^2 H}{\partial e_3^2} = 2\mu_1, \quad \frac{\partial^2 H}{\partial e_3 \partial \gamma} = \frac{\partial^2 H}{\partial e_3 \partial \eta} = 0,$$

$$\frac{\partial^2 H}{\partial \gamma \partial P} = \frac{\partial^2 H}{\partial \gamma \partial I_{sp}} = \frac{\partial^2 H}{\partial \gamma \partial e_1} = \frac{\partial^2 H}{\partial \gamma \partial e_2} = \frac{\partial^2 H}{\partial \gamma \partial e_3} = \frac{\partial^2 H}{\partial \gamma \partial \eta} = 0, \quad \frac{\partial^2 H}{\partial \gamma^2} = -2\mu_2,$$

$$\frac{\partial^2 H}{\partial \eta \partial P} = \frac{\partial^2 H}{\partial \eta \partial I_{sp}} = \frac{\partial^2 H}{\partial \eta \partial e_1} = \frac{\partial^2 H}{\partial \eta \partial e_2} = \frac{\partial^2 H}{\partial \eta \partial e_3} = \frac{\partial^2 H}{\partial \eta \partial \gamma} = 0, \quad \frac{\partial^2 H}{\partial \eta^2} = -2\mu_3.$$

The main minors of the matrix are the determinants of the matrices obtained by excluding the 1-st row and the first column, then 1-st row and the 2-nd column, etc. The determinant of the matrix A is [94]

$$det[H_{uu}] = h_{11}C_{11} + h_{12}C_{12} + ... + h_{17}C_{17},$$

where $h_{1,i}$ the element of the 1-st row and the i-th column of H_{uu}, and C_{1i} is the $1i$-th co-factor. Note that the determinant of a matrix containing a row or a column consisting of only zeros, is equal to zero.

Consider the elements of the matrix H_{uu}. If $\mu_2 = 0$, $\mu_3 \neq 0$, or $\mu_2 \neq 0$, $\mu_3 = 0$, then for the last or the previous row one can have $h_{6j} = \frac{\partial^2 H}{\partial \gamma \partial u_j} = 0$, $(j = 1, ..., 7)$ or $h_{7j} = \frac{\partial^2 H}{\partial \eta \partial u_j} = 0$, $(j = 1, ..., 7)$. This will mean that all main minors of H_{uu} are zeros, which indicates extremality of the trajectories that correspond to the cases: $\gamma \neq 0,\ \eta = 0,\ \gamma = 0,\ \eta \neq 0$.

If $\mu_2 = 0,\ \mu_3 \neq 0$, then from Eq. (2.123) and (2.124) it follows that $\gamma \neq 0,\ \eta = 0$, which corresponds to the case when $0 < P < P_{max}$, $I_{sp} = I_{sp,min}$ or $I_{sp} = I_{sp,max}$, that is to the motions with constant I_{sp} and variable P. If $\mu_2 \neq 0,\ \mu_3 = 0$, then from Eq. (2.123) and (2.124) possible to obtain $\gamma = 0$, $\eta \neq 0$, which means that $P = P_{max}$ or $P = 0$ and $I_{sp,min} < I_{sp} < I_{sp,max}$, that is the motion with variable I_{sp} and maximum P. In both cases $H_{uu} = 0$.

If $\mu_2 \neq 0,\ \mu_3 \neq 0$, then from the elements of H_{uu} it follows that the strict positiveness of the main minors of this matrix is provided by appropriate determination of $\mu_1, \mu_2, \mu_3, \lambda_1, \lambda_2, \lambda_3, \lambda_7, e_1, e_2, e_3$. If $\lambda_1, \lambda_2, \lambda_3, \lambda_7$ can be found by integration of the canonical system, then $\mu_1, \mu_2, \mu_3, e_1, e_2, e_3$ are found from Eqs. (2.118)–(2.122), including the condition $e_1^2 + e_2^2 + e_3^2 = 1$. Note that the general solution of the canonical system corresponding to the case when $\mu_2 \neq 0,\ \mu_3 \neq 0$, are remaining unknown. From these conditions, Eqs. (2.123) and (2.124) one can obtain $\gamma = 0,\ \eta = 0$, which mean the motion with $P = P_{max}$, $I_{sp} = I_{sp,min}$, or $I_{sp} = I_{sp,max}$. Consequently, the following theorem has been proved:

Theorem 2.2. *In the variation problem described in Eqs. (2.111)–(2.115) with constraints on power and specific impulse, the optimal arcs are those on which* $P = P_{max}$ *and* $I_{sp} = I_{sp,min}$, *or* $I_{sp} = I_{sp,max}$.

In the case of a motion with $P = P_{max}$, $I_{sp} = I_{sp,min}$, or $I_{sp} = I_{sp,max}$, the mass-flow rate is constant, and the mass is a linear function of time. One can conclude that among the extremals of the variational problem the Legendre-Clebsch condition, $H_{uu} > 0$, can be satisfied only by those thrust arcs on which P and I_{sp} are constants. Satisfaction of this condition allows to distinguish the thrust arcs which are candidates for optimality, but at the same time this limits the domain of admissible values of the parameters of the problem. For example, Theorem 2.2 excludes an extremal with variable I_{sp} from optimal trajectory. In this monograph, both optimal and extremal arcs are considered. On the other hand, the problem considered here is a model problem of space flight dynamics as the effects of the secondary celestial bodies, atmospheric forces, solar pressure and others are not considered here. In this sense, all extremal and optimal trajectories in the context of this problem can be considered as an approximation to the actual trajectory, and can serve as the reference trajectories in the guidance problem. The fact that the tests of the sufficient conditions, in particular, the

Legendre-Clebsch condition imposes limitations to the domain of the admissible parameters of the problem, and the problem being considered here is a model problem yields to the conclusion that separation of optimal trajectories among all extremals of the problem does not represent a practical interest and is not preferred in finding the reference trajectories. Therefore, the main attention will be given to the determination of extremal trajectories, that is those solutions which satisfy all necessary conditions of optimality, the transversality conditions, and the boundary conditions.

2.17.5 Canonical System of Equations

The equations of motion in the context of the optimal control problem, Eqs. (2.111) and (2.117), taking into account Eqs. (2.120)–(2.122) have the following form [7], [59], [60], [62], [67]:

$$\begin{aligned} \dot{\mathbf{v}} &= \frac{2P}{I_{sp}gm}\frac{\boldsymbol{\lambda}}{\lambda} - \frac{\mu}{r^3}\mathbf{r}, \\ \dot{\mathbf{r}} &= \mathbf{v}, \\ \dot{m} &= -\frac{2P}{I_{sp}^2 g^2}, \end{aligned} \tag{2.132}$$

with the Pontryagin function

$$H = -\frac{\mu}{r^3}(\boldsymbol{\lambda}\mathbf{r}) + (\boldsymbol{\lambda}_r\mathbf{v}) + \chi\frac{2P}{I_{sp}^2 g^2}, \tag{2.133}$$

where $\chi = \frac{\lambda c}{m}I_{sp}g - \lambda_m$ is the switching function, $c = I_{sp}g$. Note that Eqs. (2.133) is called the Pontryagin function, not Hamiltonian, because this expression explicitly includes the control variables I_{sp} and P. It is called Hamiltonian if this expression includes the state and costate variables, not the control variables. If the state vector-function is $\mathbf{x} = \mathbf{x}(\mathbf{v},\ \mathbf{r},\ m)$, then the costate equations of the problem are

$$\begin{aligned} \dot{\boldsymbol{\lambda}}_v &= -\left(\frac{\partial H}{\partial \mathbf{v}}\right)^T = -\boldsymbol{\lambda}_r, \\ \dot{\boldsymbol{\lambda}}_r &= -\left(\frac{\partial H}{\partial \mathbf{r}}\right)^T = \frac{\mu}{r^3}\boldsymbol{\lambda} - 3\frac{\mu}{r^5}(\boldsymbol{\lambda}\mathbf{r})\mathbf{r}, \\ \dot{\lambda}_m &= -\frac{\partial H}{\partial m} = \frac{2P}{I_{sp}gm^2}m^2\lambda. \end{aligned} \tag{2.134}$$

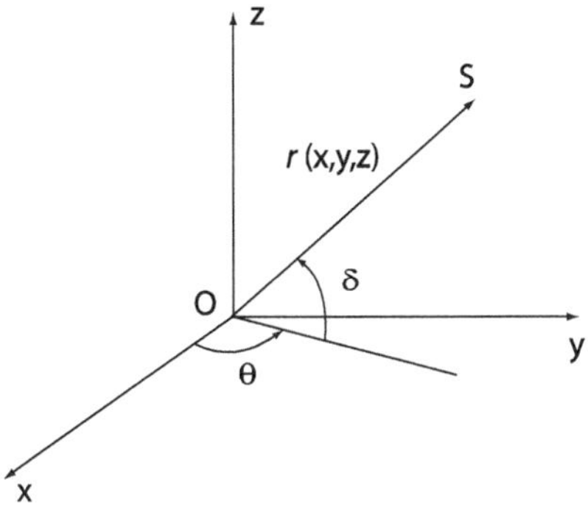

Figure 2.4 Spherical coordinate system.

2.17.6 Equations in Spherical Coordinates

Analysis of Eqs. (2.134) show that the investigation of the trajectories in the spherical system with spherical coordinates $r,\ \theta,\ \delta$ and with the origin at the center of gravitational attraction, is more convenient in obtaining the analytical solutions, and provides insight to the canonical equations, their first integrals and invariant relationships compared to the analysis in the Cartesian or other coordinates (see Fig. 2.4) [63], [139], [140]. Eqs. (2.132) and (2.134) can be rewritten in a scalar form as a canonical system [66], [141], [142]:

$$\begin{aligned}
\dot{v}_1 &= \frac{2P}{I_{sp}gm}\frac{\lambda_1}{\lambda} - \frac{\mu}{r^2} + \frac{v_2^2}{r} + \frac{v_3^2}{r}, \\
\dot{v}_2 &= \frac{2P}{I_{sp}gm}\frac{\lambda_2}{\lambda} - \frac{v_1 v_2}{r} + \frac{v_2 v_3}{r} tg\delta, \\
\dot{v}_3 &= \frac{2P}{I_{sp}gm}\frac{\lambda_3}{\lambda} - \frac{v_2^2}{r} tg\delta - \frac{v_1 v_3}{r}, \\
\dot{r} &= v_1, \\
\dot{\theta} &= \frac{v_2}{r\cos\delta}, \\
\dot{\delta} &= \frac{v_3}{r}, \\
\dot{m} &= -\frac{2P}{I_{sp}^2 g^2}, \\
\dot{\lambda}_1 &= \lambda_2\frac{v_2}{r} + \lambda_3\frac{v_3}{r} - \lambda_4, \\
\dot{\lambda}_2 &= -2\lambda_1\frac{v_2}{r} - \lambda_2\frac{v_3}{r} tg\delta + \lambda_2\frac{v_1}{r} + 2\lambda_3\frac{v_2}{r} tg\delta - \frac{\lambda_5}{r\cos\delta},
\end{aligned} \tag{2.135}$$

$$\begin{aligned}
\dot{\lambda}_3 &= -2\lambda_1\frac{v_3}{r} - \lambda_2\frac{v_2}{r}tg\delta + \lambda_3\frac{v_1}{r} - \frac{\lambda_6}{r},\\
\dot{\lambda}_4 &= \lambda_1\left(\frac{v_2^2}{r^2} + \frac{v_3^2}{r^2} - 2\frac{\mu}{r^3}\right) + \lambda_2\left(\frac{v_2 v_3}{r^2}tg\delta - \frac{v_1 v_2}{r^2}\right) -\\
&\quad \lambda_3\left(\frac{v_1 v_3}{r^2} + \frac{v_2^2}{r^2}tg\delta\right) - \lambda_5\frac{v_2}{r^2\cos\delta} + \lambda_6\frac{v_3}{r^2},\\
\dot{\lambda}_5 &= 0,\\
\dot{\lambda}_6 &= \frac{\lambda_3 v_2^2 - \lambda_2 v_2 v_3 - \lambda_5 v_2 \sin\delta}{r\cos^2\delta},\\
\dot{\lambda}_7 &= \frac{2P}{I_{sp}gm^2}\lambda
\end{aligned}$$

with the Pontryagin function, defined in Eq. (2.133) and which can be rewritten as [66]:

$$H = \lambda_1\left(\frac{2P}{I_{sp}gm}\frac{\lambda_1}{\lambda} - \frac{\mu}{r^2} + \frac{v_2^2}{r} + \frac{v_3^2}{r}\right) + \lambda_2\left(\frac{2P}{I_{sp}gm}\frac{\lambda_2}{\lambda} - \frac{v_1 v_2}{r} + \frac{v_2 v_3}{r}tg\delta\right) +$$
$$\lambda_3\left(\frac{2P}{I_{sp}gm}\frac{\lambda_3}{\lambda} - \frac{v_2^2}{r}tg\delta - \frac{v_1 v_3}{r}\right) + \lambda_4 v_1 + \lambda_5\frac{v_2}{r\cos\delta} + \lambda_6\frac{v_3}{r} - \lambda_7\frac{2P}{I_{sp}^2 g^2}. \quad (2.136)$$

The vectors in Eqs. (2.134) have the following components in the spherical system r, θ and δ with the origin at the center of gravitational attraction [63], [66]:

$$\mathbf{r}(r,0,0), \quad \mathbf{v}(v_1,v_2,v_3), \quad \boldsymbol{\lambda}_v(\lambda_1,\lambda_2,\lambda_3), \quad \lambda_m = \lambda_7,$$
$$\boldsymbol{\lambda}_r\left(\lambda_4; \quad \lambda_1\frac{v_2}{r} + \lambda_2\frac{v_3 tg\delta - v_1}{r} - \lambda_3\frac{v_2}{r}tg\delta + \frac{\lambda_5}{r\cos\delta}; \quad \frac{\lambda_1 v_3 - \lambda_3 v_1 + \lambda_6}{r}\right).$$

2.18 METHODOLOGY OF ANALYTICAL DETERMINATION OF OPTIMAL AND EXTREMAL TRAJECTORIES

- The costate equations and the conditions of local maximality are derived in the context of the variation problem stated above.
- The variation problem is reduced to the integration of the canonical system of equations for each admissible thrust arc by excluding the controls from the equations of motion and the costate equations utilizing the conditions of local maximality and the Weierstrass condition regardless the functional of the problem. These procedures allow us to determine the possibility of presence of various thrust arcs on optimal trajectories

and provide a possibility to determine the family of admissible trajectories.

- The canonical equations allow us to use various methods of integration of analytical mechanics applicable to Hamiltonian systems in order to obtain the first integrals, invariant relationships and analytical solutions. These solutions are referred to as the extremals of the problem.
- The analytical solutions obtained for each thrust arc are tested to see if they satisfy the Legendre-Clebsch condition, which express the positive semi-definiteness of the determinant of the matrix consisting of the second order partial derivatives of the Pontryagin function with respect to the controls. After the exclusion of controls from the equations of motion and costate equations, the Pontryagin function is converted into the Hamiltonian.
- The positive semi-definiteness of the second differential of the extended functional leads to an auxiliary problem of analytical optimization which is associated with the construction and solution of the auxiliary differential equations of the problem and, in particular, the Riccati matrix equation.
- The equations of the auxiliary problem are solved utilizing the analytical methods. These solutions are then used to determine the possible corner points on the extremals. This procedures allows us to further determine the finiteness of the Riccati matrix equation.
- The number of integration constants which are present in the solutions for the radius vector, velocity vector and mass is analyzed.
- The number of thrust arcs is determined depending on the number of integration constants. Consequently, the sequence of thrust arcs on extremal trajectory is determined.
- The continuity conditions for the radius and velocity vectors, primer vector and its derivative, and the condition that the switching function is zero at each corner (switching) point and the condition of transversality. These conditions represent a system with equal number of nonlinear algebraic equations and unknowns.
- If the system of the nonlinear algebraic continuity equations is solvable, then it allows us to determine the coordinates of the switching points, the values of integration constants and the unknown parameters, for example, the values of the thrust angle at the switching points.
- The analytical solutions with the integration constants and parameters of the problem mentioned above describe the extremal trajectory of the

maneuver, and all possible functional relationships between the problem parameters and their behavior during the maneuver.

- As a result of the procedures mentioned above, the trajectory constructed represent a reference or nominal trajectory which can be used in the maneuver and in solving the guidance problem.

Summary of this Chapter

1. In this chapter, the variation problem, the statement of which can be considered as an alternative to the Lawden's problem of flight dynamics, has been formulated.
2. In the context of this problem and by utilizing the sufficient conditions of optimality, it has been shown that an optimal trajectory may consist of zero thrust arc and the arcs with maximum power and constant specific impulse.
3. The fact that the satisfaction of the sufficient conditions limits the domain of admissible values of the parameters and that the problem considered here is a model problem allow us to conclude that the testing of the sufficient conditions and separation of optimal trajectories among the admissible trajectories is not meaningful from the practical point of view.
4. It has also been shown that an extremal trajectory may include the arcs of minimum thrust, intermediate thrust and maximum thrust. Depending on the ratio of thrust to weight and the specific impulse, these arcs can be considered as low thrust or high thrust arcs.
5. The auxiliary variation problem, the solution of which allows us to determine the presence of the conjugate points on the thrust arcs, has been formulated.
6. The classification of extremal thrust arcs has been proposed.
7. The methodology of analytical determination of optimal and extremal trajectories based on the analysis of the necessary conditions of optimality, the continuity conditions and on the determination of the trajectory structure has been proposed. This methodology serves as an instrument of separation of extremal trajectories from all possible and admissible trajectories. The extremal trajectories represent or can be considered as the reference trajectories and used in solving the guidance problem.

CHAPTER 3

Motion With Constant Power and Variable Specific Impulse

3.1 CANONICAL EQUATIONS AND FIRST INTEGRALS

As it was shown in the previous sections, if $\gamma = 0,\ \mu_2 \neq 0,\ \eta \neq 0,\ \mu_3 = 0$, then a motion with low thrust with variable I_{sp}, in which $P = P_{max}$ and $I_{sp_{min}} < I_{sp} < I_{sp_{max}}$. It can be shown that the corresponding equations of motion can be given in a canonical form [26]:

$$
\begin{aligned}
\dot{\mathbf{v}} &= \frac{2P}{I_{sp}gm}\boldsymbol{\lambda} - \frac{\mu}{r^3}\mathbf{r}, \\
\dot{\mathbf{r}} &= \mathbf{v}, \\
\dot{m} &= -\frac{2P}{I_{sp}^2 g^2}, \\
\dot{\boldsymbol{\lambda}} &= -\boldsymbol{\lambda}_r, \\
\dot{\boldsymbol{\lambda}}_r &= \frac{\mu}{r^3}\boldsymbol{\lambda} - 3\frac{\mu}{r^5}(\boldsymbol{\lambda}\mathbf{r})\mathbf{r}, \\
\dot{\lambda}_7 &= \frac{2P}{I_{sp}gm^2}\lambda.
\end{aligned}
\tag{3.1}
$$

If the functional of the form $\int \frac{1}{2}a_t^2\,dt$, then Hamiltonian of the problem has a form:

$$
H = \frac{2P}{I_{sp}gm}\lambda^2 - \frac{\mu}{r^3}\boldsymbol{\lambda}\mathbf{r} + \boldsymbol{\lambda}_r\mathbf{v} - \lambda_7\frac{2P}{I_{sp}^2 g^2} + \frac{a_t^2}{2}.
$$

It can also be shown that for Eqs. (3.1) the following first integrals take place [26]:

$$
\begin{aligned}
-\frac{\mu}{r^3}\boldsymbol{\lambda}\mathbf{r} + \boldsymbol{\lambda}_r\mathbf{v} + \frac{P}{b}\lambda^2 &= C, \\
\boldsymbol{\lambda}\mathbf{v} - 2\mathbf{r}\boldsymbol{\lambda}_r - 5\frac{P}{2b}\int \lambda^2 dt &= -3Ct + C_1, \\
\lambda_5 &= C_2, \\
\frac{2b}{\lambda m} &= I_{sp}g, \\
\lambda_7 m^2 &= b,
\end{aligned}
\tag{3.2}
$$

Analytical Solutions for Extremal Space Trajectories
DOI: http://dx.doi.org/10.1016/B978-0-12-814058-1.00003-5

$$\frac{\lambda}{m} - 2\frac{\lambda_7}{I_{sp}g} = 0,$$
$$\mathbf{a} = \frac{P}{b}\boldsymbol{\lambda},$$

where C, C_1, C_2, b are the integration constants. There exist some classes of solutions which can be used in the construction of interplanetary trajectories, in solving the guidance and navigation problems (3.1).

The following theorem can be proven.

Theorem 3.1. *Let the canonical equations of the problem in (2.111)–(2.115) are given by Eqs. (3.1). Then, if the functional (2.115) does not explicitly depend on the polar angle, then in the case of motion with $P = P_{max}$ and variable I_{sp} ($I_{sp,min} < I_{sp} < I_{sp,max}$), Eqs. (3.1) admit at least two families of the extremals: families of spiral and circular trajectories, which can be described in elementary functions.*

The proof of this theorem is presented below in two parts.

3.2 CIRCULAR TRAJECTORIES

In this subsection the first part of the proof of Theorem 3.1 which is associated with the family of circular trajectories in the context of the problem (2.111)–(2.115) is presented. It can be shown that Eqs. (3.1) have the following solutions describing a circular motion with P_{max} and variable I_{sp} [145].

Consider the constraint $r = r_0$ and associated Lagrange multiplier μ_1 in the context of the problem (2.111)–(2.115). For simplicity, consider also a planar motion. By introducing a polar coordinate system $O\ r\ \theta$ with the origin at the center of gravitation, and accepting that $v_1 = 0$, one can reduce Eqs. (2.135) to the following form:

$$0 = \frac{P}{b}\lambda_1 - \frac{\mu}{r^2} + \frac{v_2^2}{r}, \tag{3.3}$$
$$\dot{v}_2 = \frac{P}{b}\lambda_2, \tag{3.4}$$
$$\dot{r} = 0, \tag{3.5}$$
$$\dot{\theta} = \frac{v_2}{r}, \tag{3.6}$$
$$\dot{m} = -\frac{2P}{I_{sp}^2 g^2}, \tag{3.7}$$
$$\dot{\lambda}_1 = \lambda_2\frac{v_2}{r} - \lambda_4 + \mu_1, \tag{3.8}$$

$$\dot{\lambda}_2 = -2\lambda_1 \frac{v_2}{r} - \frac{\lambda_5}{r}, \tag{3.9}$$

$$\dot{\lambda}_4 = \lambda_1 (\frac{v_2^2}{r} - 2\frac{\mu}{r^2}) + \lambda_5 \frac{v_2}{r^2}, \tag{3.10}$$

$$\dot{\lambda}_5 = 0, \tag{3.11}$$

$$\dot{\lambda}_7 = \frac{P}{b}\frac{\lambda^2}{m}, \tag{3.12}$$

with the Hamiltonian

$$H = -\lambda_1 \frac{\mu}{r^2} + \lambda_1 \frac{v_2^2}{r} + \lambda_5 \frac{v_2}{r} + \frac{P}{2b}\lambda^2 = C, \tag{3.13}$$

where the fourth integral of Eqs. (3.2) is utilized and

$$\mathbf{r}(r, 0), \quad \mathbf{v}(v_1, v_2), \quad \boldsymbol{\lambda}(\lambda_1, \lambda_2), \quad \boldsymbol{\lambda}_r\left(\lambda_4 - \mu_1, \lambda_1 \frac{v_2}{r} + \frac{\lambda_5}{r}\right).$$

From Eq. (3.3) one can obtain

$$v_2^2 = \frac{\mu}{r} - \lambda_1 r \frac{P}{b}, \tag{3.14}$$

from which it follows that

$$\frac{dv_2}{d\lambda_1} = -\frac{Pr}{2bg}, \tag{3.15}$$

where

$$g = \sqrt{\frac{\mu}{r} - \lambda_1 r \frac{P}{b}}.$$

Based on Eq. (3.14), other unknowns can also be expressed in terms of λ_1. Indeed, using Eqs. (3.4) and (3.15), one can write

$$\dot{v}_2 = \frac{dv_2}{d\lambda_1}\frac{d\lambda_1}{dt} = \frac{P}{b}\lambda_2$$

or

$$\frac{d\lambda_1}{dt} = -2\frac{\lambda_2 g}{r}. \tag{3.16}$$

Also, Eqs. (3.14) and (3.13) yield

$$\lambda^2 = \frac{2b}{Pr}\left(Cr + \lambda_1^2 \frac{Pr}{b} - \lambda_5 g\right). \tag{3.17}$$

This expression and $\lambda_2^2 = \lambda^2 - \lambda_1^2$ allow us to obtain

$$\lambda_2^2 = \frac{b}{Pr}\left(2Cr + \lambda_1^2\frac{Pr}{b} - 2\lambda_5 g\right). \tag{3.18}$$

Consequently, from Eq. (3.16) it follows that

$$t = -\frac{r}{2}\int_{\lambda_{10}}^{\lambda_{11}} \frac{d\lambda_1}{g\sqrt{\frac{b}{Pr}\left(2Cr + \lambda_1^2\frac{Pr}{b} - 2\lambda_5 g\right)}} + t_0, \tag{3.19}$$

where t_0 is an integration constant. Also, Eqs. (3.8) and (3.16) yield

$$\lambda_4 - \mu = 3\frac{3}{r}\lambda_2 v_2. \tag{3.20}$$

Substitution of Eq. (3.20) into (3.2) gives the formulae for the mass:

$$m = \frac{5b^2}{P}\left(3Ct - 5\lambda_2(\lambda_1)v_2(\lambda_1) - C_1 + \frac{5b^2}{Pm_0}\right)^{-1}, \tag{3.21}$$

where $\lambda_2(\lambda_1)$ and $v_2(\lambda_1)$ are computed using Eq. (3.14) and (3.18), and m_0 is the integration constant. Once the formulas for λ and m are determined, I_{sp} and λ_7 can be determined using its integral given in Eqs. (3.2):

$$I_{sp}(\lambda_1) = \frac{2b}{g_0\lambda(\lambda_1)m(\lambda_1)} \tag{3.22}$$

and

$$\lambda_7 = \frac{b}{m^2(\lambda_1)}. \tag{3.23}$$

Note that Eq. (3.17) can be used to find the thrust angle, φ in terms of λ_1:

$$\sin\varphi = \frac{\lambda_1}{\sqrt{\frac{2b}{Pr}\left(Cr + \lambda_1^2\frac{Pr}{b} - \lambda_5 g\right)}}. \tag{3.24}$$

Using

$$\dot{\theta} = \frac{d\theta}{d\lambda_1}\frac{d\lambda_1}{dt} = \frac{\sqrt{g}}{r}$$

and also, Eqs. (3.16) and (3.17), one can have a quadrature for θ:

$$\theta = -\frac{1}{2}\int_{\lambda_{10}}^{\lambda_1} \frac{d\lambda_1}{\sqrt{\frac{2b}{Pr}\left(Cr + \lambda_1^2\frac{Pr}{2b} - \lambda_5 g\right)}} + \theta_0, \tag{3.25}$$

where θ_0 is the integration constant, and λ_{10} and λ_{11} are the initial and final values of λ_1. Similarly, Eqs. (3.10) and (3.16) allow one to find the quadrature for λ_4:

$$\lambda_4 = \mp \int_{\lambda_{10}}^{\lambda_1} \frac{C - \lambda_1 \frac{\mu}{r^2} - \frac{P}{2b}\lambda^2}{g\lambda_2} d\lambda_1 + \lambda_{40}, \tag{3.26}$$

where λ_{40} is the integration constant, and λ is determined from Eq. (3.17). Consequently, from Eqs. (3.8) one can obtain

$$\mu_1 = -3\lambda_2 \frac{g}{r} + \int \frac{C - \lambda_1 \frac{\mu}{r^2} - \frac{P}{2b}\lambda^2}{g\lambda_2} d\lambda_1 + \lambda_{40}. \tag{3.27}$$

Note that so far, λ_1 has been used as the independent variable. At this point, if one assumes that the low-thrust motion with P_{max} and variable I_{sp} is considered, and the performance index is given in the form $J = 1/2 \int a_t^2 dt$, where a_t is the thrust acceleration, then based on the second and the last expressions of Eq. (3.2), one can obtain

$$J = \frac{P}{5b}(3Ct - \frac{5}{2} r \frac{d\lambda_1}{dt} - C_1), \tag{3.28}$$

where $\frac{d\lambda_1}{dt}$ is computed using Eq. (3.16). As a summary of the derivations given above, the solutions to Eqs. (3.1) in the case of a planar circular motion with P_{max} and variable I_{sp} can be presented in the form:

$$\begin{aligned}
r &= const, \\
v_2^2 &= g^2, \\
\theta &= \frac{1}{2} \int_{\lambda_{10}}^{\lambda_1} \frac{d\lambda_1}{\sqrt{\frac{2b}{Pr}(Cr + \lambda_1^2 \frac{Pr}{2b} - \lambda_5 g)}} + \theta_0, \\
m &= \frac{5b^2}{P} \left(3Ct - 5\lambda_2(\lambda_1) v_2(\lambda_1) - C_1 + \frac{5b^2}{Pm_0} \right)^{-1}, \\
\lambda &= \sqrt{\frac{2b}{Pr}(Cr + \lambda_1^2 \frac{Pr}{b} - \lambda_5 g)}, \\
\sin\varphi &= \frac{\lambda_1}{\sqrt{\frac{2b}{Pr}(Cr + \lambda_1^2 \frac{Pr}{b} - \lambda_5 g)}}, \\
\lambda_2 &= \lambda \cos\varphi, \\
\lambda_4 &= \int \frac{C - \lambda_1 \frac{\mu}{r^2} - \frac{P}{2b}\lambda^2}{g\lambda_2(\lambda_1)} d\lambda_1 + \lambda_{40},
\end{aligned} \tag{3.29}$$

$$
\begin{aligned}
\lambda_5 &= \lambda_{50},\\
\lambda_7 &= \frac{b}{m^2(\lambda_1)},\\
I_{sp} &= \frac{2b}{g_0\lambda(\lambda_1)m(\lambda_1)},\\
t &= -\frac{r}{2}\int_{\lambda_{10}}^{\lambda_1}\frac{d\lambda_1}{g\sqrt{\frac{b}{Pr}\left(2Cr+\lambda_1^2\frac{Pr}{b}-2\lambda_5 g\right)}}+t_0,\\
\mu_3 &= -3\lambda_2(\lambda_1)\frac{g}{r}+\int\frac{C-\lambda_1\frac{\mu}{r^2}-\frac{P}{2b}\lambda^2}{g\lambda_2(\lambda_1)}d\lambda_1+\lambda_0,\\
J &= \frac{P}{5b}(3Ct-\frac{5}{2}r\frac{d\lambda_1}{dt}-C_1),
\end{aligned}
$$

where

$$
g=\sqrt{\frac{\mu}{r}-\lambda_1 r\frac{P}{b}},
$$

$C,\ b,\ r,\ \theta_0,\ m_0,\ \lambda_{40},\ \lambda_{50},\ t_0,\ \lambda_0,\ C_1$ are the integration constants, $P=P_{max}$ and J is the functional of the problem.

3.2.1 Case When Final Polar Angle Is Not Fixed

If the polar angle θ is not fixed, then $\lambda_5=\frac{\partial J}{\partial\theta_1}=0$. Consequently, $\lambda_5=\lambda_{50}=0$ on the trajectory. Consider the case when $\lambda_5=0$. First, consider the quadrature of Eqs. (3.29) for the polar angle, θ. It can be rewritten as

$$
\theta=-\frac{1}{2}\int_{\lambda_{10}}^{\lambda_1}\frac{d\lambda_1}{\sqrt{\frac{2b}{P}C+\lambda_1^2}}+\theta_0,
$$

which represents a table integral and can be integrated in the final form:

$$
\theta=-\frac{1}{2}\ln\left(2\sqrt{\frac{2b}{P}C+\lambda_1^2}+2\lambda_1\right)+\theta_0(\lambda_{10}). \tag{3.30}
$$

The formulas for λ and $\sin\varphi$ are reduced to the forms:

$$
\lambda=\sqrt{\frac{2b}{Pr}C+2\lambda_1^2}, \tag{3.31}
$$

$$
\sin\varphi=\frac{\lambda_1}{\sqrt{\frac{2b}{Pr}C+2\lambda_1^2}}. \tag{3.32}
$$

Now, we derive the solution for the time. Accepting

$$\frac{2bC}{P} = a, \qquad \sqrt{\frac{\mu}{r} - \lambda_1 r \frac{P}{b}} = x,$$

it can be shown that

$$\lambda_1 = \frac{b}{Pr}\left(\frac{\mu}{r} - x^2\right), \qquad d\lambda_1 = -2\frac{b}{Pr}xdx$$

and by substituting these expressions into Eq. (3.19), one can obtain that

$$t = \frac{b}{P}\int \frac{dx}{a + \frac{b^2}{P^2r^2}\left(\frac{\mu}{r} - x^2\right)^2} + t_0,$$

or

$$t = \frac{b}{P}\int \frac{dx}{x^4 + a_1x^2 + b_1} + t_0, \tag{3.33}$$

where

$$a_1 = -2\frac{\mu}{r}, \qquad b_1 = c_1 + \frac{\mu^2}{r^2}, \qquad c_1 = a\frac{P^2r^2}{b^2},$$

with the condition that $\mu/r > \sqrt{-c_1}$ and $c_1 < 0$. Eq. (3.33) can be simplified as

$$t = r\int \frac{dx}{(p_1 - x^2)(p_2 - x^2)} + t_0,$$

which can be rewritten as

$$t = \frac{r}{p_2}\int_{\sqrt{p_1}z_0}^{\sqrt{p_1}z} \frac{r}{\sqrt{p_2}} \frac{dz}{(1 - z^2)(1 - k^2z^2)} + t_0,$$

or

$$t = \frac{r}{p_2}F(z, k) + t_0, \tag{3.34}$$

where

$$p_1 = \frac{mu}{r} - \sqrt{c_1}, \quad p_2 = \frac{mu}{r} + \sqrt{c_1}, \quad z_0 = \frac{x_0}{\sqrt{p_1}}, \quad z = \frac{x}{\sqrt{p_1}},$$

$$F(z, k) = \int_0^{\sqrt{p_1}z} \frac{r}{\sqrt{p_2}} \frac{dz}{(1 - z^2)(1 - k^2z^2)}, \qquad k = \sqrt{\frac{p_1}{p_2}}.$$

Here Eq. (3.34) is known as an elliptic integral of the first kind in the Legendre normal form, and k is called the "modulus" of this integral [146].

Consequently, in the case, when $\lambda_5 = 0$, the variables θ, λ, φ, and t of Eqs. (3.29) can be computed according to Eqs. (3.30), (3.31), (3.32), and (3.34). The formulas for all other variables of Eqs. (3.29) remain the same.

3.2.2 Case When Flight Time and Final Polar Angle Are Not Fixed

If the flight time is not fixed, then $C = \frac{\partial J}{\partial t} = 0$. If the polar angle θ is not fixed, then $\lambda_5 = \frac{\partial J}{\partial \theta_1} = 0$. Consider the case when $C = 0$ and $\lambda_5 = 0$. It can be easily shown that in this case there exist the solutions with $J = m_0 - m_f$ depending on the polar angle θ. Indeed, from Eq. (3.4) it follows that

$$\dot{\lambda}_2 = -2\lambda_1 \frac{v_2}{r}. \tag{3.35}$$

Also, from Eqs. (3.3) and (3.13) one can obtain

$$\frac{P}{b}\lambda_1 = \frac{\mu}{r^2} - \frac{v_2^2}{r}, \qquad \lambda_1 \frac{\mu}{r^2} = \lambda_1 \frac{v_2^2}{r} + \frac{P}{2b}\lambda^2 = 0.$$

These two expressions yield $\lambda_1 = \pm\lambda_2$ or $\sin\varphi = \pm\cos\varphi$ or

$$\varphi = \pi/4 - \pi/2k, \qquad k = 0, 1, 2, \ldots. \tag{3.36}$$

Using $\dot{\theta} = v_2/r$, Eq. (3.35) can be rewritten in the form:

$$\dot{\lambda}_2 = \dot{\lambda}\cos\varphi = \mp 2\lambda \sin\varphi \dot{\theta}$$

or

$$\frac{\dot{\lambda}}{\lambda} = \mp 2\dot{\theta},$$

which yields the first integral:

$$\lambda = C_2 \exp(\mp 2\theta), \tag{3.37}$$

where C_2 is the integration constant. In this case, from Eq. (3.4) it follows that

$$\frac{dv_2}{dt} = \frac{dv_2}{d\theta}\frac{d\theta}{dt} = \pm \frac{P}{b} \sin\varphi C_2 \exp(\mp 2\theta)$$

or

$$v_2^2 = \frac{Pr}{b} \sin\varphi C_2 \exp(\mp 2\theta) + C_3, \tag{3.38}$$

where C_3 is the integration constant. It can be shown that from Eqs. (3.6) and (3.38) it follows that

$$\frac{d\theta}{dt} = \frac{v_2}{r} = \pm\frac{\sqrt{a_2 \exp(\mp 2\theta) + C_3 + \sqrt{C_3}}}{r} \tag{3.39}$$

which can be integrated in the form:

$$t = \frac{r}{2\sqrt{C_3}} \ln\left(\frac{x + \sqrt{C_3}}{x + \sqrt{C_3}}\right) + t_0, \tag{3.40}$$

where

$$a_1 = \sin\varphi, \quad a_2 = \frac{Pr}{b} a_1 C_2, \quad x = a_2 \exp(\mp 2\theta) + C_3$$

and t_0 is the integration constant. Eq. (3.40) can be used to find θ in the form:

$$\theta = \mp\frac{1}{2}\ln\left[\frac{C_3}{a_2}\left(\frac{(\exp(\tau)+1)^2}{(\exp(\tau)-1)^2} - 1\right)\right], \tag{3.41}$$

where $\tau = \frac{2\sqrt{C_3}}{r}(t - t_0)$. Similarly, it can be shown that from Eqs. (3.7) and (3.39) one can obtain

$$m = \left[\frac{PrC_2^2}{a_2^2 b^2}\left(\frac{v_2^3}{3} - C_3 v_2\right) + \frac{1}{m_0}\right]^{-1} \tag{3.42}$$

Based on the integral for I_{sp} of Eqs. (3.2), and utilizing (3.37), it is possible to derive the formulae for this variable:

$$I_{sp} = \frac{2Pr}{g_0 a_2^2 b}\frac{1}{\exp(\mp 2\theta)}\left(C_3 v_2 - \frac{v_2^3}{3}\right) + \frac{2b}{g_0 C_2 m_0}. \tag{3.43}$$

Eqs. (3.10) and (3.39) can be used to obtain the formulae for λ_4:

$$\dot{\lambda}_4 = \frac{d\lambda_4}{d\theta}\frac{d\theta}{dt} = \lambda_1\left(\frac{v_2^2}{r} - \frac{\mu}{r^2}\right) - \lambda_1\frac{\mu}{r^2} = \frac{P}{rb}a_1^2 C_2^2 \exp(\mp 4\theta) - \frac{\mu}{r}a_1 C_2 \exp(\mp 2\theta)$$

or

$$\lambda_4 = \int_{\theta_0}^{\theta_1} \frac{\exp(\mp 2\theta)\left(\frac{P}{b}a_1^2 C_2^2 \exp(\mp 2\theta) - \mu a_1 C_2\right)}{\pm a_2 \exp(\mp 2\theta) + C_3} d\theta + \lambda_{40}, \tag{3.44}$$

where λ_{40} is the integration constant. Consequently, Eq. (3.44) can be used to find μ_1 from Eq. (3.8) in the form:

$$\mu_1 = 3\cos\varphi\lambda\sqrt{a_2 \exp(\mp 2\theta) + C_3} - \lambda_4. \tag{3.45}$$

Also, it is obvious that based on the condition, $\lambda_5 = 0$, from Eq. (3.17) it follows that

$$\lambda = \frac{C_2}{a_2}(x^2 - C_3), \tag{3.46}$$

$$\lambda_1 = \pm\frac{\sqrt{2}}{2}\frac{C_2}{a_2}(x^2 - C_3), \quad \lambda_2 = \pm\frac{\sqrt{2}}{2}\frac{C_2}{a_2}(x^2 - C_3), \tag{3.47}$$

where the signs "−" and "+" are selected independently. Finally, using the integral for λ_7 of Eqs. (3.2), one can find that

$$\lambda_7 = \frac{b}{m^2}, \tag{3.48}$$

where m is computed using Eqs. (3.42). The integration constants introduced above can be computed using the initial values of the problem parameters. In summary, the formulas obtained above can be presented in the following form:

$$\begin{aligned}
r &= r_0 = const, \\
v_2^2(\theta) &= a_2 e^{\pm 2\theta} + C_3, \\
m(\theta) &= \left[\frac{1}{Pr}(\frac{v_2^3}{3} - C_3 v_2) + \frac{1}{\bar{m}_0}\right]^{-1}, \\
t(\theta) &= \frac{r}{2\sqrt{C_3}} \ln\frac{v_2 - \sqrt{C_3}}{v_2 + \sqrt{C_3}} + t_0, \\
\lambda(\theta) &= \lambda_0 e^{2\theta - 2\theta_0}, \\
I_{sp}(\theta) &= \frac{2b}{g_0 \lambda m}, \\
\lambda_1 &= -\frac{\sqrt{2}}{2}\lambda_0, \\
\lambda_2 &= \frac{\sqrt{2}}{2}\lambda_0, \\
\lambda_7(\theta) &= \frac{b}{m^2},
\end{aligned} \tag{3.49}$$

where $\lambda_0,\ a_2,\ C_3,\ t_0,\ r_0,\ \theta_0,\ m_0,\ b$ are the integration constants,

$$\begin{aligned}
a_2 &= \frac{\sqrt{2}}{2} Pr \frac{\lambda_0}{b} e^{2\theta_0}, \\
C_3 &= v_{20}^2 - \frac{\sqrt{2}}{2} Pr \frac{\lambda_0}{b} e^{2\theta_0},
\end{aligned}$$

$$\begin{aligned}
\frac{1}{m_0} &= \frac{1}{m_0} - \frac{1}{Pr}(\frac{v_{20}^3}{3} - C_3 v_{20}), \\
t_0 &= t(0) - \frac{r}{2\sqrt{C_3}} \ln \frac{v_{20} - \sqrt{C_3}}{v_{20} + \sqrt{C_3}}, \qquad (3.50) \\
bm_f^2 &= \left(\frac{1}{Pr}[\frac{v_{2f}^3 - v_{20}}{3} - C_3(v_{2f} - v_{20})] + \frac{1}{m_0} \right)^{-2}, \\
\lambda_{7f} &= -\frac{\partial J}{\partial m_f} = 1.
\end{aligned}$$

The constants r_0, λ_0, $t(0)$, θ_0 are determined depending on the initial conditions. The indexes 'o' and 'f' will correspond the initial and final values of the parameters. Eqs. (3.49) describe the circular trajectory. The spacecraft velocity can increase or decrease with respect to circular velocity depending on thrust angle which is constant. The velocity is increased if the thrust angle is equal to $-\frac{\pi}{4}$, or decreased if the thrust angle is equal to $\frac{\pi}{4}$. When the velocity is changed from v_0 (circular velocity) to v_f, then the polar angle will increase from θ_0 to

$$\theta_f = \frac{1}{2}\left[\frac{v_f^2 - v_{20}^2}{\frac{\sqrt{2}}{2} Pr \frac{\lambda_0}{b} e^{2\theta_0}} + 1 \right].$$

Analysis of $I_{sp}(\theta)$ and $m(\theta)$ has shown that the change in I_{sp} from $I_{sp,max}$ to $I_{sp,min}$ corresponds to a maximum final mass. This completes the first part of Theorem 3.1.

The results described above can be used in the problems of escape or transfer to a parking orbit. The possibilities of reaching a parabolic (or hyperbolic) speed of escape from a local circular orbit depend on the initial conditions and characteristics of the propulsion system. The analysis of the solutions obtained also show that I_{sp} will increase when the polar angle is increased and the velocity is increased exponentially.

Now, using the integral of Eqs. (3.2),

$$\frac{2b}{\lambda m} = I_{sp} g,$$

the equation for the mass takes the form:

$$\dot{m} = \frac{dm}{d\lambda_1}\frac{\lambda_1}{dt} = -\frac{P}{2b}\lambda^2 m^2.$$

This equation can be rewritten in the form:

$$\int \frac{dm}{m^2} = -\frac{P}{2b} \frac{\lambda^2}{\frac{d\lambda_1}{dt}} d\lambda_1,$$

and integrated as

$$\frac{1}{m} = \frac{1}{m_0} - \frac{P}{2b} \int \frac{\frac{2b}{Pr}\left(Cr + \lambda_1^2 \frac{Pr}{b} - \lambda_5 g\right)}{\frac{2}{r} g \sqrt{\frac{b}{Pr}\left(2Cr + \lambda_1^2 \frac{Pr}{b} - 2\lambda_5 g\right)}} d\lambda_1. \tag{3.51}$$

3.3 SPIRAL TRAJECTORIES

In this subsection the second part of Theorem 3.1 which is associated with the family of spiral arcs in the context of the problem (2.111)–(2.115) is presented. If the polar angle is not fixed and the functional of the problem does not depend on the final polar angle, then from the transversality condition [26]

$$\lambda_{51} = -\frac{\partial J}{\partial \theta_1}$$

it follows that $\lambda_{51} = C_2 = 0$. The corresponding analytical solutions can be obtained from Eqs. (3.2) and (3.56), and will be presented below in a form convenient for practical applications [26], [68], [143], [144], [147].

Consider again a planar motion with P_{max} and variable I_{sp} in the polar coordinate system, $Or\theta$. Thus, Eqs. (3.1) can be reduced to the form:

$$\begin{aligned}
\dot{v}_1 &= \frac{\bar{P}}{b}\lambda_1 - \frac{\mu}{r^2} + \frac{v_2^2}{r}, \\
\dot{v}_2 &= \frac{\bar{P}}{b}\lambda_2 - \frac{v_1 v_2}{r}, \\
\dot{r} &= v_1, \\
\dot{\theta} &= \frac{v_2}{r}, \\
\dot{m} &= -\frac{\bar{P}}{2b^2} m^2 \lambda^2, \\
\dot{\lambda}_1 &= \lambda_2 \frac{v_2}{r} - \lambda_4, \\
\dot{\lambda}_2 &= -2\lambda_1 \frac{v_2}{r} + \lambda_2 \frac{v_1}{r} - \frac{\lambda_5}{r}, \\
\dot{\lambda}_4 &= \lambda_1\left(\frac{v_2^2}{r^2} - 2\frac{\mu}{r^3}\right) - \frac{v_1 v_2}{r^2}\lambda_2 - \lambda_5 \frac{v_2}{r^2},
\end{aligned} \tag{3.52}$$

$$\dot{\lambda}_5 = 0,$$
$$\dot{\lambda}_7 = \frac{\bar{P}}{bm}\lambda^2,$$

where

$$H = \lambda_1(\frac{\bar{P}}{b}\lambda_1 - \frac{\mu}{r^2} + \frac{v_2^2}{r}) + \lambda_2(\frac{\bar{P}}{b}\lambda_2 - \frac{v_1 v_2}{r}) + \lambda_4 v_1 + \lambda_5 \frac{v_2}{r} - \lambda_7 \frac{\bar{P}}{2b^2} m^2 \lambda_v^2. \tag{3.53}$$

The integrals of Eqs. (3.52) are presented in Eqs. (3.2). One more integral can be obtained by using the Poisson brackets [26]. Let's denote the integral of Hamiltonian by ψ_1:

$$\psi_1 = H = -\frac{\mu}{r^3}\boldsymbol{\lambda}\mathbf{r} + \boldsymbol{\lambda}_r\mathbf{v} - \frac{P}{b}\lambda^2 = C_1. \tag{3.54}$$

Besides that, the second integral in Eqs. (3.2), taking into account the equality

$$\int \frac{a_t^2}{2P} dt = \frac{P}{2b^2}\int \lambda^2 dt = \frac{1}{m} - \frac{1}{m_0},$$

can be rewritten as ψ_2 in the form:

$$\psi_2 = \boldsymbol{\lambda}\mathbf{v} - 2\mathbf{r}\boldsymbol{\lambda}_r - \frac{5b}{m} - \frac{5b}{m_0} + 3Ct = C_1. \tag{3.55}$$

By using the Poisson brackets one can obtain that

$$[\psi_1, \psi_2] = (\frac{\partial \psi_1}{\partial \boldsymbol{\lambda}}\frac{\partial \psi_2}{\partial \mathbf{v}} - \frac{\partial \psi_1}{\partial \mathbf{v}}\frac{\partial \psi_2}{\partial \boldsymbol{\lambda}}) + (\frac{\partial \psi_1}{\partial \boldsymbol{\lambda}_r}\frac{\partial \psi_2}{\partial \mathbf{r}} - \frac{\partial \psi_1}{\partial \mathbf{r}}\frac{\partial \psi_2}{\partial \boldsymbol{\lambda}_r}) =$$
$$-3\frac{\mu}{r^3}\boldsymbol{\lambda}\mathbf{r} - 3\boldsymbol{\lambda}_r\mathbf{v} + \frac{P}{b}\lambda^2 = d = const,$$

or

$$\lambda^2 = \frac{2b}{5P}(d + 3C) = const. \tag{3.56}$$

Using the last expression of Eqs. (3.2) it can be shown that Eq. (3.56) means that the magnitude of thrust acceleration is constant. It can be shown that this case corresponds to a motion in a uniform gravity field. One can now obtain invariant relationships, $\dot{\lambda} = 0$, $\ddot{\lambda} = 0$, $\dddot{\lambda} = 0$ as a direct consequence of Eq. (3.56). These three relationships can be rewritten in the form:

$$\lambda_1\lambda_4 + \lambda_1\lambda_2\frac{v_2}{r} - \lambda_2^2\frac{v_1}{r} + \frac{\lambda_2\lambda_5}{r} = 0, \tag{3.57}$$

$$\lambda_4^2 + (\lambda_1 \frac{v_2}{r} - \lambda_2 \frac{v_1}{r} + \frac{\lambda_5}{r})^2 = \lambda_v^2 \frac{\mu}{r^3} - 3\lambda_1^2 \frac{\mu}{r^3}, \quad (3.58)$$

$$(\lambda_v^2 - 5\lambda_1^2)v_1 + 2v_1(\lambda_1 v_1 + \lambda_2 v_2) - 4\lambda_1 \lambda_4 r = 0. \quad (3.59)$$

Eqs. (3.53) and (3.57)–(3.59) allow us to obtain the following formulas in terms of the thrust angle, φ, the between local horizon and thrust direction:

$$r^2 = -3\mu \frac{\lambda_v}{C_3} \sin^3 \varphi, \quad (3.60)$$

$$v_1 = \frac{3 \sin 2\varphi}{5 \sin^2 \varphi - 3} s, \quad (3.61)$$

$$v_2 = \frac{(3 - \sin^2 \varphi)}{5 \sin^2 \varphi - 3} s, \quad (3.62)$$

$$\lambda_4 = -\frac{\lambda_v \cos \varphi}{r} s, \quad (3.63)$$

where

$$\frac{C_3}{\lambda} = \frac{C}{\lambda} - \frac{P}{2}\frac{\lambda}{b} \quad \text{and} \quad s = \pm \sqrt{\frac{\mu}{r}(1 - 3 \sin^2 \varphi)}$$

and it is assumed that $\sin^2 \varphi < 1/3$. Note that if $\sin \varphi < 0$, then $C_3 > 0$ or vice versa. The sign of s is selected depending on the problem statement. For example, if $\sin 2\varphi > 0$ and the range is decreasing, then $v_1 < 0$, and hence s must be positive. Furthermore, the equation for mass, m of Eqs. (3.52) can be easily integrated to yield

$$m = \left(\frac{1}{m_0} + \frac{P\lambda^2}{2b^2} t \right)^{-1}, \quad (3.64)$$

where m_0 is the integration constant. Consequently, the specific impulse, I_{sp} can be determined from the corresponding integral of Eqs. (3.2) for this variable:

$$I_{sp} = \frac{2b}{g\lambda m}, \quad (3.65)$$

where m is computed using Eqs. (3.64). As b and λ are constants, the relationship between mass and specific impulse is represented by the ratios:

$$\frac{m_0}{m_f} = \frac{I_{sp,f}}{I_{sp,0}}, \quad (3.66)$$

where "0" and "f" indicate the initial and final values of the variables. Assuming φ as an independent variable and utilizing the expressions

$$\frac{d\theta}{dt}=\frac{d\theta}{d\varphi}\frac{d\varphi}{dt}=\frac{v_2}{r},$$

and the second integral of Eqs. (3.2), one can find that

$$\theta=\frac{1}{4}(\frac{3}{\tan\varphi_0}+\varphi_0)-\frac{1}{4}(\frac{3}{\tan\varphi}+\varphi)+\theta_0, \tag{3.67}$$

$$t=\frac{1}{t_c}\left[\frac{3s\cos\varphi(1-5\sin^2\varphi)}{3-5\sin^2\varphi}-\frac{C_1}{\lambda_v}\right], \tag{3.68}$$

where

$$\frac{C_1}{\lambda}=\frac{3\cos\varphi_0 s_0(1-5\sin^2\varphi_0)}{3-5\sin^2\varphi_0}, \qquad t_c=\frac{5}{2}\frac{P\lambda}{b}-3\frac{C_3}{\lambda}.$$

The Lagrange multipliers, λ_1, λ_2, and λ_7 are found using similar formulas presented in the previous subsection. In summary, the analytical solutions for a planar motion with a constant thrust acceleration magnitude are given as follows:

$$r = \frac{\mu}{d^2}F_1, \tag{3.69}$$

$$v_1 = dF_2, \tag{3.70}$$

$$v_2 = dF_3, \tag{3.71}$$

$$\theta = \frac{1}{4}(\frac{3}{\tan\varphi_0}+\varphi_0)-\frac{1}{4}(\frac{3}{\tan\varphi}+\varphi)+\theta_0, \tag{3.72}$$

$$\frac{1}{m} = \frac{1}{m_0}+\frac{P_{max}\lambda^2}{2b^2}t, \tag{3.73}$$

$$I_{sp} = \frac{2b}{g\lambda m}, \tag{3.74}$$

$$\frac{m_0}{m_f} = \frac{I_{sp,f}}{I_{sp,0}}, \tag{3.75}$$

$$t = \frac{1}{t_c}\left[\frac{3zk(1-5s)}{3-5s}-\frac{C_1}{\lambda}\right]+t_0, \tag{3.76}$$

$$\lambda_1 = \sqrt{s}\lambda, \tag{3.77}$$

$$\lambda_2 = k\lambda, \tag{3.78}$$

$$\lambda_4 = -\frac{k}{r}\sqrt{\frac{\mu}{r}}(1-3s)\lambda, \tag{3.79}$$

$$\lambda_5 = 0, \tag{3.80}$$

$$\lambda_7 = \frac{b}{m^2}, \tag{3.81}$$

where

$$s = \sin^2 \varphi, \qquad k = \cos\varphi, \qquad z = \sqrt{\frac{\mu}{r}(1-3s)},$$

$$F_1(s) = s^{\frac{3}{4}},$$

$$F_2(s) = \frac{6sk}{3-5s}\sqrt{\frac{1-3s}{s^{\frac{3}{4}}}},$$

$$F_3(s) = \frac{3-s}{3-5s}\sqrt{\frac{1-3s}{s^{\frac{3}{4}}}}.$$

The integration constants can be determined as:

$$a = \left(\frac{\mu\alpha}{3}\right)^{\frac{1}{4}},$$

$$\alpha = \frac{P}{2}\frac{\lambda}{b} - \frac{C}{\lambda},$$

$$\frac{C_1}{\lambda} = \frac{3k_0 - z_0(1-5s_0)}{3-5z_0},$$

$$\frac{\bar{C}}{\lambda} = \frac{C}{\lambda} - \frac{P}{2}\frac{\lambda}{b},$$

$$t_c = \frac{5}{2}\frac{P\lambda}{b} - 3\frac{\bar{C}}{\lambda}.$$

It can be shown that for small angles of low thrust ($\varphi \sim 0$) one can obtain the following equality:

$$\tan\psi \approx 2\varphi.$$

As it can be seen from these expressions, the behavior of the angle ψ is similar to the same angle in the first class of low thrust solutions.

Eqs. (3.69)–(3.81) describe extremal motion on spiral trajectory around the center of attraction. The equations for r and θ of Eqs. (3.69)–(3.81) are considered to be the parametric equations of the trajectory. The corresponding dependencies for r and θ are presented in Fig. 3.1 and 3.2. The typical examples of the spiral trajectories are given in Fig. 3.3. This completes the second part of the proof of Theorem 3.1. In Eqs. (3.69)–(3.81), the quantity r is increased from $r_0 = 0$ to $r_f = \infty$, whereas θ is increased

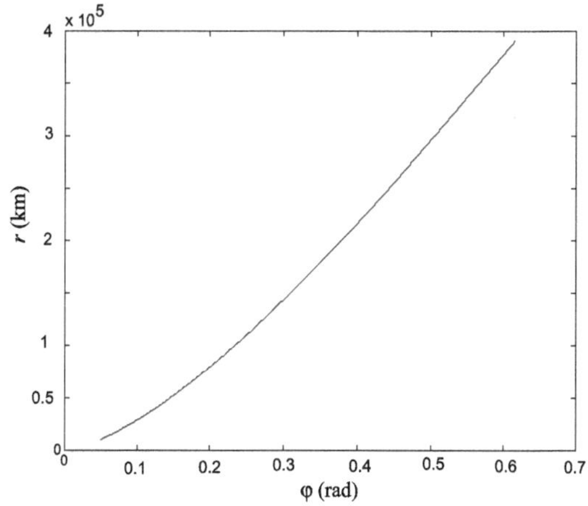

Figure 3.1 Dependency between φ and r.

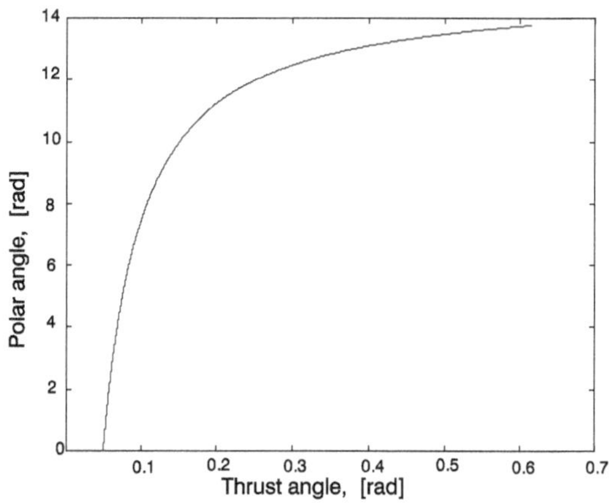

Figure 3.2 Dependency between φ and θ.

from $-\infty$ to

$$\theta_f = \frac{1}{4}(\frac{3}{\tan\varphi_f} + \varphi_f) - \frac{1}{4}(\frac{3}{\tan\varphi_f} + \varphi_f) + \theta_0,$$

and the velocity is decreased from

$$v_0 = \sqrt{\frac{9\sin^2 2\varphi_0 + (3 - sin^2\varphi_0)^2}{(5\sin^2\varphi_0 - 3)^2}}\sqrt{\frac{\mu}{r}_0(1 - 3\sin^3\varphi_0)},$$

to $v_f = 0$, and the angle φ is increased from $\varphi_0 = 0$ to $\varphi_f = \arcsin(\sqrt{\frac{1}{3}})$. Such a behavior of the variables corresponds to an escape of a spacecraft along a spiral trajectory, where the motion has a counterclockwise direction. Besides that, the quantity r is decreased from $r_0 = \infty$ to $r_f = 0$, and the polar angle θ is increased from

$$\theta_0 = \frac{1}{4}(\frac{3}{\tan\varphi_f} + \varphi_f) - \frac{1}{4}(\frac{3}{\tan\varphi_f} + \varphi_f) + \theta_0,$$

to $+\infty$, and the velocity $\vec{v}$ is increased from $v_0 = 0$ to

$$v_0 = \sqrt{\frac{9\sin^2 2\varphi_f + (3 - sin^2\varphi_f)^2}{(5\sin^2\varphi_f - 3)^2}}\sqrt{\frac{\mu}{r}}_0(1 - 3\sin^3\varphi_f),$$

whereas the angle φ will increase from $\varphi_0 = pi - \arcsin(\frac{1}{\sqrt{3}})$ (or decrease from $\pi + \arcsin(\frac{1}{\sqrt{3}})$) to $\varphi_f = 0$ the corresponding trajectory is a spiral arc along which the spacecraft will approach the center of attraction in a counterclockwise direction. The analogical results can be obtained when the thrust angle is replaced by $\varphi + \pi$. The typical examples of such spiral arcs are illustrated in Fig. 3.3 depending on the initial and final thrust angles. In particular, the various time dependencies of the thrust angle for a capture maneuver with duration from 5 to 30 days are illustrated in Fig. 3.4. Besides that from the equations for mass and specific impulse of the system (3.69)–(3.81) it can be seen that I_{sp} be increased $I_{sp,0} \neq I_{sp,max}$ and $I_{sp,f} \neq I_{sp,min}$. The ratio of $I_{sp,f}$ to $I_{sp,0}$ is inverse proportional to the initial mass to the final mass. Note that the thrust angle φ must be non-zero during the motion with low thrust. Consider now the existence of the conjugate points on the thrust arcs analyzed in this section by testing the Jacobi condition. For these tests the equations of the chapter 2 and the solutions of the auxiliary variation problem will be utilized. The integration constants in the solutions obtained are: d, θ_0, φ_0, m_0, λ, b, C_1, t_c, t_0. As in the previous section, consider the following notation.

$$\alpha_1 = d,\ \alpha_2 = \theta_0,\ \alpha_3 = \varphi_0, \alpha_4 = m_0,\ \alpha_5 = \lambda,\ \alpha_6 = b,\ \alpha_7 = C_1,\ \alpha_8 = t_c,\ \alpha_9 = t_0$$

The solutions of the auxiliary variation problem are written in the form:

$$\begin{aligned}
\eta_1 &= \sum R_i \frac{\partial r}{\partial \alpha_i} = R_1 \frac{\partial r}{\partial \alpha_1} = -2\frac{\mu}{d^3}F_1(s),\\
\eta_2 &= \sum V_i \frac{\partial v_1}{\partial \alpha_i} = V_1 \frac{\partial v_1}{\partial \alpha_1} = F_2(s),
\end{aligned}$$

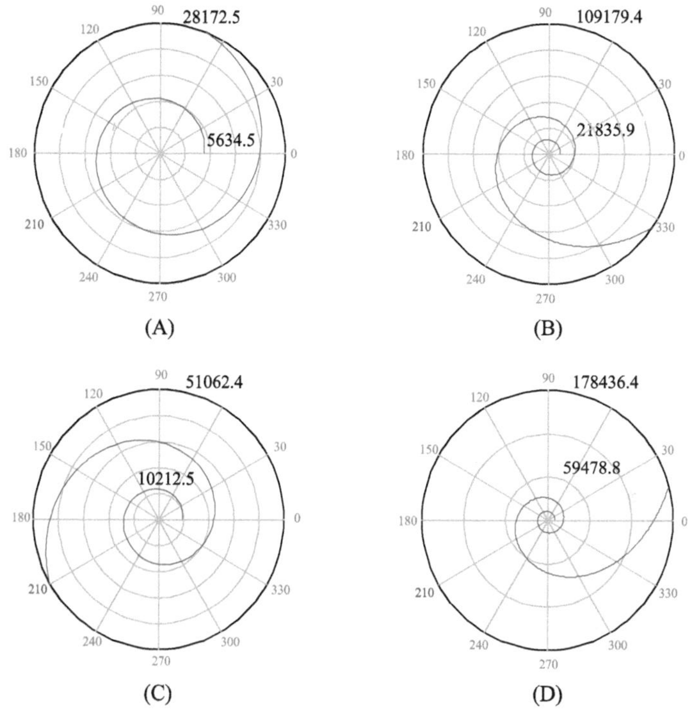

Figure 3.3 Spiral trajectories for $\theta_1 = 0$, $\omega_1 = 0$, $\varphi_1 = 0.05$, $\varphi_2 = 0.101$, 0.15, 0.25, and 0.35. (A) $\theta_1 = 0$, $\omega_1 = 0$, $\varphi_1 = 0.05$, $\varphi_2 = 0.101$, (B) $\theta_1 = 0$, $\omega_1 = 0$, $\varphi_1 = 0.05$, $\varphi_2 = 0.15$, (C) $\theta_1 = 0$, $\omega_1 = 0$, $\varphi_1 = 0.05$, $\varphi_2 = 0.25$, (D) $\theta_1 = 0$, $\omega_1 = 0$, $\varphi_1 = 0.05$, $\varphi_2 = 0.35$.

$$
\begin{aligned}
\eta_3 &= \sum W_i \frac{\partial v_2}{\partial \alpha_i} = W_1 \frac{\partial v_2}{\partial \alpha_1} = F_3(s), \\
\eta_4 &= \sum \Theta_i \frac{\partial \theta}{\partial \alpha_i} = \Theta_2 \frac{\partial \theta}{\partial \theta_0} + \Theta_3 \frac{\partial \theta}{\partial \varphi_0} = \Theta_2 + \Theta_3 (-\frac{3}{\sin^2 \varphi_0} + 1), \\
\eta_5 &= \sum M_i \frac{\partial m}{\partial \alpha_i} = \\
& M_4 \frac{\partial m}{\partial m_0} + M_5 \frac{\partial m}{\partial \lambda} + M_6 \frac{\partial m}{\partial b} + M_7 \frac{\partial m}{\partial C_1} + M_8 \frac{\partial m}{\partial t_c} + M_9 \frac{\partial m}{\partial t_0} = \\
& -M_4 \frac{1}{m_0^2} + M_5 \frac{P\lambda^2}{b^2} t - M_6 \frac{P\lambda^2}{b^3} t - M_7 \frac{1}{t_c \lambda} \\
& -M_8 \frac{1}{t_c^2} [\frac{3zk(1-5s)}{3-5s} - \frac{C_1}{\lambda}] + M_9,
\end{aligned}
$$

where R_1, V_1, W_1 are equated to 1. The angle φ is considered as an independent variable. Taking into account the form of functions F_1, F_2, F_3,

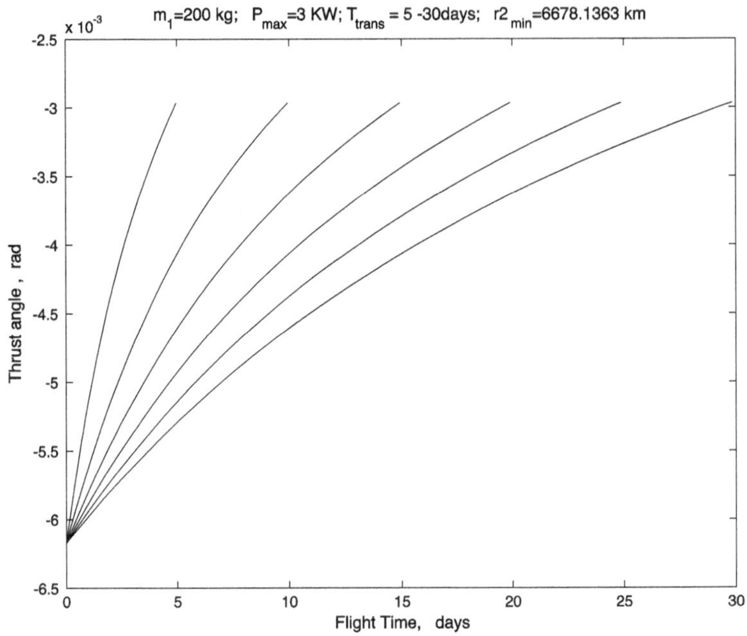

Figure 3.4 Thrust angle vs. flight time.

one can show that η_1, η_2, η_3 can not be equal to zero simultaneously, although η_4 and η_5 can be zeros at certain values of the constants. Consequently, the thrust arcs considered in this section do not contain conjugate points.

In summary, in this chapter, the analytical solutions for the circular and spiral trajectories with low thrust, maximum power and variable specific impulse have been obtained and analyzed. These thrust arcs can find applications in the problems of escape and capture on a given parking orbit. It has been shown that these trajectories do not contain conjugate points.

3.4 MITIGATION OF RADIATION DOSE WHEN PASSING THROUGH THE EARTH RADIATION BELT

In this section the strategy of minimizing the fuel consumption and mitigation of total accumulation of the radiation dose when passing through the Earth radiation belt is presented. It can be shown that the thrust arcs with variable specific impulse allow us to construct a transfer trajectory from the boundaries of the Earth sphere of influence (for instance, when returning from Mars) to a given Earth parking orbit. For practical purposes the parking orbit is considered as a weak elliptical orbit with altitude of 300 km. On this trajectory the spacecraft passes through the Earth radiation belt [144]. From this point of view, in the case of manned missions it is important to conduct analysis on determining an optimal trajectory with the purpose of mitigating the radiation dose. As is known, on altitudes from 1000 km to 14,000 km, the radiation dose is significant compared to the dose on other altitudes. Besides that, between the altitudes of 800 km to 1000 km, the probability of collision with other bodies of various origin or space debris is very high. For mitigation of the radiation dose and decreasing the probability of collision with space debris it seems reasonable to divide the trajectory into two parts. The first part includes the trajectory of descent from the Earth sphere of influence until a certain altitude below 800 km, and the second part of the trajectory is a transfer trajectory from this altitude to the given parking orbit. It has been shown that the radiation dose accumulated on a given altitude explicitly depends on an instant change of dose $D(t)$ on a given altitude, area $A(t)$ that is formed by a curve of change of dose and the axis of the altitude. This dependency is given as

$$A(t) = \int_{t_0}^{t_1} D(t)\,dt,$$

where

$$D(t) = a_1 r(\varphi(t))^2 + a_2 r(\varphi) + a_3, \qquad m = 11, 12$$

and a_1, a_2, a_3 are some constants. The curve of an instant change of dose is determined as

$$D(x) = \frac{dS(x)}{dx},$$

where x is the distance from the Earth surface to spacecraft, $S(x)$ is the area formed by the curve of change of dose and the axis of the altitude. It has been shown that this area denoted below as S_Σ, can be determined by using the equality

$$S_\Sigma = S_1 + S_2 + S_3,$$

where S_1 and S_2 represent the areas which correspond to linear relationships of the curve of dose on altitude, and S_3 corresponds to semi-parabolic part of the curve. The detailed derivation of the formulas of determination of these areas is presented in the Ref. [144]. Approximate relationship of the

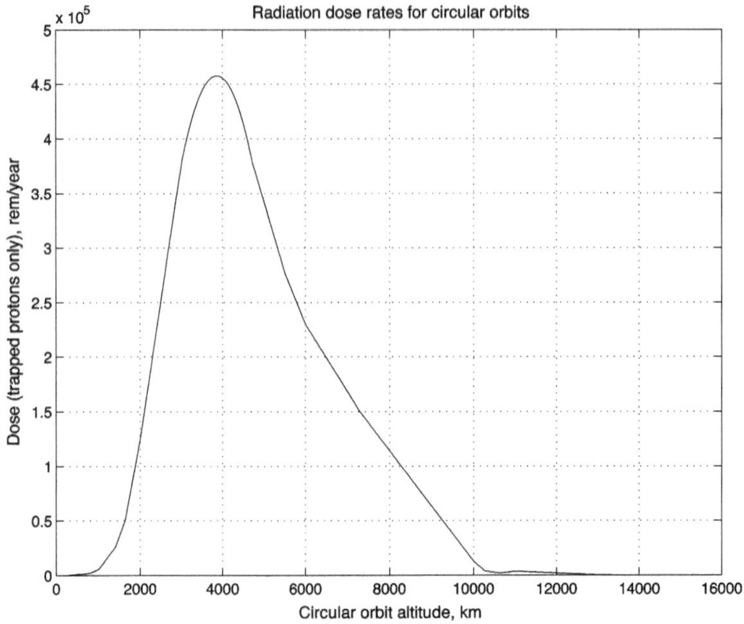

Figure 3.5 Approximate relationship of the dose magnitude from altitude (JSC).

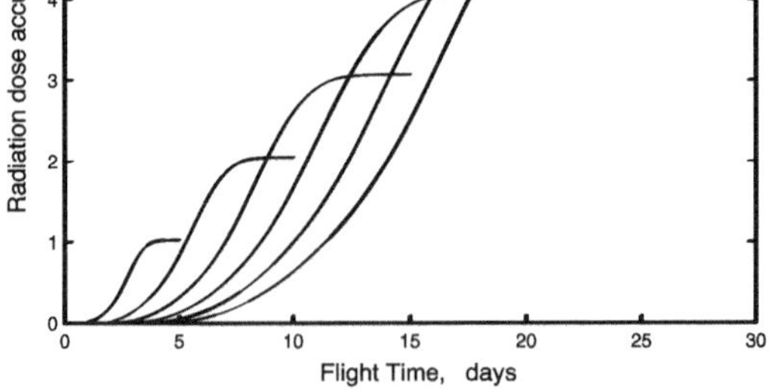

Figure 3.6 Radiation accumulation depending on time.

magnitude of the dose from altitude has been proposed by NASA JSC and illustrated in Fig. 3.5. The radiation dose and its accumulation depending on time of passing through the radiation belt are illustrated in Fig. 3.6. The

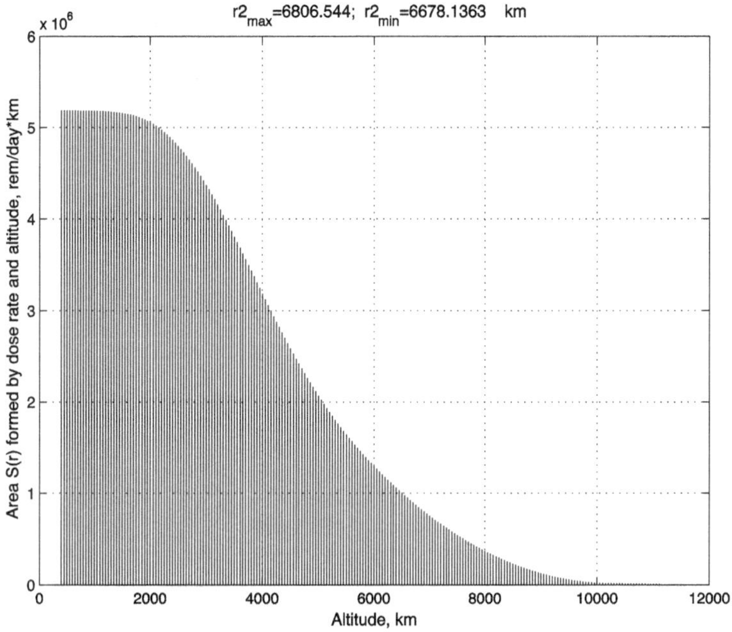

Figure 3.7 Integral of radiation dose vs. altitude.

total dose depending on the altitude is given in Fig. 3.7. It has been shown that the shortest passes through the radiation zone leads to a fast decrease of radiation accumulation. Such transfer trajectories require significant fuel consumption. Nevertheless, the total amount of fuel can be decreased by increasing the time of transfer to the parking orbit. It can also be proved that the longer transfers are more preferable from the economical stand point.

CHAPTER 4

Motion With Variable Power and Constant Specific Impulse

4.1 FIRST INTEGRALS AND INVARIANT RELATIONSHIPS

According to the analysis of the stationary conditions and Weierstrass condition, and the classification of the admissible thrust arcs, the case of motion with variable power and constant specific impulse corresponds to IT arcs, where $P < P_{max}$, $I_{sp} = const$, and $\chi = \frac{c}{m}\lambda - \lambda_7 = 0$ on a non-zero time interval, and $c = I_{sp}g_0$ and $\lambda = |\boldsymbol{\lambda}|$. In this case Eqs. (2.134) with Hamiltonian

$$H = -\frac{\mu}{r^3}(\boldsymbol{\lambda}\mathbf{r}) + (\boldsymbol{\lambda}_r\mathbf{v}) \tag{4.1}$$

has the following first integrals [63], [151]:

$$-\frac{\mu}{r^3}\boldsymbol{\lambda}\mathbf{r} + \boldsymbol{\lambda}_r\mathbf{v} = C, \tag{4.2}$$

$$[\mathbf{v}\boldsymbol{\lambda}] + [\mathbf{r}\boldsymbol{\lambda}_r] = \mathbf{K}(A_1, A_2, C_2), \tag{4.3}$$

$$\boldsymbol{\lambda}\mathbf{v} - 2\mathbf{r}\boldsymbol{\lambda}_r - c\lambda \ln\frac{m_0}{m} + 3Ct = C_1, \tag{4.4}$$

$$\lambda_7 m = C_3, \tag{4.5}$$

$$\lambda_1^2 + \lambda_2^2 + \lambda_3^2 = \lambda^2, \tag{4.6}$$

where C, C_1, C_2, C_3 $(= c\lambda)$, $\mathbf{K}(A_1, A_2, C_2)$, λ are the integration constants. Taking into account the components of all vectors, lets rewrite Eqs. (4.2)–(4.4) in the form [62]:

$$-\lambda_1\frac{\mu}{r^2} + \lambda_1\frac{v_2^2}{r} + \lambda_1\frac{v_3^2}{r} - \lambda_2\frac{v_1v_2}{r} + \lambda_2\frac{v_2v_3}{r}tg\delta -$$

$$-\lambda_3\frac{v_2^2}{r}tg\delta - \lambda_3\frac{v_1v_3}{r} + \lambda_4 v_1 + \lambda_5\frac{v_2}{r\cos\delta} + \lambda_6\frac{v_3}{r} = C, \tag{4.7}$$

$$\lambda_3 v_2\frac{\cos\varphi}{\cos\theta} - \lambda_2 v_3\frac{\cos\varphi}{\cos\theta} - \lambda_5\cos\varphi\, tg\theta + \lambda_6\sin\varphi = A_1, \tag{4.8}$$

$$\lambda_3 v_2\frac{\sin\varphi}{\cos\theta} - \lambda_2 v_3\frac{\sin\varphi}{\cos\theta} - \lambda_5\sin\varphi\, tg\theta - \lambda_6\cos\varphi = A_2, \tag{4.9}$$

$$\lambda_5 = C_2, \tag{4.10}$$

Analytical Solutions for Extremal Space Trajectories
DOI: http://dx.doi.org/10.1016/B978-0-12-814058-1.00004-7

$$\lambda_1 v_1 + \lambda_2 v_2 + \lambda_3 v_3 - 2\lambda_4 r - c\lambda \ln\frac{m_0}{m} + 3Ct = C_1. \tag{4.11}$$

In the case of a planar motion, Eqs. (4.5), (4.6), and (4.7)–(4.11) are reduced to the form:

$$\begin{aligned}
&\lambda_1(-\frac{\mu}{r^2} + \frac{v_2^2}{r}) - \lambda_{v_2}\frac{v_1 v_2}{r} + \lambda_1 v_1 + \lambda_5 \frac{v_2}{r} = C,\\
&\lambda_1 v_1 + \lambda_2 v_2 - 2\lambda_4 r - c\lambda \ln\frac{m_0}{m} + 3Ct = C_1,\\
&\lambda_5 = C_2,\\
&m\lambda_7 = C_3,\\
&\sqrt{\lambda_1^2 + \lambda_2^2} = \lambda,
\end{aligned} \tag{4.12}$$

where $C_3 = \lambda$. It is known that the invariant relationships are very important in studying the canonical systems [67]. On the thrust arcs being considered the following condition must be satisfied: $\chi = 0$, that is all time derivatives of χ must be zero over non-zero time interval [65]. For a planar case, this leads to the following invariant relationships for IT arcs (see Fig. 2.3):

$$\lambda_1\lambda_4 + \lambda_1\lambda_2\frac{v_2}{r} + \lambda_2^2\frac{v_1}{r} + \lambda_5\frac{\lambda_2}{r} = 0, \tag{4.13}$$

$$\lambda_4^2 + (\lambda_1\frac{v_2}{r} - \lambda_2\frac{v_1}{r} + \frac{\lambda_5}{r})^2 = \lambda^2\frac{\mu}{r^3} - 3\lambda_1^2\frac{\mu}{r^3}, \tag{4.14}$$

$$(\lambda^2 - 5\lambda_1^2)v_1 + 2v_1(\lambda_1 v_1 + \lambda_2 v_2) - 4\lambda_1\lambda_4 r = 0, \tag{4.15}$$

$$cm\lambda_1 r[3\lambda^2 - 5\lambda_1^2] + M\lambda\,[\Lambda_1 + \Lambda_2\Lambda_3 + \Lambda_4] = 0, \tag{4.16}$$

where

$$\begin{aligned}
\Lambda_1 &= (\lambda^2 - 5\lambda_1^2)(v^2 - \frac{\mu}{r}),\\
\Lambda_2 &= \frac{3\lambda_1\lambda_2 v_2 + (\lambda_1^2 - 2\lambda_2^2)v_1 + 2\lambda_2\lambda_5}{\lambda_1},\\
\Lambda_3 &= \frac{4\lambda_1\lambda_2 v_2 + (\lambda_1^2 - 2\lambda_2^2)v_1 + 2\lambda_2\lambda_5}{\lambda_1} - \lambda_1 v_1,\\
\Lambda_4 &= v_1(2\lambda^2 v_1 - 5\lambda_1(\lambda_1 v_1 + \lambda_2 v_2)),
\end{aligned}$$

4.2 SPHERICAL TRAJECTORIES

Theorem 4.1. *Let Eqs. (4.2)–(4.5) are the first integrals of the problem (2.111)–(2.115). If the final time is fixed,* $(C \neq 0)$*, then in the case of motion* *with variable power, ($P < P_{max}$) and constant specific impulse, ($I_{sp} = const$), there exists at least one family of extremals, representing the three dimensional spherical trajectories which are described in elementary functions and a quadrature.*

Proof of Theorem 4.1. Taking into account the components of the vectors $\boldsymbol{\lambda}$ and $\boldsymbol{\lambda}_r$, Eqs. (4.13) and (4.14) are rewritten as [148]:

$$I_1 = \boldsymbol{\lambda}\boldsymbol{\lambda}_r = 0, \qquad I_2 = \lambda_r^2 + 3\frac{\mu}{r^5}(\boldsymbol{\lambda}\mathbf{r})^2 - \frac{\mu}{r^3}\lambda^2 = 0. \tag{4.17}$$

Now, form the Poisson brackets to obtain [64]:

$$(I_1, I_2) = 9\frac{\mu}{r^5}(\boldsymbol{\lambda}\mathbf{r})\lambda^2 - 15\frac{\mu}{r^7}(\boldsymbol{\lambda}\mathbf{r})^3 = K_1 \tag{4.18}$$

$$(I_1, (I_1, I_2)) = 105\frac{\mu}{r^9}(\boldsymbol{\lambda}\mathbf{r})^4 - 90\frac{\mu}{r^7}(\boldsymbol{\lambda}\mathbf{r})^2\lambda^2 + 9\frac{\mu}{r^5}\lambda^2 = K_2, \tag{4.19}$$

where K_1, K_2 are the integration constants. Eq. (4.19) is considered as a bi-quadratic equation with respect to $\boldsymbol{\lambda}\mathbf{r}$. The solution of the equation has the form:

$$\boldsymbol{\lambda}\mathbf{r} = r\sqrt{\frac{3}{7}\lambda^2 + \sqrt{\frac{K_2 r^5}{105\mu} + \frac{24}{245\lambda^4}}}. \tag{4.20}$$

Substituting Eq. (4.20) into Eq. (4.18), one can obtain the 16-th order nonlinear equation with respect to r:

$$K_1^4 r^{16} - \frac{15}{343}K_2^3\mu r^{15} - \frac{18}{49}K_1^2 K_2\mu\lambda^2 r^{13} - \frac{828}{343}K_2^2\mu^2\lambda^4 r^{10} + \frac{8460}{343}K_1^2\mu^2\lambda^6 r^8 - \frac{324}{2401}\mu^3\lambda^8\left(\frac{1409}{5}K_2 r^5 + \frac{1749091}{49}\mu\lambda^4\right) = 0.$$

As is known, if the coefficient of a highest degree and free term of an equation with real coefficients have various signs, then this equation has at least one positive root [135]. In this case the roots of the equation above obtained are $r_i = const$, $(i \leq 16)$. This corresponds to a spherical motion, that is the relationships are satisfied at $r = const$. Consider the IT arcs, which lie on a spherical surface with radius $r = const$. On IT arcs the following equality takes place:

$$\chi^{(3)} = 6\frac{\mu\beta}{m\lambda^2 r^5}(\boldsymbol{\lambda}\mathbf{r}) + (2\dot{\boldsymbol{\lambda}}\mathbf{r} + \boldsymbol{\lambda}\dot{\mathbf{r}}) = 0.$$

Then the following cases are possible: 1. $\boldsymbol{\lambda}\mathbf{r} = 0$, $-2\boldsymbol{\lambda}_r\mathbf{r} + \boldsymbol{\lambda}\mathbf{v} \neq 0$; 2. $\boldsymbol{\lambda}\mathbf{r} \neq 0$, $-2\boldsymbol{\lambda}_r\mathbf{r} + \boldsymbol{\lambda}\mathbf{v} \neq 0$; 3. $\boldsymbol{\lambda}\mathbf{r} = 0$, $-2\boldsymbol{\lambda}_r\mathbf{r} + \boldsymbol{\lambda}\mathbf{v} = 0$. Consider the following cases:

1. Let

$$\boldsymbol{\lambda}\mathbf{r}=0,\ -2\boldsymbol{\lambda}_r\mathbf{r}+\boldsymbol{\lambda}\mathbf{v}\neq 0. \tag{4.21}$$

In this case, from Eqs. (2.135) one can obtain the following equations:

$$\begin{aligned}
0 &= \frac{c\beta}{m}\frac{\lambda_1}{\lambda}-\frac{\mu}{r^2}+\frac{v_2^2}{r}+\frac{v_3^2}{r},\\
\dot{v_2} &= \frac{c\beta}{m}\frac{\lambda_2}{\lambda}+\frac{v_2v_3}{r}tg\delta,\\
\dot{v_3} &= \frac{c\beta}{m}\frac{\lambda_3}{\lambda}-\frac{v_2^2}{r}tg\delta,
\end{aligned} \tag{4.22}$$

and

$$\begin{aligned}
\dot{\lambda_1} &= \lambda_2\frac{v_2}{r}+\lambda_3\frac{v_3}{r}-\lambda_4,\\
\dot{\lambda_2} &= -2\lambda_1\frac{v_2}{r}-\lambda_2\frac{v_3}{r}tg\delta+2\lambda_3\frac{v_2}{r}tg\delta-\frac{\lambda_5}{r\cos\delta},\\
\dot{\lambda_3} &= -2\lambda_1\frac{v_3}{r}-\lambda_2\frac{v_2}{r}tg\delta-\frac{\lambda_6}{r}.
\end{aligned} \tag{4.23}$$

From the first equation of Eq. (4.21) it follows that $\lambda_1=0$. Then the first equations of Eqs. (4.22) and (4.23) take the following form:

$$v_2^2+v_3^2=\frac{\mu}{r}, \qquad \lambda_2v_2+\lambda_3v_3=\lambda_4r. \tag{4.24}$$

Multiplying the second equation of Eqs. (4.22) by v_2, and the third equation by v_3, and adding them together, one can obtain

$$v_2\dot{v_2}+v_3\dot{v_3}=\frac{c\beta}{m\lambda}(\lambda_2v_2+\lambda_3v_3). \tag{4.25}$$

From Eqs. (4.24), (4.25) it follows that $\beta=0$ or $(\lambda_2v_2+\lambda_3v_3)=0$ or both are equal to zero.

Let $\beta\neq 0$, $(\lambda_2v_2+\lambda_3v_3)=0$. Then due to Eqs. (4.24) we have $r\neq 0$, $\lambda_4=0$. Now take the projections of the equation for the primer vector

$$\ddot{\boldsymbol{\lambda}}=3\frac{\mu}{r^5}(\boldsymbol{\lambda}\mathbf{r})\mathbf{r}-\frac{\mu}{r^3}\boldsymbol{\lambda}$$

on the directions of the radius vector and with Eqs. (4.24) one can obtain $\beta=0$ [118]. But this contradicts to the assumption that $m\neq 0$. The similar result may be obtained in the case 3.

2. Let

$$\boldsymbol{\lambda}\mathbf{r} \neq 0, -2\boldsymbol{\lambda}_r\mathbf{r} + \boldsymbol{\lambda}\mathbf{v} \neq 0. \tag{4.26}$$

Due to Eqs. (4.20) we have $\lambda_1 = const$. From Eqs. (4.26) we can obtain

$$-2\lambda_4 r + \lambda_2 v_2 + \lambda_3 v_3 = 0. \tag{4.27}$$

Then from Eqs. (4.24) and (4.27) it follows that

$$\lambda_4 = 0, \qquad \lambda_2 v_2 + \lambda_3 v_3 = 0. \tag{4.28}$$

Then the projection of the primer vector equation on the radius vector's direction is

$$\frac{c\beta}{m}/\lambda/ = 3\frac{\mu}{r^2}\lambda_1. \tag{4.29}$$

If $\lambda_1 > 0$, then we have $\frac{c\beta}{m}\lambda = 3\frac{\mu}{r^2}\lambda_1$; if $\lambda_1 < 0$ then $-\frac{c\beta}{m}\lambda = 3\frac{\mu}{r^2}\lambda_1$. Based on the equation for λ_4, and Eqs. (4.28), one can derive the following expression for the Hamiltonian:

$$C = \frac{\mu}{r^2}\lambda_1. \tag{4.30}$$

Here λ_1 and C will have the same signs. Integrating Eq. (4.29), and using the condition that the switching function is zero, we can obtain the formulas for mass and its multiplier:

$$m = m_0 e^{-3\frac{\mu\lambda_1}{r^2 c\lambda}}, \qquad \lambda_7 = \frac{c\lambda}{m_0} e^{3\frac{\mu\lambda_1}{r^2 c\lambda}}. \tag{4.31}$$

Other solutions for the spherical arcs are obtained as follows. Differentiating the expression

$$\lambda_2^2 + \lambda_3^2 = \lambda^2 - \lambda_1^2 = const \tag{4.32}$$

with Eq. (4.29) we have

$$v_3 = \lambda_6 N_1, \tag{4.33}$$

where

$$N_1 = \frac{-3\lambda_1^2 + \lambda^2}{\lambda_1(3\lambda_1^2 + \lambda^2)}. \tag{4.34}$$

Multiplying Eq. (4.8) by $\sin\varphi$, and Eq. (4.9) by $\cos\varphi$ and subtracting the second equation from the first one, we have

$$\lambda_6 = a\sin(\varphi - \alpha), \tag{4.35}$$

where a and α are the integration constants, which are associated with c_1, c_2 and $tg\alpha = \frac{c_2}{c_1}$, $a = \sqrt{c_1^2 + c_2^2}$. Substituting Eqs. (4.35) into Eq. (4.33), and also taking into account the equality

$$v_2^2 + v_3^2 = \frac{\mu}{r} - 3\frac{\mu}{r}\frac{\lambda_1^2}{\lambda^2}, \tag{4.36}$$

obtained by using the equation for v_1 of Eqs. (2.135) and relationships Eqs. (4.30), one will have

$$v_3 = N_1 a \sin(\varphi - \alpha), \tag{4.37}$$

$$v_2 = \sqrt{N_2 - N_1^2 a^2 \sin^2(\varphi - \alpha)}, \tag{4.38}$$

where

$$N_2 = \mu\frac{-3\lambda_1^2 + \lambda^2}{r\lambda^2}.$$

Then Eqs. (4.28) and (4.32) allow us to find the primer vector components:

$$\lambda_2 = N_1 a\frac{\sqrt{\lambda^2 - \lambda_1^2}}{\sqrt{N_2}}\sin(\varphi - \alpha), \tag{4.39}$$

$$\lambda_3 = \frac{\sqrt{\lambda^2 - \lambda_1^2}}{\sqrt{N_2}}\sqrt{N_2 - N_1^2 a^2 \sin^2(\varphi - \alpha)}. \tag{4.40}$$

To determine the relationship between φ and θ, differentiate Eq. (4.33):

$$\frac{v_2^2}{r}tg\theta - \lambda_5\frac{N_1 v_2}{r\cos\theta}tg\theta + \lambda_3(\frac{N_1(v_2^2 + v_3^2)}{r\cos^2\theta} - \frac{c\beta}{m\lambda}) = 0.$$

Solving this equation for v_2 and comparing it with Eq. (4.38), one can have

$$\tau[N_2\tau + \lambda_5\frac{N_1\sqrt{N_2}}{\cos\theta}tg\theta + \sqrt{\lambda^2 - \lambda_1^2}(\frac{N_1(v_2^2 + v_3^2)}{r\cos^2\theta} - \frac{rcm}{M\lambda})\cot\theta] = 0,$$

where

$$\tau = \sqrt{1 - \frac{N_1^2}{N_2^2}a^2\sin^2(\varphi - \alpha)}. \tag{4.41}$$

The following cases take place: 1. If $\tau = 0$, then

$$\varphi = \arcsin\sqrt{\frac{N_2}{N_1^2 a^2}} + \alpha = const = \varphi_1.$$

From Eqs. (2.135) and (4.28) we obtain

$$\lambda_2 = 0, \tag{4.42}$$

and also $\lambda_1^2 = \lambda^2$. In this case on the right side of Eq. (4.29) there will be a negative quantity that is not acceptable. From this it follows that the case (4.42) is not admissible. Therefore, the case (1) does not give any solution for IT arcs.

2) If $\tau \neq 0$, then

$$N_2\tau + \lambda_5 \frac{N_1\sqrt{N_2}}{\cos\theta} tg\theta + \sqrt{\lambda^2 - \lambda_1^2}\left(\frac{N_1(v_2^2 + v_3^2)}{r\cos^2\theta} - \frac{rc\beta}{m\lambda}\right)\cot\theta = 0.$$

From this and from Eq. (4.41) one can have the expression for the angle φ in terms of θ:

$$\varphi - \alpha = \arcsin\left[\frac{N_2}{a}\sqrt{\frac{\mu}{r}}N_3\right]. \tag{4.43}$$

Here

$$N_3 = \sqrt{\frac{s_1\sin^2 2\theta - (s_2 + s_3\cos^2\theta + s_4\sin\theta)^2}{s_5\sin^2 2\theta}}.$$

Consequently, the variables $v_2, v_3, \lambda_2, \lambda_3$ can also be expressed via θ using Eq. (4.43). Substituting Eq. (4.43) into Eq. (4.37), one can obtain

$$v_3 = \sqrt{\frac{\mu}{r}}N_3, \tag{4.44}$$

where

$$s_1 = \frac{\lambda_1^2}{4}(-9\lambda_1^4 + \lambda^4)^2, \qquad s_2 = \sqrt{\lambda^2 - \lambda_1^2}(-9\lambda_1^4 + \lambda^4)$$

$$s_3 = -3\lambda_1^2\sqrt{\lambda^2 - \lambda_1^2}(3\lambda_1^2 + \lambda^2), \tag{4.45}$$

$$s_4 = \lambda\lambda_5\frac{r}{\mu}(-3\lambda_1^2 + \lambda^2)^{\frac{3}{2}}, \qquad s_5 = \lambda_1^2\lambda^2\frac{(-9\lambda_1^4 + \lambda^4)}{4(-3\lambda_1^2 + \lambda^2)}.$$

Substituting Eqs. (4.44) into the equation of (2.135) for θ, we can obtain:

$$t - t_0 = \sqrt{\frac{r^3}{\mu}}\int\frac{d\theta}{N_3}, \tag{4.46}$$

where t_0 is the integration constant.

Theorem 4.1 has been proved. □

To determine the boundary values of φ and θ compute the time derivative of (4.37) and compare with the third equation of Eqs. (4.22):

$$N_1 a \cos(\varphi - \alpha) \frac{v_2}{r \cos \theta} = \lambda_3 \frac{c\beta}{m\lambda} - \frac{v_2^2}{r} tg\theta.$$

From this and from Eqs. (4.29), (4.38) and (4.40) for $\varphi = 0$ one can obtain

$$\sqrt{N_2} N_1 a \cos(\varphi - \alpha) = 3\mu \frac{\sqrt{\lambda^2 - \lambda_1^2}}{r^2 \lambda^2}.$$

Consequently, when θ is changed from $\theta_1 = 0$ to some θ_2, the angle φ is changed from

$$\varphi_1 = \arcsin \sqrt{\frac{N_4}{N_5 N_1^2 a^2}} \quad to \quad \varphi_2 = \arcsin \frac{N_2}{a} \sqrt{\frac{\mu}{r} N_3}$$

Here

$$N_4 = 36\mu^2 \lambda_1^2 (\lambda^2 - \lambda_1^2) - N_2^2 a^2 r^4 \lambda^4, \quad N_5 = 36\mu^2 \lambda_1^2 (\lambda^2 - \lambda_1^2) - N_2 r^4 \lambda^4.$$

Denote by γ the angle between the velocity vector and the coordinate axis that shows the direction of φ, and η denotes the angle between the projections of the thrust vector on the tangent plane to the sphere and the coordinate axis for φ. Then these angles are

$$tg\gamma = \frac{v_2}{v_3}, \qquad tg\eta = \frac{\lambda_2}{\lambda_3}. \tag{4.47}$$

From Eq. (4.38) it follows that $\gamma = \eta + \frac{\pi}{2}$. If $N_1^2 a^2 = N_2$, then Eq. (4.47) gives

$$\gamma = \theta - \theta_0 + \frac{\pi}{2}, \qquad \eta = \theta - \theta_0. \tag{4.48}$$

It is seen from (4.48) that on all IT arcs lying on a spherical surface, the thrust direction remains perpendicular to the direction of motion, and in the case when $N_1^2 a^2 = N_2$, it is inclined to a constant angular distance from the angle θ in the plane where $\varphi = 0$. So, if the flight time is fixed, then the spherical IT arcs are described by Eqs. (4.31), (4.35), (4.37), (4.40), (4.43), (4.45), and (4.46), which allow us to determine $v_2, v_3, \varphi, M, \lambda_2, \lambda_3, \lambda_6, \lambda_7, t, \eta, \gamma$ depending on the independent variable θ and can describe any spherical trajectories.

Existence of Conjugate Points

To determine the existence of the conjugate points, consider the derivatives of $r, \varphi, v_1, v_2, v_3, m$ with respect to the constants a, α, C, r_0, m_0:

$$\eta_{11} = \frac{\partial r}{\partial a} = 0, \quad \eta_{12} = \frac{\partial r}{\partial \alpha} = 0, \quad \eta_{13} = \frac{\partial r}{\partial C} = 0,$$

$$\eta_{14} = \frac{\partial}{\partial r_0} = 0, \quad \eta_{15} = \frac{\partial r}{\partial m_0} = 0,$$

$$\eta_{21} = \frac{\partial \varphi}{\partial a} = \frac{-\frac{N_2}{a^2}\sqrt{\frac{\mu}{r}}N_3}{\sqrt{1 - [\frac{N_2}{a^2}\sqrt{\frac{\mu}{r}}N_3]^2}}, \quad \eta_{22} = \frac{\partial \varphi}{\partial \alpha} = 1, \quad \eta_{23} = \eta_{24} = \eta_{25} = 0,$$

$$\eta_{31} = \frac{\partial v_2}{\partial a} = 0, \quad \eta_{32} = \eta_{33} = \eta_{34} = \eta_{35} = 0,$$

$$\eta_{41} = \frac{\partial v_2}{\partial a} = \frac{-2N_1^2 a^2 \sin^2(\varphi - \alpha)}{\sqrt{N_2 - N_1^2 a^2 \sin^2(\varphi - \alpha)}},$$

$$\eta_{42} = \frac{\partial v_2}{\partial \alpha} = \frac{2N_1^2 a^2 \sin(\varphi - \alpha)\cos(\varphi - \alpha)}{\sqrt{N_2 - N_1^2 a^2 \sin^2(\varphi - \alpha)}}, \quad \eta_{43} = \eta_{44} = \eta_{45} = 0,$$

$$\eta_{51} = \frac{\partial v_3}{\partial a} = N_1 \sin(\varphi - \alpha), \quad \eta_{52} = \frac{\partial v_3}{\partial \alpha} = -N_1 \cos(\varphi - \alpha),$$

$$\eta_{53} = \eta_{54} = \eta_{55} = 0,$$

$$\eta_{61} = \eta_{62} = \eta_{63} = 0, \quad \eta_{64} = \frac{\partial m}{\partial r_0} = m_0 \exp(-\frac{\mu}{r_0^2 c}t)(-\frac{2\mu}{r_0^2 c}t),$$

$$\eta_{65} = \frac{\partial m}{\partial m_0} = \exp(-\frac{\mu}{r_0^2 c}t).$$

The expressions for η_3 on a non-zero time interval contradict to the assumption that the solutions of the auxiliary equations are not zeros on interval $t_0 < t' < t_1$. Consequently, the spherical IT arcs do not contain the conjugate points.

4.3 CIRCULAR TRAJECTORIES

Note that for various constants λ_1, a, α in Eq. (2.21), there may exist particular solutions. For example, if the minimization of the characteristic velocity is considered ($\lambda = 1$), with conditions $\lambda_1 = \frac{1}{3}$ and $a = 0$, then from Eqs. (4.31), (4.35), and (4.37)–(4.40) one can obtain the following particular solutions:

$$v_1 = 0, \qquad v_2 = \sqrt{\frac{2}{3}\frac{\mu}{r}}, \qquad v_3 = 0,$$
$$r = r_0, \qquad \varphi = \sqrt{\frac{2\mu}{3r^3\cos^3\theta}}t + \varphi_0, \qquad \theta = \theta_0, \tag{4.49}$$
$$\lambda_2 = 0, \quad \lambda_3 = \frac{2\sqrt{2}}{3}, \quad \lambda_4 = 0, \quad \lambda_5 = 0, \quad \lambda_6 = 0.$$

If $a = -\frac{2}{3}\frac{\sqrt{2\mu}}{r_0}$ with $\lambda_1 = \frac{1}{3}$, then from the equations above mentioned we can obtain

$$v_1 = 0, \quad v_2 = \sqrt{\frac{2\mu}{3r_0}(1 - 3\sin^2(\varphi - \alpha)}, \quad v_3 = -\sqrt{\frac{2\mu}{r_0}}\sin(\varphi - \alpha),$$
$$\theta = \arcsin(\frac{1}{\sqrt{3}}\sin\nu), \quad \nu = \sqrt{\frac{6\mu}{r}}(t - t_0), \quad \varphi = \frac{\sqrt{1 - 3\sin^2\theta}}{\sqrt{3}\sin\theta} + \alpha \tag{4.50}$$
$$\lambda_2 = -\frac{2\sqrt{2}}{\sqrt{3}}\sin(\varphi - \alpha), \quad \lambda_3 = -\frac{2\sqrt{2}}{\sqrt{3}}\sqrt{\frac{1}{3} - \sin^2(\varphi - \alpha)}, \quad \lambda_4 = 0,$$
$$\lambda_5 = const, \quad \lambda_6 == -\frac{2}{3}\sqrt{\frac{2\mu}{r}}\sin(\varphi - \alpha).$$

The following expressions are valid for the solutions (4.49), (4.50):

$$m = m_0 e^{-\frac{\mu}{r_0^3 c}t}, \qquad \lambda_7 = \frac{c\lambda}{m_0}e^{\frac{\mu}{r_0^3 c}t}. \tag{4.51}$$

Note that Eqs. (4.49)–(4.51) have been obtained by utilizing the Levi-Civita's method of determining the particular solutions [62], [139]. From Eq. (4.29) it follows that λ_1 can accept the negative values. In the case when ($\lambda_1 < 0$), this means that the spherical trajectories obtained above can satisfy the necessary conditions of optimality. From Eq. (4.29) it also follows that there may exist non-optimal spherical trajectories, which are consistent with the conclusion of Ref. [86], although in this work the constraints are imposed on the thrust magnitude.

4.4 EXTREMALS FOR MANEUVERS WITH FREE FINAL TIME

In the case of motion with variable power and constant specific impulse, the system Eqs. (2.134) with Hamiltonian

$$H = -\frac{\mu}{r^3}(\boldsymbol{\lambda}\mathbf{r}) + (\boldsymbol{\lambda}_r\mathbf{v}) \tag{4.52}$$

admits following first integrals [150]:

$$\lambda_1(-\frac{\mu}{r^2}+\frac{v_2^2}{r})-\lambda_2\frac{v_1v_2}{r}+\lambda_4v_1+\lambda_5\frac{v_2}{r} = C, \tag{4.53}$$

$$\lambda_1v_1+\lambda_2v_2-2\lambda_4r-c\lambda\ln\frac{m_0}{m}+3Ct = C_1, \tag{4.54}$$

$$\lambda_5 = C_2, \tag{4.55}$$

$$m\lambda_7 = C_3, \tag{4.56}$$

$$\sqrt{\lambda_1^2+\lambda_2^2} = \lambda, \tag{4.57}$$

and the following invariant relationships for IT arcs:

$$\lambda_1\lambda_4+\lambda_1\lambda_2\frac{v_2}{r}+\lambda_2^2\frac{v_1}{r}+\lambda_5\frac{\lambda_2}{r}=0, \tag{4.58}$$

$$\lambda_4^2+(\lambda_1\frac{v_2}{r}-\lambda_2\frac{v_1}{r}+\frac{\lambda_5}{r})^2=\lambda^2\frac{\mu}{r^3}-3\lambda_1^2\frac{\mu}{r^3}, \tag{4.59}$$

$$(\lambda^2-5\lambda_1^2)v_1+2\lambda_1(\lambda_1v_1+\lambda_2v_2)-4\lambda_1\lambda_4r=0, \tag{4.60}$$

$$cm\lambda_1r[3\lambda^2-5\lambda_1^2]+M\lambda[(\lambda^2-5\lambda_1^2)(v^2-\frac{\mu}{r})+$$
$$[\frac{3\lambda_1\lambda_2v_2+(\lambda_1^2-2\lambda_2^2)v_1+2\lambda_2\lambda_5}{\lambda_1}][[\frac{4\lambda_1\lambda_2v_2+(\lambda_1^2-2\lambda_2^2)v_1+2\lambda_2\lambda_5}{\lambda_1}]$$
$$-\lambda_1v_1]+v_1(2\lambda^2v_1-5\lambda_1(\lambda_1v_1+\lambda_2v_2))]=0. \tag{4.61}$$

If the final polar angle is fixed, and the flight time is free, then $C_2\neq 0$ and $C=0$ [3], [143]. From Eqs. (4.119) it follows that

$$r=\frac{9\mu\lambda^2}{C_2^2}\frac{s^6}{1-3s^2}, \tag{4.62}$$

which is similar to the Lawden's formula for $\lambda=1$. Here $s=\sin\varphi$, $\lambda_5=C_2$. It can be shown that in the case being considered that integrals in Eqs. (4.52)–(4.56) and invariant relationships in Eqs. (4.58)–(4.61) allow us to obtain two classes of solutions which are presented below [67], [63], [53], [152], [153]. Prove the following theorem.

Theorem 4.2. *Let in the problem (2.111)–(2.115) the canonical system Eqs. (2.134) admits the first integrals (4.53)–(4.57). Then, if the final time is not fixed, ($C=0$), and the final polar angle is fixed, then in the case of motion with variable power, ($P<P_{max}$), and constant specific impulse, ($I_{sp}=const$), there exist at least two classes of planar spiral extremals, one of which represents alternative class of Lawden spirals. The solutions for both classes are expressed in terms*

of elementary functions and quadratures. In analytical solutions for these classes the magnitudes of radius vectors are described by the same formula, and the other solutions have the same structure, but are described by various functions depending on the thrust angle.

The proof of this theorem consists of two parts:

First part of the Proof of Theorem 4.2. Rewrite Eqs. (4.118) in the form:

$$\lambda_4 = \pm\frac{\lambda_2}{\lambda}\sqrt{\frac{\mu}{r^3}(\lambda^2 - 3\lambda_1^2)},$$

$$\lambda_1\frac{v_2}{r} - \lambda_2\frac{v_1}{r} + \frac{C_2}{r} = \mp\frac{\lambda_1}{\lambda}\sqrt{\frac{\mu}{r^3}(\lambda^2 - 3\lambda_1^2)},$$

or

$$\frac{\lambda_4}{\lambda} = \pm k\sqrt{\frac{\mu}{r^3}(1 - 3s^2)}, \tag{4.63}$$

$$s\frac{v_2}{r} - k\frac{v_1}{r} + \frac{n}{r} = \mp s\sqrt{\frac{\mu}{r^3}(1 - 3s^2)}, \tag{4.64}$$

where

$$s = \sin\varphi, \quad k = \cos\varphi, \quad \lambda_1 = \lambda s, \quad \lambda_2 = \lambda k, \quad n = \pm\frac{C_2}{\lambda}. \tag{4.65}$$

Substituting Eq. (4.63) and (4.64) in (4.53), one can have [67]

$$\frac{C^2}{\lambda_v^2}r^4 + 6\frac{C}{\lambda_v}\mu s^3 r^2 - n^2\mu(1 - 3^2)r + 9\mu^2 s^6 = 0. \tag{4.66}$$

Some solutions of this equation in the case of arbitrary C and n have been obtained in Ref. [61] and [67]. It will be shown below that using (4.63), (4.64), (4.66) and other first integrals and invariant relationships, we obtain the solutions of Eqs. (2.134) in terms of functions of φ (see Fig. 4.15). If $n \neq 0$ and $C = 0$, then from Eqs. (4.66) it follows that

$$r = 9\frac{\mu}{n^2}\frac{s^6}{1 - 3s^2}. \tag{4.67}$$

Using Eqs. (4.67) and (4.65), it is possible to rewrite Eqs. (4.64) and (4.60) in the form:

$$sv_2 - kv_1 \quad = \quad \pm n\frac{(1 - 3s^2)}{3s^2} - n,$$

$$sv_1 + kv_2 = \mp 2n\frac{k(1-3s^2)}{3s^3} - \frac{1-5s^2}{2s}v_1. \tag{4.68}$$

The velocity components v_1 and v_2 can be found from Eqs. (4.68) in the form of functions of φ. Below the first and second classes of extremals of Eqs. (2.134) can be found by selecting the corresponding sign on the right sides of Eqs. (4.63) and (4.68).

4.4.1 First Class of Extremals – Lawden Spirals

By selecting the sign on the right sides of Eqs. (4.63) and (4.68), one can obtain [150]

$$\begin{aligned} sv_2 - kv_1 &= -n\frac{(1-3s^2)}{3s^2} - n, \\ sv_1 + kv_2 &= 2n\frac{k(1-3s^2)}{3s^3} - \frac{1-5s^2}{2s}v_1. \end{aligned} \tag{4.69}$$

Eqs. (4.69) allow us to find v_1 and v_2 in the form:

$$\begin{aligned} v_1 &= 2n\frac{k(1-2s^2)}{s^2(3-5s^2)}, \\ v_2 &= \frac{n}{3}\frac{3-13s^2+12s^4}{s^3(3-5s^2)}. \end{aligned} \tag{4.70}$$

Eqs. (4.67) and (4.70) can be rewritten in a compact form:

$$\begin{aligned} r &= \frac{\mu}{n^2}L_1(s), \\ v_1 &= nL_2(s), \\ v_2 &= nL_3(s), \end{aligned} \tag{4.71}$$

where

$$L_1(s) = 9\frac{s^6}{1-3s^2}, \quad L_2(s) = 2\frac{k(1-2s^2)}{s^2(3-5s^2)}, \quad L_3(s) = \frac{3-13s^2+12s^4}{3s^3(3-5s^2)}.$$

Note that the solutions (4.71) are valid if $\sin^2\varphi < \frac{1}{3}$, or when $|\varphi| < 0.6155$. For a capture maneuver, which will be considered below, one can have $\frac{dr}{dt} < 0$ at any instant of time, and consequently, $n < 0$, as $L_2(s) > 0$ during the entire maneuver.

Substituting r, v_1, and v_2 in Eqs. (4.54), we obtain:

$$m = m_0 \exp h(\varphi), \tag{4.72}$$

where

$$h(\varphi) = \frac{1}{c}\left[n\frac{k(10s^4 - 7s^2 + 1)}{s^3(3 - 5s^2)} + d\right],$$

and $d = \frac{C_1}{\lambda} = const$. Using $\dot{r} = v_1$, the expressions (4.71) and

$$\frac{dr}{dt} = \frac{dr}{d\varphi}\frac{d\varphi}{dt},$$

one can find time t as a function of φ in quadratures:

$$t = t_0 + 27\frac{\mu}{n^3}\int_{\varphi_0}^{\varphi} \frac{s^7(3 - 5s^2)}{(1 - 3s^2)^2}d\varphi, \tag{4.73}$$

which can be integrated to obtain

$$t = t_0 + 27\frac{\mu}{n^3}\left(-\frac{1}{9}k^5 + \frac{32}{81}k^3 - \frac{5}{9}k - \frac{k}{81(3k^2 - 2)} - \frac{2\sqrt{2}}{81\sqrt{3}}\ln\left|\frac{3k - \sqrt{6}}{3k + \sqrt{6}}\right|\right)\Bigg|_{\varphi_0}^{\varphi}. \tag{4.74}$$

Complete integration of Eq. (4.73) and derivation of Eq. (4.74) are presented in the next subsection. The equations for θ can be obtained using $\dot{\theta} = \frac{v_2}{r}$, the formula for v_2 of Eqs. (4.71) and

$$\frac{d\theta}{dt} = \frac{d\theta}{d\varphi}\frac{d\varphi}{dt},$$

in the following form:

$$\frac{d\theta}{d\varphi} = \frac{3 - 13s^2 + 12s^4}{s^2(3 - 5s^2)} = \frac{3}{s^2} - 4,$$

by integrating of which, we can find (see Fig. 4.15)

$$\theta = \theta_0 - 3\cot\varphi - 4\varphi. \tag{4.75}$$

Note that as θ_2 is fixed, the transversality condition is not applicable. The constants $C_1, C_2, \lambda_v, t_0, \theta_0$ can be found by using the boundary conditions. From Eq. (4.72) we find

$$\frac{dm}{d\varphi} = n\frac{m_i}{c}\frac{(1 - 3s^2)(9 - 25s^2 + 20s^4)}{s^4(3 - 5s^2)^2}. \tag{4.76}$$

By accepting

$$\frac{dt}{d\varphi} = 27\frac{\mu}{n^3}\frac{s^7(3-5s^2)}{(1-3s^2)^2},$$

it can be found that

$$\beta = -\frac{\frac{dm}{d\varphi}}{\frac{dt}{d\varphi}} = -\frac{n^4}{27\mu}\frac{m}{c}\frac{(1-3s^2)^3(9-25s^2+20s^4)}{s^{11}(3-5s^2)^3}. \tag{4.77}$$

Consequently, the thrust acceleration is found as function of thrust angle φ in the form:

$$a_t = \frac{c\beta}{m} = \frac{n^4}{27\mu}\frac{(1-3s^2)^3(9-25s^2+20s^4)}{s^{11}(3-5s^2)^3}. \tag{4.78}$$

If the functional of the problem is given in the form of integral from the thrust acceleration squared, then

$$J = \frac{1}{2}\int a_t^2\,dt = \frac{1}{54}\frac{n^4}{\mu}\int g(\varphi)\frac{dt}{d\varphi}d\varphi = \frac{1}{2n}\int\frac{(1-3)(9-25s+20s^2)}{s^2(3-5s)}d\varphi. \tag{4.79}$$

As it can be seen from Eq. (4.78), the thrust acceleration depends on thrust angle and does not depend on the propulsion parameters. The corresponding Lagrange multipliers can be computed using

$$\begin{aligned}
\lambda_1 &= \lambda s,\\
\lambda_2 &= \lambda k,\\
\lambda_4 &= \lambda\frac{n^3}{27\mu}\frac{k(1-3s^2)}{s^9},\\
\lambda_5 &= \lambda n,\\
\lambda_7 &= \lambda\frac{c}{m},
\end{aligned} \tag{4.80}$$

where $C=0,\ C_2=\lambda n,\ C_3=\lambda c$. The integration constants $d,\ n,\ \lambda,\ t_0,\ \theta_0$ can be found depending on the initial and final conditions.

Eqs. (4.72) and (4.75) are the parametric equations describe spiral trajectories around the center of attraction depending on the values of the integration constants and the thrust angle. Examples of such spirals are illustrated in Fig. 4.1. This figure shows that the shape of a spiral trajectory and the revolutions around the center of attraction are highly sensitive to the changes in the final thrust angle.

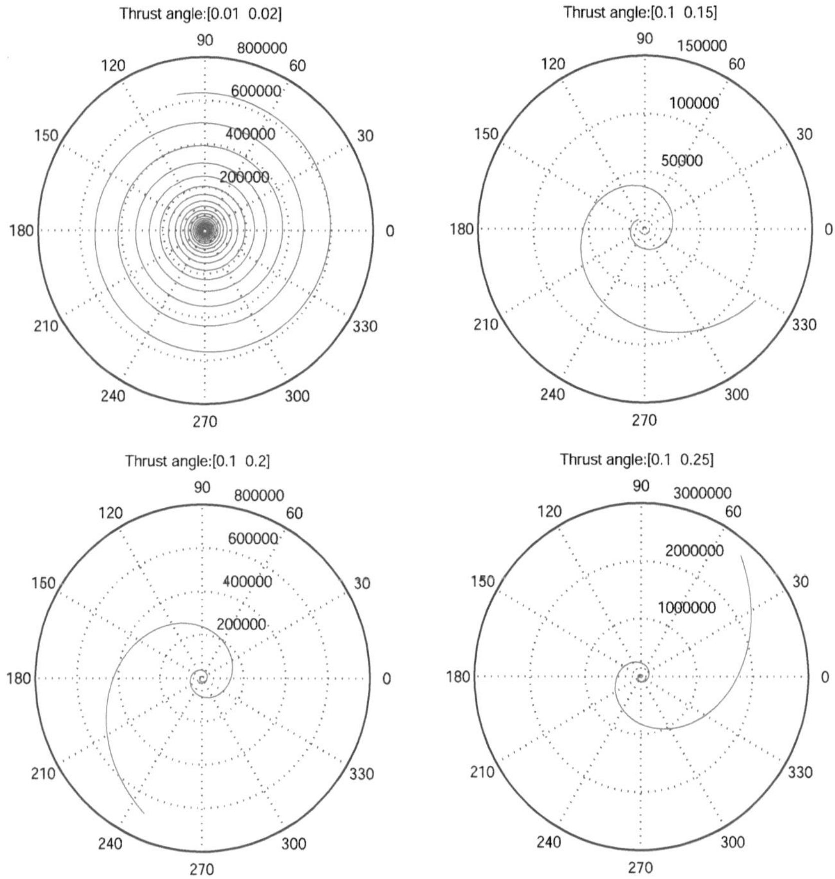

Figure 4.1 Examples of spirals, the first class of extremals (Lawden spirals) depending on final thrust angle.

The first part of the proof of Theorem 4.2 is completed. □

The second part of the proof is presented in the next subsection. It can be shown that the solutions obtained above are similar to Lawden spirals for IT arcs [3]. The Lawden spirals have been obtained by using the first integrals [3], [130], by introducing the new transformations of the variables [54], and also by testing the second variation of the functional [154]. The difference in obtaining the Lawden's solutions in this work is associated with the analysis of invariant relationships and the first integral, which explicitly contains mass and time. The advantage of using such approach is that it allows us to obtain all possible extremals of the problem. As it can

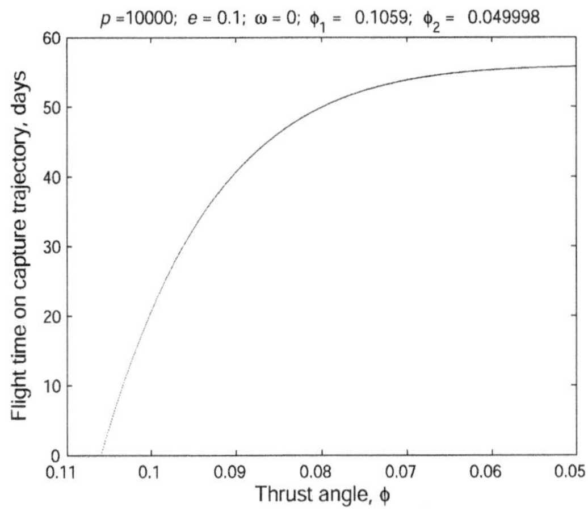

Figure 4.2 Flight duration vs. thrust angle.

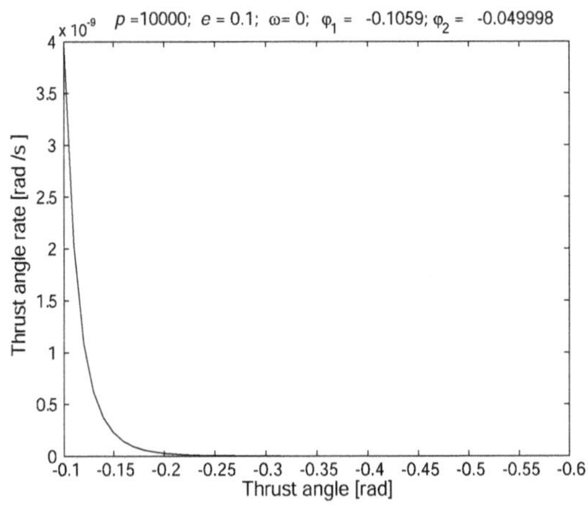

Figure 4.3 Rate of change of thrust angle vs. thrust angle.

be seen from Eq. (4.73), the time is expressed as a quadrature of the thrust angle. This dependency is illustrated in Figs. 4.2 and 4.3. The magnitude of radius vector is increased from r_0 to ∞, when the thrust angle is increased from $\varphi_0 \neq 0$ to $\varphi_f = \arcsin \frac{1}{\sqrt{3}} = 0.61548$ or decreased from $\varphi_0 \neq 0$ to $\varphi_f = -0.61548$. The spacecraft approaches the center of attraction from ∞ to $r_f (r_0 > r_f)$, while the thrust angle is decreased from $|\varphi_0| = 0.61548$ to $|\varphi_f|$, $(\varphi_f \neq 0, |\varphi_0| > |\varphi_f|)$. The corresponding segment can be used in the capture problem. This problem will be considered in the next section.

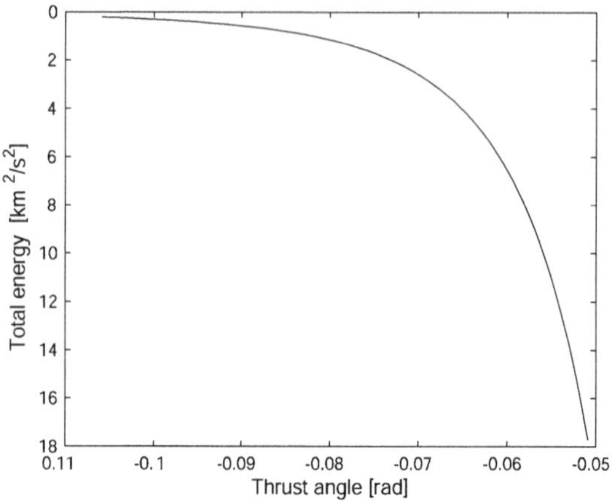

Figure 4.4 Total energy vs. thrust angle.

Direction of motion. The dependency between the angles ψ and φ can be found from (4.71) in the following form (see Fig. 4.15):

$$\tan\psi = \frac{6\sin\varphi\cos\varphi(1-2s)}{3-13s+12s^2}. \tag{4.81}$$

For small thrust angles one can have $\sin\varphi \approx \varphi$ and $\cos\varphi \approx 1$. Consequently,

$$\tan\psi \approx 2\varphi.$$

Energy. As is known, the escape and capture maneuvers require corresponding analysis of energy. An escape occurs when the spacecraft energy changes from a negative to a positive value. The opposite change in the sign may correspond to a capture when an appropriate design of the maneuver is in place. It can be shown that the energy is a function of thrust angle:

$$E = \frac{n^2}{18}\frac{36sk^2(1-2s)^2 + (3-13s+12s^2)^2 - 2(3-5s)^2(1-3s)}{s^3(3-5s)^2} \tag{4.82}$$

and as it is seen, the energy is not an explicit function of propulsion characteristics (see Fig. 4.4). By equating this function to zero, one can obtain $|\varphi_E| = 0.48805023$. This angle may determine an instant of escape or capture.

4.4.2 Time Solution for Lawden Spirals

Here the complete integration of Eq. (4.73) and derivation of the analytical solution for time of flight on Lawden spirals are presented (see Eq. (4.74)). Consider the integral part of Eq. (4.73):

$$\int_{\varphi_0}^{\varphi} \frac{s^7(3-5s^2)}{(1-3s^2)^2}\, d\varphi, \tag{4.83}$$

where $s = \sin\varphi$. By accepting $sd\varphi = -d(\cos\varphi) = -dk$, where $k = \cos\varphi$, one can show that

$$\int \frac{s^7(3-5s^2)}{(1-3s^2)^2} d\varphi = -\int \frac{(1-k^2)^3(5(1-k^2)-3)dk}{(1-3(1-k^2))^2} = \\ -\int \frac{(5k^8-17k^6+21k^4-11k^2+2)dk}{9k^4-12k^2+4}. \tag{4.84}$$

It is known that if $F(k)$ and $f(k)$ are polynomials, in which the order of $F(k)$ is higher than $f(k)$, then

$$\int \frac{F(k)}{f(k)} dk = \int E(k)dk + \int \frac{\psi(k)}{f(k)} dk$$

where $E(k)$ is a polynomial and $\psi(k)$ is the polynomial, whose order is less than that of $f(k)$. Consequently, the numerator of the ratio under the integral is rewritten in the form:

$$(9k^4-12k^2+4)(\frac{5}{9}k^4+ak^2+b)+ck^2+d = \\ 5k^8+(9a-\frac{5}{9}\times 12)k^6+(9b-12a+\frac{5}{9}\times 4)k^4 \\ +(-12b+4a+c)k^2+(4b+d).$$

By equating the corresponding coefficients before the respective orders of k, one can obtain the system

$$9a-\frac{5}{9}\times 12=-17, \quad 9b-12a+\frac{5}{9}\times 4=21, \\ -12b+4a+c=-11, \quad 4b+d=2,$$

which can be solved to obtain

$$a=-\frac{31}{27}, \quad b=\frac{5}{9}, \quad c=\frac{7}{27}, \quad d=-\frac{2}{9}.$$

Eq. (4.84) can be reduced to the form:

$$\int \frac{(5k^8 - 17k^6 + 21k^4 - 11k^2 + 2)dk}{9k^4 - 12k^2 + 4} =$$
$$\int (\frac{5}{9}k^4 + ak^2 + b)dk + \int \frac{(ck^2 + d)dk}{9k^4 - 12k^2 + 4} =$$
$$\frac{1}{9}k^5 + \frac{a}{3}k^3 + bk + \int \frac{(ck^2 + d)dk}{9k^4 - 12k^2 + 4}. \tag{4.85}$$

The integral expression in the second term can be simplified into separate fractions as

$$\int \frac{(ck^2 + d)dk}{9k^4 - 12k^2 + 4} = \int \frac{A_1 dk}{(3k^2 - 2)^2} + \int \frac{A_2 dk}{(3k^2 - 2)} = \int \frac{A_1 + A_2(3k^2 - 2)}{(3k^2 - 2)^2} dk. \tag{4.86}$$

By equating the coefficients of the respective terms,

$$A_1 - 2A_2 = d, \quad 3A_2 = c,$$

one can obtain that $A_1 = d + 2/3c$ and $A_2 = c/3$. From the table of indefinite integrals, one can obtain that

$$\int \frac{dk}{(A + Ck^2)} = \frac{1}{2\sqrt{-AC}} \ln \left| \frac{Ck - \sqrt{-AC}}{Ck + \sqrt{-AC}} \right|,$$

and

$$\int \frac{(Ndk)}{(A + Ck^2)^p} = \frac{Nk}{2A(p-1)(A + Ck^2)^{p-1}} + \frac{2p - 3}{2A(p-1)} \int (A + Ck^2)^{p-1} dk,$$

where p is an integer and $AC < 0$. For the integrals being considered, $p = 2$, $A = -2$, and $C = 3$. These two formulas allow one to reduce the integrals in Eq. (4.86) to the following forms:

$$\int \frac{dk}{(3k^2 - 2)} = \frac{1}{2\sqrt{6}} \ln \left| \frac{Ck - \sqrt{-AC}}{Ck + \sqrt{-AC}} \right|$$
$$\int \frac{dk}{(3k^2 - 2)^2} = -\frac{k}{4(3k^2 - 2)} + \frac{1}{4} \int \frac{dk}{(3k^2 - 2)}$$
$$= -\frac{k}{4(3k^2 - 2)} - \frac{1}{8\sqrt{6}} \ln \left| \frac{Ck - \sqrt{-AC}}{Ck + \sqrt{-AC}} \right|.$$

By substituting these expressions into Eq. (4.86), one can obtain

$$\int \frac{(ck^2+d)dk}{9k^4-12k^2+4} = \int \frac{A_1 dk}{(3k^2-2)^2} + \int \frac{A_2 dk}{(3k^2-2)} =$$
$$-(d+\frac{2c}{3})\left(\frac{k}{4(3k^2-2)} + \frac{1}{8\sqrt{6}}\ln\left|\frac{Ck-\sqrt{-AC}}{Ck+\sqrt{-AC}}\right|\right)$$
$$+\frac{c}{6\sqrt{6}}\ln\left|\frac{Ck-\sqrt{-AC}}{Ck+\sqrt{-AC}}\right| =$$
$$= -(d+\frac{2c}{3})\frac{k}{4(3k^2-2)} - \left((d+\frac{2c}{3})\frac{1}{8\sqrt{6}} + \frac{c}{6\sqrt{6}}\right)\ln\left|\frac{Ck-\sqrt{-AC}}{Ck+\sqrt{-AC}}\right|. \tag{4.87}$$

Similarly, subsequent substitution Eq. (4.87) into (4.85), and then the resulting expression into (4.84) with using the values of $A,\ C,\ a,\ b,\ c,$ and d obtained above, yields

$$\int \frac{s^7(3-5s^2)}{(1-3s^2)^2}d\varphi = \frac{1}{9}k^5 + \frac{a}{3}k^3 + bk + \int \frac{(ck^2+d)dk}{9k^4-12k^2+4} =$$
$$\frac{1}{9}k^5 + \frac{a}{3}k^3 + bk - (d+\frac{2c}{3})\frac{k}{4(3k^2-2)}$$
$$-\left((d+\frac{2c}{3})\frac{1}{8\sqrt{6}} + \frac{c}{6\sqrt{6}}\right)\ln\left|\frac{Ck-\sqrt{-AC}}{Ck+\sqrt{-AC}}\right| =$$
$$-\frac{1}{9}k^5 + \frac{31}{81}k^3 - \frac{5k}{9} - \frac{k}{81(3k^2-2)} - \frac{2\sqrt{2}}{81\sqrt{3}}\ln\left|\frac{3k-\sqrt{6}}{3k+\sqrt{6}}\right|. \tag{4.88}$$

Consequently, by substituting Eq. (4.88) into (4.73), and using the boundary values of φ, one can obtain a complete for of the time solution (see Eq. (4.74)) for the Lawden spirals (e.g. the first class of extremals for free final time):

$$t = t_0 + 27\frac{\mu}{n^3}\left(-\frac{1}{9}k^5 + \frac{32}{81}k^3 - \frac{5}{9}k - \frac{k}{81(3k^2-2)} - \frac{2\sqrt{2}}{81\sqrt{3}}\ln\left|\frac{3k-\sqrt{6}}{3k+\sqrt{6}}\right|\right)_{\varphi_0}^{\varphi}. \tag{4.89}$$

This formulae for time is consistent with the solution for time given in Ref. [3].

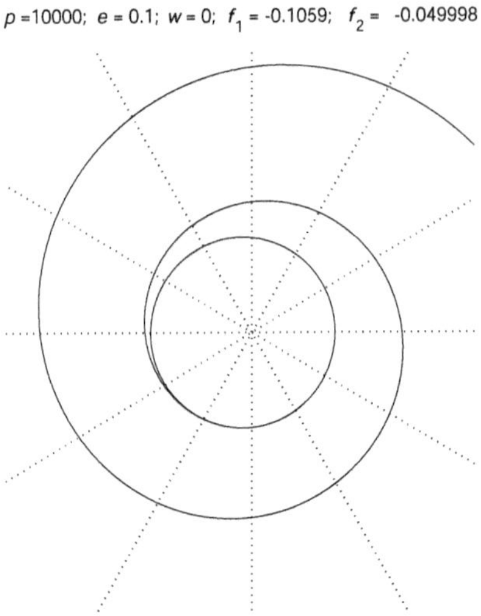

Figure 4.5 Transfer trajectory.

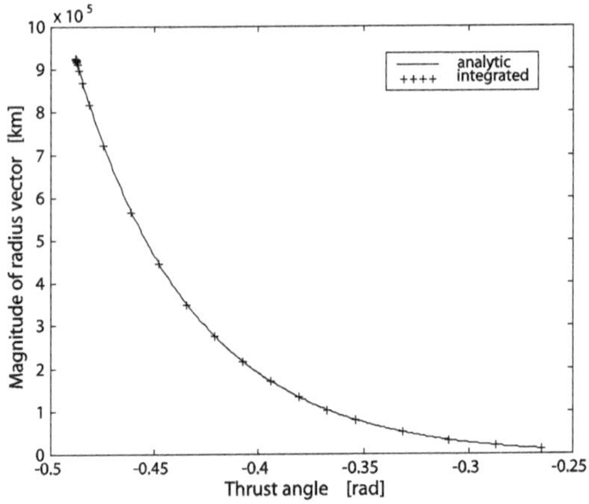

Figure 4.6 Dependency between φ and r.

4.4.3 Comparison With Numerical Integration

The computations have shown that the equations of motion can be integrated numerically if the thrust angle is replaced by using the analytical solutions. From Figs. 4.6–4.9 it can be concluded that the behaviors of r, θ, v_1, v_2, obtained from the analytical solutions and from the numerical integration are the same.

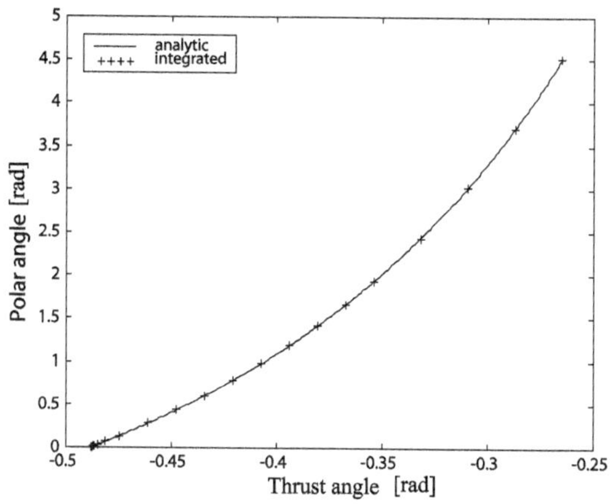

Figure 4.7 Dependency between φ and θ.

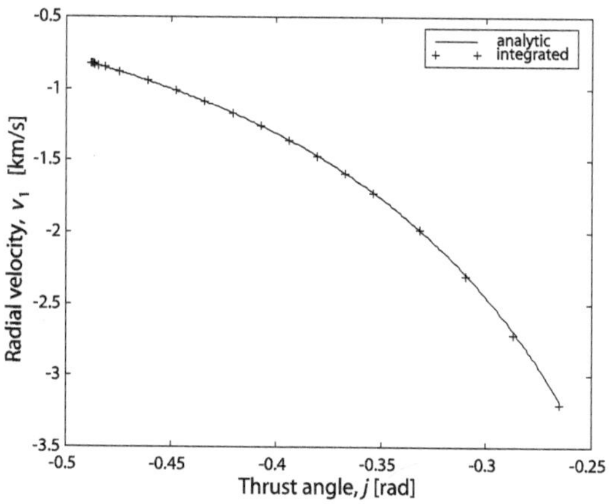

Figure 4.8 Dependency between φ and v_1.

To determine whether the solutions obtained above contain conjugate points, one can use the equations presented in Chapter 2 and the solutions of the auxiliary variation problem. The integration constants are $n, m_0, C_1, C_2, C_3, \theta_0, \lambda$. Denoting them as

$$\alpha_1 = n,\ \alpha_2 = m_0,\ \alpha_3 = C_1,\ \alpha_4 = C_2,\ \alpha_5 = C_3,\ \alpha_6 = \theta_0,\ \alpha_7 = \lambda_v,$$

and using the superposition principle, we can rewrite the solutions of the auxiliary variation problem in the form:

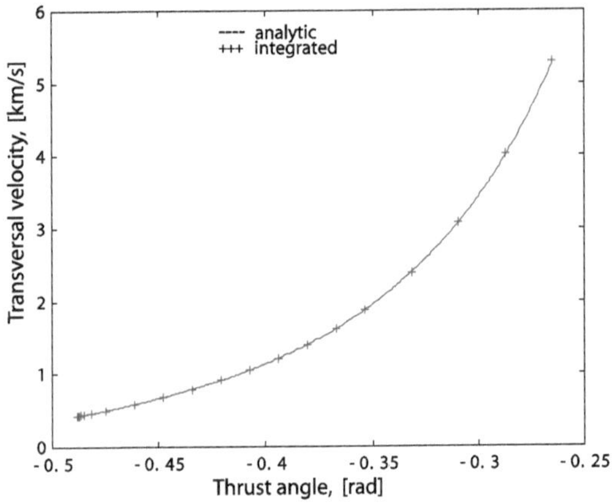

Figure 4.9 Dependency between φ and v_2.

$$\eta_1 = \sum R_i \frac{\partial r}{\partial \alpha_i} = R_1 \frac{\partial r}{\partial \alpha_1} = -2\frac{\mu}{n^3} L_1(s),$$

$$\eta_2 = \sum V_i \frac{\partial v_1}{\partial \alpha_i} = V_1 \frac{\partial v_1}{\partial \alpha_1} = L_2(s),$$

$$\eta_3 = \sum W_i \frac{\partial v_2}{\partial \alpha_i} = W_1 \frac{\partial v_2}{\partial \alpha_1} = L_3(s),$$

$$\eta_4 = \sum M_i \frac{\partial m}{\partial \alpha_i} = M_1\ m_0\ \exp(h(s))[-\frac{1}{c}[k\frac{10s^2 - 7s + 1}{s^3(3-5s)}]]+$$

$$M_2\ \exp(h(s)) + M_3\ m_0\ \exp(h(s))[-\frac{1}{c}],$$

$$\eta_5 = \sum \Theta_i \frac{\partial \theta}{\partial \alpha_i} = \Theta_6 \frac{\partial \theta}{\partial \theta_0} = 1,$$

where the constants R_1, V_1, W_1, Θ_6 are equal to one, as the other terms of summation in the corresponding solutions are zero. As it was shown above, a conjugate point exists on an extremal if $\eta_i = 0,\ (i = 1, ..., 5)$ at $\varphi = \varphi_0$ and $\varphi = \varphi',\ (\varphi_0 < \varphi' \leq \varphi_1)$. It can be seen that $\eta_i,\ (i = 1, ..., 5)$ are not zeros simultaneously, from which it follows that the trajectory does not contain the conjugate points.

4.4.4 Second Class of Extremals

Second part of the proof of Theorem 4.2. By selecting the second sign on the right hand side of Eqs. (4.63) and (4.68), we obtain

$$\begin{aligned} sv_2 - kv_1 &= n\frac{(1-3s^2)}{3s^2} - n, \\ sv_1 + kv_2 &= -2n\frac{k(1-3s^2)}{3s^3} - \frac{1-5s^2}{2s}v_1. \end{aligned} \tag{4.90}$$

Eqs. (4.90) allows us to find v_1 and v_2 in the form:

$$\begin{aligned} v_1 &= 2n\frac{k(-1+4s^2)}{s^2(3-5s^2)}, \\ v_2 &= \frac{n}{3}\frac{(6s^4+7s^2-3)}{s^3(3-5s^2)}. \end{aligned}$$

The expressions in Eqs. (4.67) and (4.91) can be rewritten as

$$\begin{aligned} r &= \frac{\mu}{n^2}W_1(s), \\ v_r &= nW_2(s), \\ v_\theta &= nW_3(s), \end{aligned} \tag{4.91}$$

where

$$W_1(s) = 9\frac{s^6}{1-3s^2}, \quad W_2(s) = 2\frac{k(-1+4s^2)}{s^2(3-5s^2)}, \quad W_3(s) = \frac{1}{3}\frac{(6s^4+7s^2-3)}{s^3(3-5s^2)}.$$

As it can be seen from these formulas, both extremals are described by the same formula for the magnitude of the radius vector, $W_1 = L_1$. The remaining variables of the second class of extremals can be found similar to the first class of extremals. It can be shown that these variables can be determined by the formulas:

$$\begin{aligned} m &= m_0 \exp z(\varphi), \\ t &= t_0 + 27\frac{\mu}{n^3}\int_{\varphi_0}^{\varphi} \frac{s^7(1-2s^2)(3-5s^2)}{(-1+4s^2)(1-3s^2)^2}d\varphi, \\ \theta &= \theta_0 + \int_{\varphi_0}^{\varphi} \frac{(1-2s^2)(3-5s^2)(6s^4+7s^2-3)}{s^2(-1+4s^2)(1-3s^2)^2}d\varphi, \\ \beta &= -\frac{n^4}{81\mu}\frac{m}{c}\frac{(-1+4s^2)(1-3s^2)^2(170s^6-258s^4+53s^2-9)}{s^{11}(1-2s^2)(3-5s^2)^3}, \\ a_t &= -\frac{n^4}{81\mu}\frac{(-1+4s^2)(1-3s^2)^2(170s^6-258s^4+53s^2-9)}{s^{11}(1-2s^2)(3-5s^2)^3}, \\ z(\varphi) &= \frac{1}{c}\left[\frac{nk(1-10s^2)(1+3s^2)}{3s^3(3-5s^2)} + d\right], \quad d = \frac{C_1}{\lambda} = const. \end{aligned} \tag{4.92}$$

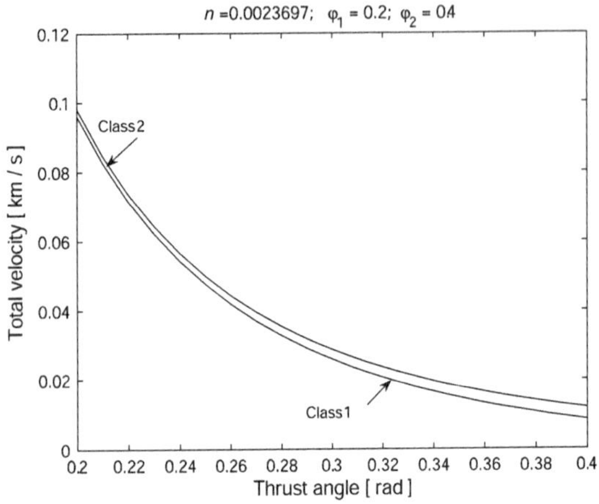

Figure 4.10 Velocity vs. thrust angle.

It can be shown that the formulas for the Lagrange multipliers are not different from those in the first class of extremals (see (4.80)). The computations of the direction of motion, analysis of energy as function of thrust angle, and comparison with numerical integration ca also be conducted similarly as in the first class of extremals. This completes the proof of Theorem 4.2. □

The analysis of mass, rate of change of thrust angle, and velocity show that there exists a small difference between these classes, although the fuel consumption is more effective in the case of the first class of extremals (see Figs. 4.10, 4.11, and 4.12). Despite this difference, both classes of extremals can be used in the guidance problems as reference trajectories.

4.4.5 Example of Transfer to a Given Elliptical Orbit

In this subsection we consider a capture maneuver for a spacecraft approaching the Earth gravitational field via the extremals of first class (Lawden's spirals). It is assumed that the spacecraft enters the gravitational field when the magnitude of spacecraft's position vector is equal to or less than the radius of Earth sphere of influence, $(r_0 \leq r_{si})$, where $r_{si} = 9.2482 * 10.^5$ km and when the spacecraft's total energy becomes negative. The final conditions are assumed satisfied if the spacecraft reaches the given final (parking) orbit with semi-latus rectum, p, eccentricity e, and angular distance of perigee, ω. Consider a case when the maneuver time is not

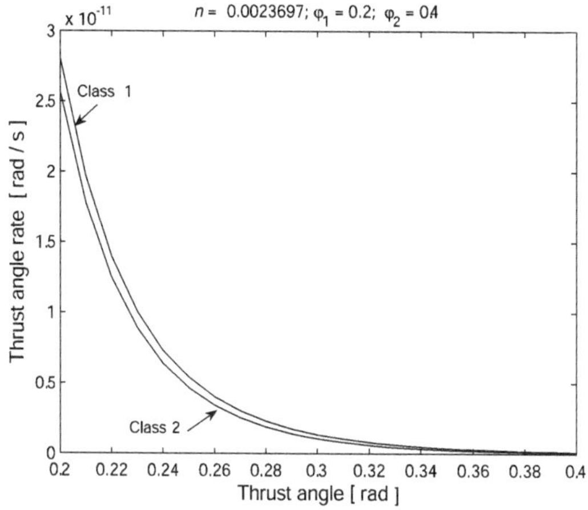

Figure 4.11 Rate of thrust angle vs. thrust angle.

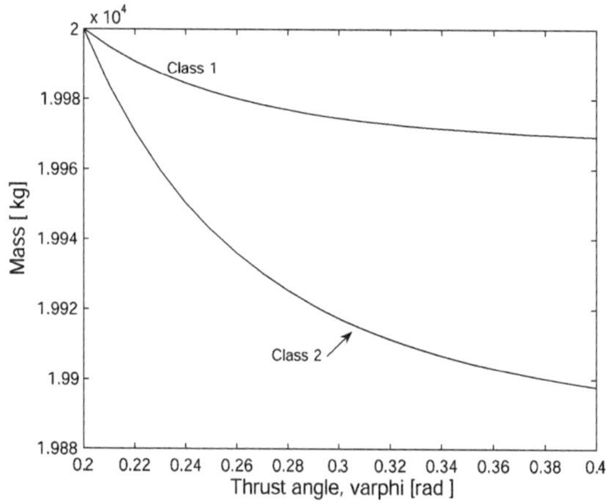

Figure 4.12 Mass vs. thrust angle.

fixed, $(C=0)$, and the final polar angle has a certain fixed value subject to determination, $(n\neq 0)$, and the duration of flight in this orbit is considered arbitrary. The spacecraft's engine is on all the time during the maneuver. In this case, there exist two junction points to be determined. Below the method of application of analytical solutions to the design of extremals of the maneuvers will be utilized. The details of this method will be given in Chapter 7 of this monograph. The values of variables given at the first and second junction points (or junctions) will be denoted by indexes "1" and "2" respectively. The continuity conditions at the first junction can be written in the form:

$$\frac{\mu}{n^2}L_1(s_1) = r_{si}, \tag{4.93}$$

$$nL_2(s_1) = v_{r1}, \tag{4.94}$$

$$nL_3(s_1) = v_{\theta 1}, \tag{4.95}$$

$$\theta_1 = \theta_0 - 3\cot\varphi_1 - 4\varphi_1, \tag{4.96}$$

where

$$s_1 = \sin\varphi_1, \quad k_1 = \cos\varphi_1.$$

At the second junction these conditions are as follows:

$$\frac{\mu}{n^2}L_1(s_2) = \frac{p}{1+e\cos f_2}, \tag{4.97}$$

$$nL_2(s_2) = \frac{\mu}{p}e\sin f_2, \tag{4.98}$$

$$nL_3(s_2) = \frac{\mu}{p}(1+e\cos f_2), \tag{4.99}$$

$$\theta_2 = \theta_0 - 3\cot\varphi_2 - 4\varphi_2 = f_2 + \omega, \tag{4.100}$$

where

$$s_2 = \sin\varphi_2, \qquad k_2 = \cos\varphi_2.$$

Eqs. (4.93)–(4.100) allow us to find unknowns $n, \varphi_1, \varphi_2, f_2, v_{11}, v_{21}$ depending on the magnitude of the initial position vector, r_0, and p, e and ω. Excluding $\frac{\mu}{p}$ from Eqs. (4.97) and (4.98), one can obtain

$$4n^2\frac{k_2^2}{s_2^4}\frac{(1-2s_2^2)^2}{(3-5s_2^2)^2} = n^2\frac{(1-3s_2^2)}{9s_2^6(1+e\cos f_2)}e^2\sin^2 f_2$$

or

$$\frac{e^2\sin^2 f_2}{1+e\cos f_2} = \frac{36k_2^2s_2^2(1-2s_2^2)^2}{(1-3s_2^2)(3-5s_2^2)^2}, \tag{4.101}$$

where

$$\frac{\mu}{p} = n^2\frac{(1-3s_2^2)}{9s_2^6(1+e\cos f_2)}. \tag{4.102}$$

Assuming that $1 - 3s_2^2 > 0$, from Eqs. (4.98) and (4.99) it follows that

$$\frac{e\sin f_2}{1+e\cos f_2} = \frac{6k_2 s_2(1-2s_2^2)}{3-13s_2^2+12s_2^4}. \tag{4.103}$$

Eqs. (4.101) and (4.103) can serve to compute e and f_2 depending on the final thrust angle, φ_2. From Eq. (4.101) and (4.103), after some simplifications, one obtains

$$e\cos f_2 = \frac{(3-13s_2^2+12s_2^4)^2}{(1-3s_2^2)(3-5s_2^2)^2} - 1. \tag{4.104}$$

Besides that, Eq. (4.101) can be rewritten as

$$e\sin f_2 \frac{e\sin f_2}{1+e\cos f_2} = \frac{36k_2^2 s_2^2(1-2s_2^2)^2}{(1-3s_2^2)(3-5s_2^2)^2}. \tag{4.105}$$

After substituting the right side of Eq. (4.103) to Eq. (4.105), the latter is represented in the form:

$$e\sin f_2 \frac{6k_2 s_2(1-2s_2^2)}{3-13s_2^2+12s_2^4} = \frac{36k_2^2 s_2^2(1-2s_2^2)^2}{(1-3s_2^2)(3-5s_2^2)^2}$$

or

$$e\sin f_2 = \frac{6k_2 s_2(3-13s_2^2+12s_2^4)(1-2s_2^2)}{(1-3s_2^2)(3-5s_2^2)^2}. \tag{4.106}$$

Let's introduce the following notations for simplification:

$$\begin{aligned} q_1(\varphi_2) &= 3-13s_2^2+12s_2^4, \\ q_2(\varphi_2) &= 1-3s_2^2, \\ q_3(\varphi_2) &= 1-2s_2^2, \\ q_4(\varphi_2) &= 3-5s_2^2. \end{aligned}$$

Then Eqs. (4.104) and (4.106) can be rewritten as

$$e\cos f_2 = \frac{q_1^2 - q_2 q_4^2}{q_2 q_4^2}, \tag{4.107}$$

$$e\sin f_2 = \frac{6k_2 s_2 q_1 q_3}{q_2 q_4^2}. \tag{4.108}$$

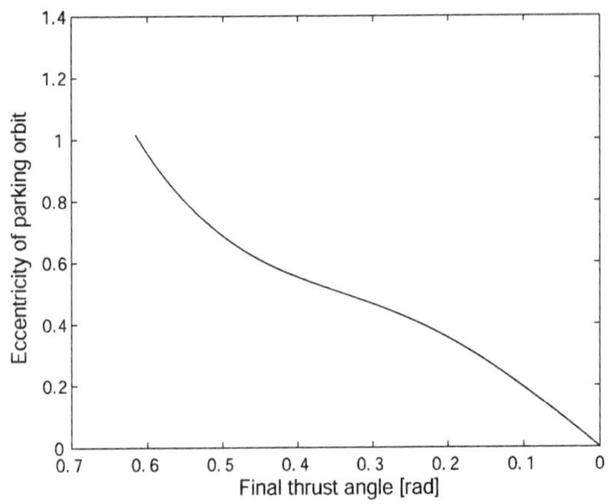

Figure 4.13 Relationship between φ_2 and *e*.

Excluding f_2 from these expressions, one can obtain equation for φ_2:

$$(q_1^2 - q_2 q_4^2)^2 + 36 k_2^2 s_2^2 q_1^2 q_3^2 - e^2 q_2^2 q_4^4 = 0. \tag{4.109}$$

This equation allows us to find φ_2 as a function of e: $\varphi_2 = \varphi_2(e)$. This functional dependency is presented in Fig. 4.13. The true anomaly, f_2 can easily be found from Eqs. (4.107) and (4.108):

$$f_2 = f_2(e) = \tan^{-1}\left[\frac{k_2 s_2 q_1 q_3}{q_1^2 - q_2 q_4^2}\right]. \tag{4.110}$$

The constant n can now be determined from Eq. (4.102) as a function of p and e:

$$n = n(p, e) = \sqrt{\frac{p}{9\mu}\frac{q_2^2 q_4^2}{s_2^6 q_1^2}}. \tag{4.111}$$

Then substituting Eq. (4.111) into (4.93), we obtain

$$9\mu \sin^6\varphi_1 - n^2 r_{si}(1 - 3\sin^2\varphi_1) = 0$$

or

$$(s_1^2)^3 + d_1 s_1^2 + d_2 = 0, \tag{4.112}$$

where

$$s_1 = \sin\varphi_1, \quad d_1 = \frac{n^2 r_{si}}{3\mu}, \quad d_2 = -\frac{n^2 r_{si}}{9\mu}.$$

By considering this equation as a cubic equation with respect to $s_1^2(=\sin^2\varphi_2)$, one can find that the discriminant of this equation is positive: $\frac{d_1^3}{27}+\frac{d_2^2}{4}>0$. This means that there exist one real and two complex roots of this equations. The real root can be given in the form:

$$s_1 = s_1(p, e, r_{si}) = \sin^2\varphi_1 = 2\sqrt{\frac{d_1}{3}}\cot 2\xi, \tag{4.113}$$

where

$$\tan\xi = (\tan\gamma)^{\frac{1}{3}}, \quad \cot 2\gamma = \sqrt{\frac{9}{4}\frac{\mu}{n^2 r_{si}}}.$$

Choose $\sin\varphi<0$ to satisfy the Robbins' condition [42]. Then from Eq. (4.113) we obtain:

$$\sin\varphi_1(p, e, r_{si}) = -2\sqrt{\frac{d_1}{3}}\cot 2\xi, \tag{4.114}$$

if $\cot 2\xi>0$, and

$$\sin\varphi_1(p, e, r_{si}) = 2\sqrt{\frac{d_1}{3}}\cot 2\xi, \tag{4.115}$$

if $\cot 2\xi<0$. As mentioned above, at any given time,

$$0<\sin^2\varphi<\frac{1}{3}.$$

Eqs. (4.94) and (4.95) can be used to compute the initial velocity components:

$$\begin{aligned} v_{11} &= n(p, e)L_2(p, e, r_{si}), \\ v_{21} &= n(p, e)L_3(p, e, r_{si}). \end{aligned} \tag{4.116}$$

The constant θ_0 in Eq. (4.96) can be determined from Eq. (4.100) as

$$\theta_0(p, e, \omega) = f_2 + \omega + 3\cot\varphi_2 + 4\varphi_2$$

This allows us to find θ_1 and θ_2 as functions of p, e, ω, and r_{si} from Eqs. (4.96) and (4.100) in the form:

$$\theta_1 = \theta_1(p, e, \omega, r_{si}),$$

$$\theta_2 = \theta_2(p, e, \omega). \quad (4.117)$$

The duration (phase), θ_x, of a flight in the final orbit can be computed using the condition that the final polar angle, $\tilde{\theta_2}$ is fixed, and from the expression:

$$\theta_x(p, e, \omega) = \tilde{\theta_2} - \theta_2(p, e, \omega).$$

As it can be seen from Eqs. (4.109), (4.111), and (4.113), the angle φ_2 is a function of eccentricity, e, whereas φ_1 is a function of e, p, and r_0. In other words, the ellipticity of the final parking orbit determines the final thrust angle, and this ellipticity and the size of the orbit determine the integration constant, n. The orientation of the final orbit does not affect φ_1, φ_2, or n, and used only to determine the integration constant of the polar angle and its boundary values. As is known, the primer vector for zero thrust arcs is determined by the formulas [3]:

$$\lambda_r = A\cos f + Be\sin f,$$

$$\lambda_\theta = -A\sin f + B(1 + e\cos f) + \frac{D - A\sin f}{1 + e\cos f},$$

where A, B, and D are the constants, and $C = 0$. It can be shown that if $(C = 0)$, then $A = 0$ [3]. If $\lambda_1 = \lambda\sin\varphi$, $\lambda_2 = \lambda_v\cos\varphi$, then the continuity conditions for the primer vector allow us to find the constants:

$$\frac{B}{\lambda} = \frac{\sin\varphi_2}{e\sin f},$$

$$\frac{D}{\lambda} = (\varphi_2 - \frac{\sin\varphi_2}{e\sin f}(1 + e\cos f_2))(1 + e\cos f_2).$$

Note that by adding the errors on the right hand sides of Eqs. (4.93)–(4.100), one can construct the equations for analysis of navigation scheme of this problem. One can conclude that the spacecraft can be transferred from the extremals of the first class (qualitatively, the second class too) to a given elliptical orbit and vice versa. This means that the extremals of this class can be used to perform capture and escape maneuvers, and in the transfers between elliptical orbits. Besides that, the form of the analytical solutions allows us to separate the analysis of the kinematic parameters from the dynamic parameters, and also to separately analyze the primer vector. Some typical spiral trajectories for various elements of the parking orbit are presented in Figs. 4.5 and 4.14.

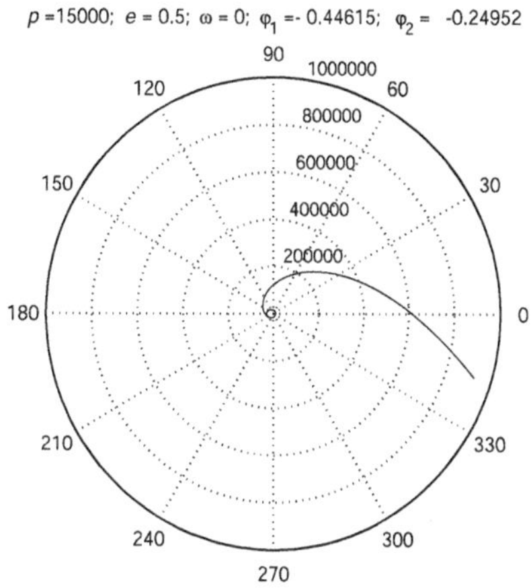

Figure 4.14 Typical transfer trajectory to parking orbit.

4.5 EXTREMALS FOR MANEUVERS WITH FIXED FINAL TIME

This section describes the derivation and analysis of six classes of analytical solutions with variable power and constant specific impulse when the final time is fixed and the functional of the problem explicitly depends on final polar angle.

Theorem 4.3. *Let Eqs. (4.2)–(4.6) are the first integrals, and Eqs. (4.13)–(4.16) are the invariant relationships in the problem described in Eqs. (2.111)–(2.115). If the final time is fixed and the functional explicitly depends on the final polar angle, then in case of motion with variable power ($P < P_{max}$) and constant specific impulse, ($I_{sp} = const$), there exist at least six classes of planar extremal trajectories, which can be described in elementary functions and quadratures. For certain values of the thrust angle these classes can satisfy the Robbins' necessary condition of optimality.*

Proof of Theorem 4.3. In the case of a planar motion, find λ_4 and the expression $(\lambda_1 v_2 - \lambda_2 v_1 + \lambda_5)$ from Eqs. (4.13) and (4.14) to obtain

$$\lambda_4 = \pm \frac{C_2^3 \lambda_2 \lambda^2 (\lambda^2 - 3\lambda_1^2)^2}{27 \mu \lambda_1^9},$$

$$\lambda_1 v_2 - \lambda_2 v_1 + C_2 = \mp \frac{C_2}{3\lambda_1^2}(\lambda^2 - 3\lambda_1^2). \tag{4.118}$$

Substituting these expressions into the first integral of Eqs. (4.12), one can obtain the equation for r [155]:

$$\frac{C^2}{\lambda_v^2} r^4 + 6\frac{C}{\lambda_v}\mu s^3 r^2 - n^2\mu(1-3^2)r + 9\mu^2 s^6 = 0, \tag{4.119}$$

where

$$C_2 = \lambda_5, \quad n = \frac{C_2}{\lambda}, \quad \lambda_1 = \lambda \sin\varphi, \quad \lambda_2 = \lambda\cos\varphi,$$

φ the angle between the thrust vector and perpendicular to $\mathbf{r}$. The case when the flight time is fixed, ($C \neq 0$), and the functional explicitly depends on the final polar angle, ($C_2 \neq 0$), is the most general case in solving Eqs. (4.119). The discriminant of this equation is

$$Q = C_2^4 \frac{\mu^4}{C^8}(3\frac{\lambda_1^2}{\lambda^2} - 1)^2 [\frac{C_2^4}{256}(3\frac{\lambda_1^2}{\lambda^2} - 1)^2 - \mu C \frac{\lambda_1^9}{\lambda^6}]. \tag{4.120}$$

The following cases can take place:

$$Q \geq 0, \quad \lambda_1 C < 0 \qquad or \qquad Q < 0, \quad \lambda_1 C > 0.$$

The form of the solutions depends on the signs of the expressions $(3s^2 - 1)$ and $(2\cos(\frac{\alpha}{3}) - 1)$.

Let $(3s^2 - 1) < 0$. If $Q \geq 0, \quad \lambda_1 C < 0$ one can have

$$r_1 = F_1 + (\frac{F}{F_1})^{\frac{1}{2}}, \tag{4.121}$$

where

$$F = \frac{\mu C_2^2}{2C^2}(1 - 3s^2), \quad F_1 = (-2\lambda s^3 \frac{\mu}{C}(1 + \frac{2}{\sec 2\alpha}))^{\frac{1}{2}}, \quad s = \sin\varphi$$

$$tg\alpha = (tg\frac{\beta}{2})^{\frac{1}{3}}, \quad /\alpha/ \leq \frac{\pi}{4}, \quad 1 + \frac{2}{\sec 2\alpha} > 0,$$

$$\sin\beta = \frac{16\lambda^3 s^9}{16\lambda^3 s^9 - \frac{C_2^4}{C}(3s^2 - 1)^2}.$$

If $1 + \frac{2}{\sec 2\alpha} \leq 0$, then the IT arcs do not have real solutions.

If $Q < 0, \lambda_1 C > 0$, then

$$r_2 = F_2 + (\frac{F}{F_2})^{\frac{1}{2}}, \tag{4.122}$$

where

$$F_2 = (-2\lambda s^3 \frac{\mu}{C}(2\cos\frac{\alpha}{3} - 1))^{\frac{1}{2}}, \quad \cos\alpha = \frac{16\lambda^3 s^9 - \frac{\lambda s^4}{8C}(1-3s^2)^2}{16\lambda^3.s^9}$$

If $2\cos\frac{\alpha}{3} - 1 \leq 0$, then

$$r_3 = F_3 + (\frac{F}{F_3})^{\frac{1}{2}}, \tag{4.123}$$

where

$$F_3 = (2\lambda s^3 \frac{\mu}{C}(1 - 2\cos(\frac{\alpha}{3} + \frac{\pi}{3})))^{\frac{1}{2}}, \quad 1 - 2\cos\frac{\alpha}{3} + \frac{\pi}{3} > 0.$$

In the case when the latter expression is not satisfied, the IT arcs do not have real solutions.

Let $3s^2 - 1 > 0$. When $Q \geq 0,\ \lambda_1 C < 0$ we have

$$r_4 = F_4 + (\frac{W}{F_4})^{\frac{1}{2}}, \tag{4.124}$$

where

$$W = \frac{C_2^2}{4\lambda^2 C^2}\sqrt{2}\mu(3s^2 - 1), \qquad F_4 = (-2\lambda s^3 \frac{\mu}{C}(\frac{2}{\sec 2\alpha} + 1))^{\frac{1}{2}},$$

$$1 + \frac{2}{\sec 2\alpha} > 0,$$

α, β, s are determined according to the formulas obtained above.

If $1 + \frac{2}{\sec 2\alpha} \leq 0$, then the IT arcs do not have real solutions.

If $3s^2 - 1 > 0,\ Q < 0,\ \lambda_1 C > 0$, then

$$r_5 = F_5 + (\frac{W}{F_5})^{\frac{1}{2}}, \tag{4.125}$$

where

$$F_5 = (2\lambda s^3 \frac{\mu}{C}(\frac{2}{\sec 2\alpha} - 1))^{\frac{1}{2}}, \quad 2\cos\frac{\alpha}{3} - 1 > 0,$$

$$\cos\alpha = \frac{\frac{\lambda s^4}{8C}(1-3s^2)^2 - 16\lambda^3 s^9}{16\lambda^3 s^9}.$$

If $2\cos\frac{\alpha}{3} - 1 \leq 0$, then

$$r_6 = F_6 + (\frac{W}{F_6})^{\frac{1}{2}}. \tag{4.126}$$

Here

$$F_6 = (2\lambda s^3 \frac{\mu}{C}(1 - 2\cos(\frac{\alpha}{3} + \frac{\pi}{3})))^{\frac{1}{2}}.$$

After determining

$$r = r(s) \tag{4.127}$$

in the form of one of the equalities in Eqs. (4.121)–(4.126), to obtain other solutions of the equations of the problem, from Eqs. (4.7), (4.10), and (4.13)–(4.15) one can have

$$\lambda_1 v_2 - \lambda_2 v_1 = \frac{3\mu\lambda_1^4 + C\lambda_1\lambda^2 r^2 + C_2^2 r}{C_2\lambda^2 r},$$

$$\lambda_4 = \frac{\lambda_2(3\mu\lambda_1^3 + Cr^2\lambda^2)}{r^2 C_2\lambda^2}, \quad \lambda_1 v_1 + \lambda_2 v_2 = 2\lambda_4 r,$$

from which by excluding λ_4, and taking into account $s = \sin\varphi$, $k = \cos\varphi$, it follows that

$$v_1 = \frac{2Ck(3w + 2C_2)}{\lambda(5s^2 - 3)}, \quad v_2 = \frac{(3 - s^2)w + 6kC_2}{\lambda s(5s^2 - 3)}. \tag{4.128}$$

Then the equations of Eqs. (2.135) for r and θ, and also expression for λ_4:

$$\lambda_4 = k\frac{\lambda - w}{\lambda s},$$

lead to

$$t = \frac{\lambda}{2}\int \frac{(5s^2 - 3)Rd\varphi}{k(3w + 2C_2)} + C_4, \quad \theta = \frac{1}{2}\int \frac{((3 - s^2)Rw + 6kC_2)d\varphi}{rsk(3w + 2C_2)} + C_5.$$

Besides that, from the integrals (4.3) and (4.4) of this chapter it follows that

$$m = m_0 e^{\frac{P}{c\lambda}}, \quad \lambda_7 = \frac{C_3}{M}. \tag{4.129}$$

Here

$$\lambda_1 = \lambda s, \quad \lambda_2 = \lambda k, \quad w = \frac{3\lambda^2 s^4 \frac{\mu}{r} + C\lambda sr + C_2^2}{C_2},$$

$$R = \frac{dr}{d\varphi}, k = \cos\varphi, \quad P = C_3 - k\frac{(15s^2 - 3)w + 8s^2 C_2}{s(s - 3)^2} + 3C(t - C_4),$$

C_4, C_5 are the integration constants. The mass-flow rate can be computed from the expression $\chi^{(4)} = 0$ in the form:

$$\beta = \frac{[10\mu s^2\lambda^2 - v_2^2 r\lambda^2(3-13s^2) - 2\lambda_4\lambda r^2(sv_1 - 3kv_2) - 4\lambda_4^2 r^3 + 6\lambda s C_2 \frac{v_2}{r}]}{cs\lambda^2 r^2(3-5s^2)}. \tag{4.130}$$

So, Eqs. (4.127)–(4.130) are the solutions to Eqs. (2.111)–(2.115).

If $s < 0,\ C > 0$ in Eqs. (4.121), (4.122), (4.124) and when $s < 0,\ C < 0$ in Eqs. (4.123), (4.125), (4.126) the solutions Eqs. (4.128)–(4.130) represent the analytical solutions of the variational problem that satisfy the necessary conditions of optimality [42]. More detailed investigation of 28 classes of trajectories can be found in Ref. [40].

More detailed analysis and studies of all possible classes of the trajectories with variable power and constant specific impulse, including those presented above and in the previous sections, are presented in Ref. [40]. Tables 4.1 and 4.2 provide the classification of these trajectories depending on the free or fixed time and final polar angle.

4.6 ROBBINS NECESSARY CONDITION

This condition means that the radial component of the primer vector must be negative. Besides that, on IT arcs the switching function has to be identically zero. For a complete test of necessary conditions of optimality, following Robbins, check the sign of thrust acceleration. Consider the projection of the first equation of the system Eqs. (2.135) on the direction of the primer vector and obtain (see Fig. 4.15) [108]:

$$\dot{u}_1 - u_2\dot{\psi} = \frac{c\beta}{m} - \frac{\mu}{r^2}s, \tag{4.131}$$

where

$$u_1 = \dot{r}\sin\varphi + r\dot{\theta}\cos\varphi = \frac{(\lambda_1 v_1 - \lambda_2 v_2)}{\lambda},$$
$$u_2 = -\dot{r}\cos\varphi + r\dot{\theta}\sin\varphi = \frac{(-\lambda_2 v_1 + \lambda_1 v_2)}{\lambda}, \quad \psi = \frac{\pi}{2} - \varphi + \theta. \tag{4.132}$$

Here $u_1,\ u_2$ are the velocity components on the primer vector's direction and its perpendicular, ψ is the angle between the primer vector and the polar axis ($\theta = 0$). In the solutions obtained above, the angle φ is considered

Table 4.1 Summary of solutions for intermediate-thrust arcs

$C_1=0,\ C_4\neq 0$ **(Class 1) [3], [61]**	$C_1=0,\ C_4\neq 0$ **(Class 2)**	$C_1\neq 0,\ C_4=0$ **(Class 3)**	$C_1\neq 0,\ C_4\neq 0$ **(Classes 4–28)**
$v=2n\frac{k(1-2s^2)}{s^2(3-5s^2)}$	$v=2n\frac{k(-1+4s^2)}{s^2(3-5s^2)}$	$v=\frac{6bsk}{3-5s^2}\sqrt{\frac{1-3s^2}{s^{\frac{3}{2}}}}$	$v=\frac{2kn}{1-3s^2}$
$w=n\frac{3-13s^2+12s^4}{3s^3(3-5s^2)}$	$w=\frac{n}{3}\frac{(6s^4+7s^2-3)}{s^3(3-5s^2)}$	$w=\frac{b(3-s^2)}{3-5s^2}\sqrt{\frac{1-3s^2}{s^{\frac{3}{2}}}}$	$w=\frac{n(1+s^2)+zs(5-s^2)}{s(1-3s^2)}$
$r=\frac{9\mu}{n^2}\frac{s^6}{1-3s^2}$	$r=\frac{9\mu}{n^2}\frac{s^6}{1-3s^2}$	$r=\frac{\mu}{b^2}s^{\frac{3}{2}}$	$r=\frac{1}{4}\left[\sqrt{\hat{\gamma}}\pm\sqrt{-\hat{\gamma}\mp\frac{2D}{A\sqrt{\hat{\gamma}}}}\right]^2$
$\theta=\theta_0-3\cot\varphi-4\varphi$	$\theta=\theta_0+\int_{\varphi_i}^{\varphi}S_5\,d\varphi$	$\theta=\theta_0-\frac{3}{4}\cot\varphi-\frac{1}{4}\varphi$	$\theta=\frac{1}{2n}\int Z_5\frac{d\hat{r}}{d\varphi}d\varphi+\theta_0$
$m=m_i\exp\left[\frac{G_1}{c_e}\right]$	$m=m_i\exp\left[\frac{S_1}{c_e}\right]$	$m=m_i\exp\left[\frac{H_1}{c_e}\right]$	$m=m_i\exp\left[\frac{Z_1}{c_e}\right]$
$\lambda_v=C_3s$	$\lambda_v=C_3s$	$\lambda_v=C_3s$	$\lambda_v=C_3s$
$\lambda_w=C_3k$	$\lambda_w=C_3k$	$\lambda_w=C_3k$	$\lambda_w=C_3k$
$\lambda_r=\frac{C_4k}{3rs^3}(1-3s^2)$	$\lambda_r=\frac{C_4k}{3rs^3}(1-3s^2)$	$\lambda_r=-C_3\frac{k}{r}z$	$\lambda_r=-C_3\frac{k}{r}z$
$\lambda_\theta=C_4$	$\lambda_\theta=C_4$	$\lambda_\theta=0$	$\lambda_\theta=C_4$
$\lambda_m=\frac{C_3c_e}{m}$	$\lambda_m=\frac{C_3c_e}{m}$	$\lambda_m=\frac{C_3c_e}{m}$	$\lambda_m=\frac{C_3c_e}{m}$
$t=t_i+27\frac{\mu}{n^3}\int G_2\,d\varphi$	$t=t_i+27\frac{\mu}{n^3}\int S_2\,d\varphi$	$t=t_i+\frac{\mu}{4b^2}\int H_2\,d\varphi$	$t=t_i+\frac{1}{2n}\int Z_2\frac{d\hat{r}}{d\varphi}d\varphi$
$\beta=\frac{n^4}{27\mu}\frac{m}{c_e}G_3$	$\beta=-\frac{m}{c_e}S_3$	$\beta=-\frac{m}{c_e}H_3$	$\beta=\frac{m}{c_eC_3}Z_3,$
$G_1=-G_4+\frac{C_2}{C_3}$	$S_1=-S_4+\frac{C_2}{C_3}$	$H_1=-H_4+3d_2t+d_1$	$Z_1=-Z_4+3C_1t-C_2$
$G_2=\frac{s^7(3-5s^2)}{(1-3s^2)^2}$	$S_2=\frac{s^7(1-2s^2)(3-5s^2)}{(-1+4s^2)(1-3s^2)^2}$	$H_2=\frac{(3-5s^2)\,d\varphi}{zs}$	$Z_2=\frac{1-3s^2}{k}$
$G_3=\frac{(1-3s^2)^3(9-25s^2+20s^4)}{s^{11}(3-5s^2)^3}$	$S_3=\frac{dS_1}{d\varphi}\frac{d\varphi}{dt}$	$H_3=\frac{dH_1}{d\varphi}\frac{d\varphi}{dt}$	$Z_3=\frac{dZ_1}{d\varphi}\frac{d\varphi}{dt}$
$G_4=n\frac{k(10s^4-7s^2+1)}{s^3(3-5s^2)}$	$S_4=sv+kw-2\frac{\lambda_r}{C_3}r$	$H_4=sv+kw-2r\frac{\lambda_r}{C_3}$	$Z_4=2kz(1-C_3)-Z_6$
	$S_5=\frac{(1-2s^2)(6s^4+7s^2-3)}{s^2(-1+4s^2)(1-3s^2)}$	$d_2=\frac{C_1}{C_3}-\frac{C_3}{2}$	$Z_5=\frac{n(1+s^2)+zs(5-s^2)}{\hat{r}sk}$
		$z=\sqrt{\frac{\mu}{r}(1-3s^2)}$	$Z_6=\frac{k}{s}\frac{(1-5s^2)}{(1-3s^2)}(n+3sz)$
			$z=\sqrt{\frac{\mu}{r}(1-3s^2)}$
			$\hat{r}=r=r_{ij},\ \hat{\gamma}=(\gamma_1)_j$
			$i=1,\ldots,4;\ j=1,\ldots,6$

Table 4.2 Solutions for y_1 in the case $C_1 \neq 0,\ C_4 \neq 0$

$C_1 s < 0$, $Q > 0,\ \forall p$ (*Cardan*)	$C_1 s < 0$ $Q > 0,\ p > 0$ (*Trig*)	$C_1 s > 0$ $Q < 0,\ p < 0$	$C_1 s > 0$ $Q \geq 0,\ p < 0$
$y_{11} = \left(\frac{1}{2}c^2 + \sqrt{Q}\right)^{\frac{1}{3}} +$ $\left(\frac{1}{2}c^2 - \sqrt{Q}\right)^{\frac{1}{3}}$, $Q = -64\mu^3 \frac{C_3^3}{C_1^4}(C_1 s)s^8 +$ $\frac{\mu^2 C_4^4}{4C_1^4}(1 - 3s^2)^2$	$y_{12} = -\frac{4}{\sqrt{3}}\sqrt{-d}\cos 2\alpha$ $\tan\alpha = \left(\tan\left(\frac{\beta}{2}\right)\right)^{\frac{1}{3}}$ $\tan\beta = -\frac{16}{\sqrt{27}}\frac{\sqrt{-d^3}}{c^2}$ $\lvert\alpha\rvert \leq \frac{\pi}{4},\ \lvert\beta\rvert \leq \frac{\pi}{2}$ $d = \frac{E}{A} < 0$	$y_{13} = \frac{4}{\sqrt{3}}\sqrt{d}\cos\frac{\alpha}{2}$ $y_{14} = -\frac{4}{\sqrt{3}}\sqrt{d}(\cos\frac{\alpha}{3} + \frac{\pi}{3})$ $y_{15} = -\frac{4}{\sqrt{3}}\sqrt{d}(\cos\frac{\alpha}{3} - \frac{\pi}{3})$ $\cos\alpha = \frac{\sqrt{27}}{16}\frac{c^2}{\sqrt{d^3}}$ $d = \frac{E}{A} > 0$	$y_{16} = -\frac{4}{\sqrt{3}}\sqrt{d}\ cosec 2\alpha$ $\tan\alpha = \left(\tan\left(\frac{\beta}{2}\right)\right)^{\frac{1}{3}}$ $\sin\beta = -\frac{16}{\sqrt{27}}\frac{\sqrt{d^3}}{c^2},\ \lvert\beta\rvert \leq \frac{\pi}{2}$ $\lvert\alpha\rvert \leq \frac{\pi}{4},\ \lvert\beta\rvert \leq \frac{\pi}{2}$ $d = \frac{E}{A} > 0$

as an independent variable, and therefore one can write the derivatives of u_1, θ, λ_1 in the form:

$$\dot{u}_1 = \frac{du_1}{d\varphi}\frac{d\varphi}{dt}, \quad \dot{\theta} = \frac{v_2}{r}, \quad \dot{\lambda}_1 = \lambda_2\frac{v_2}{r} - \lambda_4. \tag{4.133}$$

As

$$a_t = \frac{c\beta}{m} = \dot{u}_1 - u_2\dot{\psi} + \frac{\mu}{r^2}s, \tag{4.134}$$

then due to Eqs. (4.127), (4.129), (4.132), and (4.133), and accepting for generality $\lambda = 1$, from Eq. (4.131) one can obtain the expression for the thrust acceleration:

$$a_t = \frac{c\beta}{m} = \frac{du_1}{d\varphi}\frac{(\frac{v_2}{r} - \lambda_4)}{k} - \frac{\lambda_4(sv_2 - kv_1)}{k} + \frac{\mu}{r^2}s,$$

where

$$\frac{du_1}{d\varphi} = -\frac{(a_3 w + a_4 C_2 + a_5 sk\frac{dw}{d\varphi})}{(s^2(3-5s^2)^2)}, \quad \frac{dw}{d\varphi} = -\frac{a_1 r + a_2\frac{dr}{d\varphi})}{r^2 C_2}, \quad \frac{dr}{d\varphi} = \frac{v_1}{\frac{v_2}{r} - \frac{\lambda_4}{k}},$$

$$a_1 = 12\mu s^3 k + Cr^2 k + 2Csr\frac{dr}{d\varphi} + C_2^2\frac{dr}{d\varphi}, \quad a_2 = 3\mu s^4 + Csr^4 + C_2^2 r,$$

$$a_3 = -24 + 40s^4 + 114s^2k^2 + 3k^4 - 15s^2k^4,$$

$$a_4 = -12 + 20s^2 + 40s^2k^2, \qquad a_5 = 24 - 40s^2 - 3k^2 + 5s^2k^2.$$

The analysis conducted for the constants C and C_3 have shown that for the set of values of s, satisfying the conditions $s = \sin\varphi < 0$ and $(3s^2 - 1) > 0$, the following conditions are valid: $a > 0$ if $-1 < s < -0.7746$, and $a < 0$ if $-0.7746 < s < 0$. Consequently, the classes of solutions obtained above satisfy the necessary conditions of optimality if $s \in (-1.0, -0.7746)$. This completes the proof of Theorem 4.3. □

4.7 CONJUGATE POINTS

Consider now the question about the presence of the conjugate points. The constants of the solutions are C, C_2, C_3, C_4, C_5, λ, m_0. We can now compute the derivatives to obtain the solutions to the auxiliary problem:

$$\eta_1 = R_1\frac{\partial r}{\partial C} + R_2\frac{\partial r}{\partial C_2} + R_3\frac{\partial r}{\partial \lambda}, \quad \eta_2 = V_1\frac{\partial v_1}{\partial C} + V_2\frac{\partial v_1}{\partial C_2} + V_3\frac{\partial v_1}{\partial \lambda},$$

$$\eta_3 = W_1\frac{\partial v_2}{\partial C} + W_2\frac{\partial v_2}{\partial C_2} + W_3\frac{\partial v_2}{\partial \lambda},$$
$$\eta_4 = \Theta_1\frac{\partial \theta}{\partial C} + \Theta_2\frac{\partial \theta}{\partial C_2} + \Theta_3\frac{\partial \theta}{\partial C_5} + \Theta_4\frac{\partial \theta}{\partial \lambda},$$
$$\eta_5 = M_1\frac{\partial m}{\partial C} + M_2\frac{\partial m}{\partial C_2} + M_3\frac{\partial m}{\partial C_3} + M_4\frac{\partial m}{\partial C_4} + M_5\frac{\partial m}{\partial \lambda} + M_6\frac{\partial m}{\partial m_0}.$$

Utilizing the expressions for F, F_1, W, P, it can be shown that these equations have at least a 4-th order terms. This means that it is always possible to find the values of φ_0 and φ', $(\varphi_0 < \varphi' \leq \varphi_1)$, at which $\eta(\varphi_0) = 0$ and $\eta(\varphi') = 0$. Consequently, the angles φ_0 and φ' are the conjugate points.

4.8 OPTIMALITY AND APPLICABILITY OF LAWDEN SPIRALS

The analytical solutions for planar IT arcs known as Lawden spirals have been first obtained by Lawden in Ref. [3], [152], [153]. These solutions have been obtained regardless the functional of the problem and satisfy the equations of motion and costate equations. The analysis of the literature devoted to the studies of IT arcs and their optimality have shown that it is not clear yet, to what degree these arcs are optimal and not only stationary, and can they be considered as the solutions to the practical problems [42], [156], [52], [53], [59], [67]. Below some of these questions will be addressed.

Theorem 4.4. *If according to the problem (2.111)–(2.115), the final flight time is not fixed, $(C = 0)$, no constraint is imposed on the angular distance, and the functional of the problem does not explicitly depend on the final polar angle, then the Lawden spirals are degenerated into zero thrust arcs.*

Proof of Theorem 4.4. Using the equations of motion obtained by Lawden for IT arcs and Eqs. (2.135) one can show that the constant A in the formulas of the spirals, the cyclic constant λ_5 and the primer vector magnitude are related to each other by the expression:

$$\lambda_5 = -\lambda A. \tag{4.135}$$

Indeed, consider the first integral

$$u_2 + r\dot{\psi}\sin\varphi = A, \tag{4.136}$$

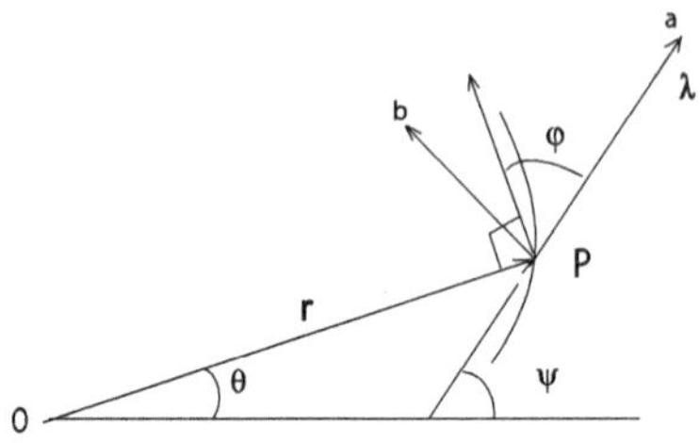

Figure 4.15 Coordinate systems.

where u_2 is the projection of the velocity on the direction perpendicular to $\boldsymbol{\lambda}$ (see Fig. 4.15):

$$u_2 = -\dot{r}\cos\varphi + r\dot{\theta}\sin\varphi, \tag{4.137}$$

where the angles φ, ψ, and θ are related by the expressions: [3]

$$\theta = \varphi + \psi - \frac{\pi}{2}. \tag{4.138}$$

As

$$tg\varphi = \frac{\lambda_1}{\lambda_2}, \quad \dot{\psi} = \dot{\theta} - \dot{\varphi}, \quad \dot{\theta} = \frac{v_2}{r}, \tag{4.139}$$

then the integral (4.136) with (4.138) and (4.139) is reduced to

$$-\lambda_2(v_1\lambda_2^2 - v_2\lambda_1\lambda_2 - r\lambda_1\lambda_4 - \lambda_2\lambda_5) = \lambda^2(\lambda_5 + \lambda A),$$

from which one can obtain Eq. (4.135). As the Lawden's solutions contain A, which is associated with λ_5, it is of interest to consider these solutions taking into account the boundary conditions and the transversality conditions. If there is no constraint on the final polar angle, the functional J of the problem does not explicitly depend on this variable, then for the final time from the transversality condition it follows that

$$\lambda_{51} = \frac{\partial J}{\partial \theta} = 0. \tag{4.140}$$

As $\lambda_5 = const$ is the integral of the canonical system Eqs. (2.135), then from (4.140) one can obtain $\lambda_5 = 0$, valid at all points of the optimal trajectory. Consequently, the constant A is also equal to zero. In this case simple analysis of the integrals (4.7), relationships (4.13)–(4.15) and Eqs. (2.135) lead to $m = 0$, that is to the degeneration of the Lawden spirals into the zero thrust arcs.

This completes the proof of Theorem 4.4. □

Until present time the Lawden spirals are studied only for optimality by utilizing the Kopp-Moyer conditions or Robbins condition, and non-optimality of the spirals has been shown [53], [54], [42], [154]. Note that the tests of the Robbins condition is complicated by the necessity of verifying that the angle ψ (hat is φ in the Lawden's coordinate system) must be chosen such that the thrust acceleration

$$f = \frac{\mu}{A^2}(\frac{1-3s^2}{3-5s^2})^3 s^{-11}(27 - 75s^2 + 60s^4),$$

is positive [42]. However, if the conditions $1 - 3s^2 > 0$ and $3 - 5s^2 > 0$, and the Robbins condition, $\sin\varphi < 0$, are satisfied, then one can obtain $f < 0$, which shows non-optimality of the spirals. But more importantly, in the previously conducted studies mentioned above, the Lawden spirals are investigated without determining whether these solutions exist for the given problem with specified boundary conditions. The three remarks given below are associated with these findings.

Remark 1. When obtaining the solutions for the spirals, it is accepted that $\lambda = 1$ [3]. Such an assumption is valid for the minimization of the characteristic velocity $J = c \ln \frac{m_0}{m_1}$ [91]. In this case the expression (4.140) will be valid, which will also lead to degeneration of the arcs being considered. Therefore, any assumption about the value of the magnitude of the primer vector without taking into account the boundary conditions and the transversality condition is not always an appropriate assumption.

The solutions for the spirals have been obtained using the first integrals and the primer vector equation [63]. Note that these solutions can also be obtained using the first integrals and invariant relationships. Indeed, from Eq. (4.120) with $C = 0$ one can obtain

$$r = \frac{9\mu\lambda_1^6}{A^2\lambda^2(\lambda^2 - 3\lambda_1^2)}.$$

One can also verify that, as it will be shown in the next subsection, the other solutions for the spirals can be obtained from (4.7), (4.13)–(4.15).

Remark 2. In Ref. [152], the Lawden spirals represent the solutions valid in the case of free time and minimization of the fuel mass. It is noted that there exists a possibility that the corresponding trajectory may be a solution to the problem of optimal escape from a circular orbit. However, in this specific case, from the transversality condition (see (4.140)) it follows that $\lambda_5 = 0$, as the functional of the problem does not depend on the final polar angle. Using (4.135) we obtain $A = 0$, and this leads to the degeneration of the thrust arcs. Consequently, in Ref. [152] the IT arcs, that is the Lawden spirals can not be included into an optimal trajectory of escape from a circular orbit.

Consequently, the spirals can be the solutions only in the case when the final polar angle is given or the functional of the problem explicitly depends on the final polar angle. In particular, if the minimization of the characteristic velocity is considered, and there is no constraint on the angular distance, the Lawden spirals can not serve as the IT arc solutions to the problem.

In summary, the following comments can be made regarding the optimality and applicability of the Lawden spirals. According to the variation problem, the IT arcs have to satisfy the canonical equations, the boundary conditions, the transversality conditions, and the conditions $\chi = \dot{\chi} = \ddot{\chi} = \chi^{(3)} = \chi^{(4)} = \ldots = 0$. The latter conditions represent the conditions of existence and optimality of the thrust arcs.

Remark 3. It is helpful to note about the Ref. [42], where Robbins studied the problem based on the necessary condition of optimality derived in the same work without analyzing the other conditions. Analyzing the formulas of this work, specifically Eqs. (10'–13'),

$$\lambda\dot{\lambda} = \boldsymbol{\lambda}\dot{\boldsymbol{\lambda}}, \tag{4.141}$$

$$\dot{\lambda}\dot{\lambda} + \lambda\ddot{\lambda} = \dot{\boldsymbol{\lambda}}\dot{\boldsymbol{\lambda}} - (\boldsymbol{\lambda}\nabla)^2 U, \tag{4.142}$$

$$3\dot{\lambda}\ddot{\lambda} + \lambda\lambda^{(3)} = -(\dot{\mathbf{r}}\nabla)(\boldsymbol{\lambda}\nabla)^2 U - 4(\boldsymbol{\lambda}\nabla)(\dot{\boldsymbol{\lambda}}\nabla)U - (\boldsymbol{\lambda}\nabla)^2\frac{\partial U}{\partial t}, \tag{4.143}$$

$$3\ddot{\lambda}\ddot{\lambda} + 4\dot{\lambda}\lambda^{(3)} + \lambda\lambda^{(4)} = -a(\boldsymbol{\lambda}\nabla)^3 U + B(\boldsymbol{\lambda}, \dot{\boldsymbol{\lambda}}, \mathbf{r}, \vec{r}, t), \tag{4.144}$$

where B is a complicated expression not considered here, it can be shown that the Robbins necessary condition partially reflects the condition $\chi^{(4)} = 0$, and the equalities obtained by equating the left hand sides of (4.141)–(4.144) to zero, partially represents the conditions $\dot{\chi} = \ddot{\chi} = \chi^{(3)} = \chi^{(4)} = 0$. Besides that, if $a_t = \frac{B}{q} > 0$, where $q = (\boldsymbol{\lambda}\nabla)^3 U$ (see Eq. (15) of the Ref. [42]), then this does not mean yet that Eqs. (4.141)–(4.143) are satisfied.

Consequently, if the solutions for IT arcs are obtained, then to determine the optimality of these solutions it is not enough to test only Robbins condition, but it is also necessary to consider other conditions of existence and optimality, boundary conditions and the transversality conditions with given functional. Only after these analysis one can make a conclusion about optimality of these arcs.

In the next subsection, the two classes of the extremals will be derived for the case when the flight time is not fixed, and the final polar angle is fixed. One of these classes represent the Lawden spirals. The purpose of deriving the Lawden spirals is to demonstrate the advantage of using the method of obtaining the solutions as it allows us to also derive other previously unknown solutions for IT arcs for arbitrary values of the constants C and C_2.

4.9 HAMILTON-JACOBI EQUATION FOR INTERMEDIATE THRUST ARCS

Obtaining analytical solutions for IT arcs employing various classical methods of analytical mechanics represents an important task in solving the variational problem being considered. In this subsection, the questions of integration of Hamilton-Jacobi equation for IT arcs are considered. It will be shown that the first integrals of the canonical system of the variation problem and some invariant relationships allow us to obtain incomplete integral of Hamilton-Jacobi equation and reduce the solution of the problem to quadratures [157]. Let us accept the following notations:

$$v_1 = q_1, \quad v_2 = q_2, \quad r = q_3, \quad \theta = q_4, \quad m = q_5.$$

Consider the free univalent transformations, $Q_i = Q_i(q, \lambda, t)$, $\wedge_i = \wedge_i(q, \lambda, t)(i = 1, ..., 5)$, which can be determined by the equations [140]:

$$\frac{\partial S(q, Q, t)}{\partial q_i} = \lambda_i, \qquad \frac{\partial S(q, Q, t)}{\partial Q_i} = -\wedge_i, (i = 1, ..., 5) \tag{4.145}$$

where the following assumptions are made for so called derivative (characteristic) function S:

$$det\left| \frac{\partial^2 S}{\partial Q_i \partial q_j} \right|^n \neq 0, \qquad (i, j = 1, ..., 5) \tag{4.146}$$

In this case, Eqs. (2.135) are reduced to a form:

$$\frac{dQ_i}{dt} = \frac{\partial K}{\partial \wedge_i}, \qquad \frac{d\wedge_i}{dt} = -\frac{\partial K}{\partial Q_i} \tag{4.147}$$

with new Hamiltonian

$$K = H + \frac{\partial S}{\partial t}.$$

If it is required that $K = 0$, then (4.147) gives integrals

$$Q_i = \alpha_i, \qquad \Lambda_i = \beta_i \tag{4.148}$$

where α_i, β_i are integration constants. Then based on Eqs. (4.145) and (4.146) the following solutions can be obtained:

$$q_i = q_i(\alpha_i, \beta_i, t), \qquad \lambda_i = \lambda_i(\alpha_i, \beta_i, t). \tag{4.149}$$

Here, taking into account (4.145), the function S must satisfy the following Hamilton-Jacobi equation:

$$\frac{\partial S}{\partial t} + H(q_i, \frac{\partial S}{\partial q_i}, t) = 0. \tag{4.150}$$

In the presence of a complete integral $S(q_i, \alpha_i, t)$ of Eq. (4.150), the solution (4.149) can be found from the expressions:

$$\frac{\partial S}{\partial q_i} = \lambda_i, \qquad \frac{\partial S}{\partial \alpha_i} = -\beta_i. \tag{4.151}$$

As the Hamiltonian,

$$H = -\lambda_1 \frac{\mu}{q_3^2} + \lambda_1 \frac{q_1^2}{q_3} - \lambda_2 \frac{q_1 q_2}{q_3} + \lambda_3 q_1 + \lambda_4 \frac{q_2}{q_3} \tag{4.152}$$

of the problem considered does not explicitly depend on time, then there exist integral $H = C$, where C is an integration constant. Assuming

$$S = -Ct + V,$$

from (4.150) taking into account (4.145), one can obtain Hamilton-Jacobi equation for IT arcs in the following form:

$$-\frac{\partial V}{\partial q_1} \frac{\mu}{q_3^2} + \frac{\partial V}{\partial q_1} \frac{q_2^2}{q_3} - \frac{\partial V}{\partial q_2} \frac{q_1 q_2}{q_3} + \frac{\partial V}{\partial q_3} q_1 + \frac{\partial V}{\partial q_4} \frac{q_2}{q_3} = C. \tag{4.153}$$

In summary, the solution of Eqs. (4.147) is reduced to the derivation of the complete integral $V(q, \alpha)$ of Eq. (4.153).

Below for integration of Eq. (4.153), consider the methods of separation of variables and Hamilton-Jacobi.

4.9.1 Non-integrability of Hamilton-Jacobi Equation by Separation of Variables

In this subsection, it will be shown that the variables in Eq. (4.153) are not separable [157].

Lemma 4.1. *In the context of the problem (2.111)–(2.115), Eq. (4.153) for IT is not integrable by the method of separation of variables.*

Proof of Lemma 4.1. The criterion of applicability of the method of separation of variables is the condition of Morera-Levi-Civita-Forbat [140]:

$$\frac{\partial H}{\partial \lambda_i}\left(\frac{\partial H}{\partial \lambda_j}\frac{\partial^2 H}{\partial q_i \partial q_j} - \frac{\partial H}{\partial q_j}\frac{\partial^2 H}{\partial q_i \partial \lambda_j}\right) = \frac{\partial H}{\partial q_i}\left(\frac{\partial H}{\partial \lambda_j}\frac{\partial^2 H}{\partial \lambda_j \partial q_j} - \frac{\partial H}{\partial q_j}\frac{\partial^2 H}{\partial \lambda_i \partial \lambda_j}\right). \quad (4.154)$$

To test Eq. (4.154), one can compute all necessary derivatives from Hamiltonian with respect to q_i and λ_i, $i = 1, ..., 5$. They have the following form:

$$\frac{\partial H}{\partial q_1} = -\lambda_2 \frac{q_2}{q_3} + \lambda_3, \qquad \frac{\partial H}{\partial \lambda_1} = -\frac{\mu}{q_3^2} + \frac{q_2^2}{q_3},$$

$$\frac{\partial H}{\partial q_2} = 2\lambda_1 \frac{q_2}{q_3} - \lambda_2 \frac{q_1}{q_3} + \frac{\lambda_4}{q_3}, \qquad \frac{\partial H}{\partial \lambda_2} = -\frac{q_1 q_2}{q_3},$$

$$\frac{\partial^2 H}{\partial q_1 \partial q_2} = -\frac{\lambda_2}{q_3}, \qquad \frac{\partial^2 H}{\partial q_1 \partial \lambda_2} = -\frac{q_2}{q_3},$$

$$\frac{\partial^2 H}{\partial \lambda_1 \partial q_2} = 2\frac{q_2}{q_3}, \qquad \frac{\partial^2 H}{\partial \lambda_1 \partial \lambda_2} = 0,$$

$$\frac{\partial^2 H}{\partial q_2 \partial q_1} = \frac{\partial^2 H}{\partial q_1 \partial q_2}, \qquad \frac{\partial^2 H}{\partial q_2 \partial \lambda_1} = \frac{\partial^2 H}{\partial \lambda_1 \partial q_2},$$

$$\frac{\partial^2 H}{\partial \lambda_2 \partial q_1} = \frac{\partial^2 H}{\partial q_1 \partial \lambda_2}, \qquad \frac{\partial^2 H}{\partial \lambda_2 \partial \lambda_1} =, \frac{\partial^2 H}{\partial \lambda_1 \partial \lambda_2} = 0,$$

$$\frac{\partial H}{\partial q_3} = 2\lambda_1 \frac{\mu}{q_3^3} - \lambda_1 \frac{q_2^2}{q_3^2} + \lambda_2 \frac{q_1 q_2}{q_3^2} - \lambda_4 \frac{q_2}{q_3^2}, \qquad \frac{\partial H}{\partial \lambda_3} = q_1,$$

$$\frac{\partial^2 H}{\partial q_1 \partial q_3} = \lambda_2 \frac{q_2}{q_3^2}, \qquad \frac{\partial^2 H}{\partial q_1 \partial \lambda_3} = 1,$$

$$\frac{\partial^2 H}{\partial \lambda_1 \partial \lambda_3} = 0, \qquad \frac{\partial^2 H}{\partial \lambda_1 \partial q_3} = 2\frac{\mu}{q_3^3} - \frac{q_2^2}{q_3^2},$$

$$\frac{\partial^2 H}{\partial q_3 \partial q_1} = \frac{\partial^2 H}{\partial q_1 \partial q_3}, \qquad \frac{\partial^2 H}{\partial q_3 \partial \lambda_1} = \frac{\partial^2 H}{\partial \lambda_1 \partial q_3} = 1, \qquad \frac{\partial^2 H}{\partial \lambda_3 \partial q_1} = \frac{\partial^2 H}{\partial q_1 \partial \lambda_3} = 1,$$

$$\frac{\partial^2 H}{\partial q_2 \partial q_3} = -2\lambda_1 \frac{q_2}{q_3^2} + \lambda_2 \frac{q_1}{q_3^2} - \frac{\lambda_4}{q_3^2},$$

$$\frac{\partial^2 H}{\partial q_2 \partial \lambda_3} = 0, \qquad \frac{\partial^2 H}{\partial \lambda_2 \partial q_3} = \frac{q_1 q_2}{q_3^2},$$

$$\frac{\partial^2 H}{\partial q_3 \partial q_2} = \frac{\partial^2 H}{\partial q_2 \partial q_3}, \qquad \frac{\partial^2 H}{\partial \lambda_3 \partial q_2} = \frac{\partial^2 H}{\partial q_2 \partial \lambda_3}, \qquad \frac{\partial^2 H}{\partial q_3 \partial \lambda_2} = \frac{\partial^2 H}{\partial \lambda_2 \partial q_3},$$

$$\frac{\partial^2 H}{\partial q_2 \partial q_4} = 0, \qquad \frac{\partial^2 H}{\partial q_2 \partial \lambda_4} = \frac{1}{q_3}, \qquad \frac{\partial H}{\partial q_4} = 0,$$

$$\frac{\partial H}{\partial q_5} = -\lambda \frac{cm}{q_5^2}, \qquad \frac{\partial H}{\partial \lambda_5} = -m,$$

$$\frac{\partial^2 H}{\partial q_1 \partial q_5} = 0, \qquad \frac{\partial^2 H}{\partial q_1 \partial \lambda_5} = 0, \qquad \frac{\partial^2 H}{\partial \lambda_1 \partial q_5} = 0,$$

$$\frac{\partial^2 H}{\partial q_5 \partial q_1} = 0, \qquad \frac{\partial^2 H}{\partial q_5 \partial \lambda_1} = 0, \qquad \frac{\partial^2 H}{\partial \lambda_5 \partial q_1} = 0,$$

$$\frac{\partial^2 H}{\partial q_3 \partial q_5} = 0, \qquad \frac{\partial^2 H}{\partial q_3 \partial \lambda_5} = 0, \qquad \frac{\partial^2 H}{\partial \lambda_3 \partial q_5} = 0,$$

$$\frac{\partial^2 H}{\partial q_2 \partial q_5} = 0, \qquad \frac{\partial^2 H}{\partial q_2 \partial \lambda_5} = 0,$$

$$\frac{\partial^2 H}{\partial \lambda_2 \partial q_5} = 0, \qquad \frac{\partial^2 H}{\partial \lambda_5 \partial q_2} = 0, \qquad \frac{\partial^2 H}{\partial \lambda_2 \partial \lambda_5} = 0.$$

It can be shown that all other derivatives of Hamiltonian for other values of i and j are equal to zero. Furthermore, substituting the function in Eq. (4.152) into Eq. (4.154), and after simplifications, one can obtain the following equalities:
at $i = 1, j = 2$:

$$\left(-\frac{\mu}{q_3} + q_2^2\right)(2\lambda_1 q_2 + \lambda_4) = -2 q_1 q_2(-\lambda_2 q_2 + \lambda_3 q_3), \tag{4.155}$$

at $i = 1, j = 3$:

$$\left(-\frac{\mu}{q_3} + q_2^2\right)\left(-2\lambda_1 \frac{\mu}{q_3} + \lambda_1 q_2^2 + \lambda_4 q_2\right) = q_1(-\lambda_2 q_2 + \lambda_3 q_3)\left(-2\frac{\mu}{q_3} - q_2^2\right) \tag{4.156}$$

In the cases when $i = 2, j = 1$ and $i = 3, j = 1$ one can obtain equalities similar to Eq. (4.155) and (4.156). Analysis have shown that for all other values of i and j one can obtain identities. From Eq. (4.155) and (4.156) we have:

$$-\frac{2 q_1 q_2}{2\lambda_1 q_2 + \lambda_4} = -\frac{q_1\left(2\frac{\mu}{q_3} - q_2^2\right)}{2\lambda_1 \frac{\mu}{q_3} - \lambda_1 q_2^2 - \lambda_4 q_2} = \frac{-\frac{\mu}{q_3} + q_2^2}{-\lambda_2 q_2 + \lambda_3 q_3}.$$

These expressions lead to the equalities of the form:

$$-\lambda_4 q_2^2 = 2\lambda_4 \frac{\mu}{q_3}, \qquad -\frac{q_1}{\lambda_1} = \frac{-\frac{\mu}{q_3} + q_2^2}{-\lambda_2 q_2 + \lambda_3 q_3} \tag{4.157}$$

or

$$q_3[\lambda_1(-\frac{\mu}{q_3^2} + \frac{q_2^2}{q_3}) - \lambda_2 \frac{q_1 q_2}{q_3} + \lambda_3 q_1] = 0. \tag{4.158}$$

From Eq. (4.157) at $-q_2^2 \neq 2\frac{\mu}{q_3}$ it follows that $\lambda_4 = 0$, and from Eq. (4.158) at $q_3 \neq 0$ and $\chi = 0$ we obtain $H = C = 0$. Consequently, one can conclude that the variables in Eq. (4.153) are separable if

$$C = 0, \qquad \lambda_4 = 0. \tag{4.159}$$

Note that on IT arcs $\chi = \frac{c}{M}\lambda - \lambda_5$ and all time derivatives are zero, and Eq. (2.135) can be used to derive invariant relationships. Excluding λ_3 from Eq. (4.145) and (4.150), and taking into account Eq. (4.153) we obtain

$$(\lambda_1 q_2 - \lambda_2 q_1)^2 + \lambda_4(\lambda_1 q_2 - \lambda_2 q_1) = C\lambda_1 q_3 + \lambda_1^2 \frac{\mu}{q_3}, \tag{4.160}$$

$$(\lambda_1 q_2 - \lambda_2 q_1 + \lambda_4)^2 = \lambda_1^2 \frac{\mu}{q_3} - 3\lambda_1^4 \frac{\mu}{\lambda^2 q_3}. \tag{4.161}$$

From Eqs. (4.160) and (4.161) it follows that $\lambda_1 = 0$, which leads to degeneration of IT arcs to ZT arcs. Consequently, if Eq. (4.159) are satisfied, then IT arcs do not exist, and therefore, separation of variables in Eq. (4.153) is not possible.

Lemma 4.1 has been proved. □

Below the application of Hamilton-Jacobi method will be considered for integration of Eq. (4.153).

4.9.2 Integration of Hamilton-Jacobi Equation for Intermediate Thrust Arcs

It is known that if the first-order partial differential equation [140]

$$\varphi\left[\frac{\partial V}{\partial q_1}, \frac{\partial V}{\partial q_2}, \ldots, \frac{\partial V}{\partial q_n}, q_1, q_2, \ldots, q_n\right] = C_1, \tag{4.162}$$

is given, then according to Hamilton-Jacobi method, the integration of this equation is associated with finding $(n-1)$ expressions that involve $\lambda_i = \frac{\partial V}{\partial q_i}$

and $q_i (i = 1, ..., n)$. Let this expressions are of the following form:

$$\varphi = H = C, \quad \varphi_1 = C_1, \quad \varphi_2 = C_2, ..., \quad \varphi_{n-1} = C_{n-1},$$

where $C, C_1, C_2, ..., C_{n-1}$ are the integration constants. In this case, these expressions can be considered using Jacobi's theory [140]. It can be proved that $(n\frac{n-1}{2})$ conditional equations,

$$(\varphi_i, \varphi_k) = \frac{\partial \varphi_i}{\partial \lambda_1}\frac{\partial \varphi_k}{\partial q_1} + \frac{\partial^2 \varphi_i}{\partial \lambda_2 \partial q_2} + ... + \frac{\partial \varphi_i}{\partial \lambda_n}\frac{\partial \varphi_k}{\partial q_n} - \frac{\partial \varphi_k}{\partial \lambda_1}\frac{\partial \varphi_i}{\partial q_1} - \frac{\partial \varphi_k}{\partial \lambda_2}\frac{\partial \varphi_i}{\partial q_2} - ... - \frac{\partial \varphi_k}{\partial \lambda_n}\frac{\partial \varphi_i}{\partial q_n} \quad (4.163)$$

are necessary and sufficient for the values of $\lambda_1, \lambda_2, ..., \lambda_n$ obtained from the equations $\varphi_i = C_i$ as functions of $q_1, q_2, ..., q_n$ make the expression $\lambda_1 dq_1 + \lambda_2 dq_2 + ...\lambda_n dq_n$ as a complete differential, and its integral

$$V = \int \lambda_1 dq_1 + \lambda_2 dq_2 + ...\lambda_n dq_n \quad (4.164)$$

can be made as a complete solution of Eq. (4.162).

Theorem 4.5. *Let in the problem (2.111)–(2.115), the $s \leq 7$ first integrals of Eqs. (2.134) with s integration constants, and some number of invariant relationships are known. Then incomplete solutions for IT arcs can be expressed in quadratures with 2s integration constants.*

Proof of Theorem 4.5. Consider the following first integrals of Eqs. (2.135):

$$\varphi = H = C, \quad \varphi_1 = \lambda_1^2 + \lambda_2^2 + \lambda_3^2 = C_1, \quad \varphi_2 = \lambda_4 = C_2, \quad (4.165)$$

and invariant relationships given in Eqs. (4.13)

$$\varphi_3 = \lambda_1\lambda_4 + \lambda_1\lambda_2\frac{q_2}{q_3} - \lambda_2^2\frac{q_1}{q_3} + \lambda_4\frac{\lambda_2}{q_3} = 0. \quad (4.166)$$

Note that as m and λ_5 are not explicitly present in Eq. (4.153) and not present in Eqs. (4.165) and (4.166), the question on determination of these variables will be addressed later in this work. The solutions of Eqs. (4.164) and (4.166) yield:

$$\lambda_1 = \lambda_1(q_1, q_2, q_3, C, C_1, C_2),$$

$$\lambda_2 = \lambda_2(q_1, q_2, q_3, C, C_1, C_2) = \sqrt{C_1 - \lambda_1^2}, \tag{4.167}$$

$$\lambda_3 = \lambda_3(q_1, q_2, q_3, C, C_1, C_2) = \frac{\sqrt{C_1 - \lambda_1^2}}{\lambda_1}\left(-\lambda_1 \frac{q_2}{q_3} + \sqrt{C_1 - \lambda_1^2}\frac{q_1}{q_3} - \frac{C_2}{q_3}\right),$$

$$\lambda_4 = C_2.$$

The function λ_1 in explicit form is given in Appendix. As the relationships Eqs. (4.167) contain three (not four) constants of integration, then the incomplete integral (according to Lehmann-File's definition [62]) of Eqs. (4.153) on the basis of Eqs. (4.164) can be written in the form:

$$V = C_3\theta + \int \lambda_1 dq_1 + \int \lambda_2 dq_2 + \int \lambda_3 dq_3, \tag{4.168}$$

and correspondingly,

$$S = -Ct + C_2\theta + \int \lambda_1 dq_1 + \int \lambda_2 dq_2 + \int \lambda_3 dq_3.$$

Consequently, Eqs. (4.164) are rewritten in the form:

$$-\beta_1 = \frac{\partial S}{\partial C_1}, \quad -\beta_2 = \frac{\partial S}{\partial C_2}, \quad -\beta_3 = \frac{\partial S}{\partial C}. \tag{4.169}$$

By solving equations $\lambda_i = \frac{\partial S}{\partial q_i}$, $(i = 1, 2, 3)$, and taking into account Eqs. (4.169) and $\frac{dq_3}{dt} = q_1$, one can find incomplete solutions of Eq. (4.145):

$$\begin{gathered} q_i = v_i(\theta, C, C_1, C_2, \beta_1, \beta_2, \beta_3), \quad i = 1, 2, 3. \\ \lambda_j = \lambda_j(\theta, C, C_1, C_2, \beta_1, \beta_2, \beta_3), \quad j = 1, 2, 3, 4. \\ t = t(\theta, C, C_1, C_2, \beta_1, \beta_2, \beta_3). \end{gathered} \tag{4.170}$$

Theorem 4.5 has been proved. □

Theorem 4.5 can be applied in the analysis of other IT arcs, and in this case, the presence of invariant relationships is not necessary.

The remaining variables, m and λ_5 are determined from the following first integrals:

$$\lambda_4 m = C_3, \quad \lambda_1 v_1 + \lambda_2 v_2 + 2\lambda_3 r + c\lambda \ln \frac{m_0}{m} + 3Ct = C_4$$

in the form:

$$m = m_0 e^{-f}, \quad \lambda_5 = \frac{C_3}{m_0} e^{f},$$
$$f = \frac{C_4 - \lambda_1 v_1 - \lambda_2 v_2 - 2\lambda_3 r - 3Ct}{c\lambda}, \quad \lambda = const.$$

It follows from what mentioned above, the application of Hamilton-Jacobi method allows us to determine the integral of this equation with six integration constants, and find all solutions of the equations of the variational problem with eight constants. Note that by a method of exclusion, the first integrals and invariant relationships allow us to obtain the solutions with six constants [67].

4.10 CLASSIFICATION OF INTERMEDIATE THRUST ARCS

Using on the analysis of the previous sections, one can classify the intermediate thrust (IT) arcs based on the fixed or free final time and given functional of the problem.

1. If the flight time is fixed, ($C \neq 0$), then there exist the circular spherical solutions, and the solutions in quadratures [62], [139].

2. If the flight time is fixed, ($C \neq 0$), and the functional of the problem explicitly depends on the final polar angle, ($\lambda_5 \neq 0$), then one can have six classes of spiral trajectories.

3. If the flight time is fixed, ($C \neq 0$), and the functional of the problem does not explicitly depend on the final polar angle, ($\lambda_5 = 0$), then the existing solutions of the problem admit two classes of spiral trajectories.

4. If the flight time is not fixed, ($C = 0$), and the functional of the problem explicitly depends on the final polar angle, ($\lambda_5 \neq 0$), then one can have two classes of spiral trajectories, one of which corresponds to Lawden's spirals.

5. If the flight time is not fixed, ($C = 0$), and the functional of the problem does not explicitly depend on the final polar angle, ($\lambda_5 = 0$), then there exist solutions in quadratures that will be studied in the next chapter of this monograph.

For certain values of the Hamiltonian constant and radial component of the primer vector, the classes of the trajectories studied above can satisfy the necessary conditions of optimality.

In this chapter, IT arcs have been investigated by using the canonical system of equations, its first integrals and invariant relations. The following solutions have been obtained for IT arcs:

1. The spherical solutions have been obtained in the case of free final time. They satisfy the necessary conditions of optimality if the Hamiltonian is positive. This class of solutions generalizes the previously known circular trajectories for IT arcs.
2. Six classes of spiral trajectories in the case when the flight time is fixed, and the functional of the problem explicitly depends on the final polar angle. These classes satisfy the necessary conditions of optimality when the Hamiltonian is positive and radial component of the primer vector is negative.
3. Two classes of spiral trajectories in the case when the flight time is fixed, and the functional of the problem does not explicitly depend on the final polar angle. These classes also satisfy the necessary conditions of optimality when the Hamiltonian is positive and radial component of the primer vector is negative.
4. Three classes of spiral trajectories in the case when the flight time is not fixed, and the functional of the problem explicitly depend on the final polar angle. One of these classes corresponds to Lawden's spirals.

Besides that, the Lawden's spirals and their optimality have been analyzed. If the fuel-minimum problem is considered, then the Lawden's spirals do not satisfy the necessary conditions of existence of IT arcs. An approach that can account for all conditions of the problem, including the functional, have been developed. The existing classes of IT arcs have been classified based on the fixed or free final time and given functional of the problem. This classification can be useful in determining the appropriate class of the solutions depending on the problem's conditions. It has also been shown that the application of Hamilton-Jacobi method allows us to determine an incomplete integral of Hamilton-Jacobi equation with six integration constants and to obtain the solutions in quadratures.

CHAPTER 5

Motion With Constant Power and Constant Specific Impulse

5.1 CANONICAL EQUATIONS AND FIRST INTEGRALS

As the power and the specific impulse are constants, the mass-flow rate, β is also constant. In particular, if a chemical propulsion system with constant exhaust velocity and limited mass-flow rate is considered, for a maximum thrust arc this constant is equal to the maximum value of mass-flow rate, $\beta = \beta_{max}$. It can be shown that the equations of extremal motion can be given in a canonical form [158]:

$$
\begin{aligned}
\dot{\mathbf{v}} &= \frac{c\beta}{m}\frac{\boldsymbol{\lambda}}{|\boldsymbol{\lambda}|} - \frac{\mu}{r^3}\mathbf{r}, \\
\dot{\mathbf{r}} &= \mathbf{v}, \\
\dot{m} &= -\beta, \\
\dot{\boldsymbol{\lambda}} &= -\boldsymbol{\lambda}_r, \\
\dot{\boldsymbol{\lambda}}_r &= \frac{\mu}{r^3}\boldsymbol{\lambda} - 3\frac{\mu}{r^5}(\boldsymbol{\lambda}\mathbf{r})\mathbf{r}, \\
\dot{\lambda}_7 &= \frac{c\beta}{m^2}|\boldsymbol{\lambda}|,
\end{aligned}
\tag{5.1}
$$

where $\beta = const$, $I_{sp} = const$, and $c = I_{sp}g$. The corresponding Hamiltonian is written as

$$H = -\frac{\mu}{r^3}\boldsymbol{\lambda}\mathbf{r} + \boldsymbol{\lambda}_r\mathbf{v} + (\frac{c}{m}\lambda - \lambda_7)\beta.$$

It can be shown that for thrust arcs with constant power and constant specific impulse, Eqs. (5.1) admit the following first integrals [26]:

$$
\begin{aligned}
-\frac{\mu}{r^3}\boldsymbol{\lambda}\mathbf{r} + \boldsymbol{\lambda}_r\mathbf{v} + (\frac{c}{m}\lambda - \lambda_7)\beta &= C, \\
\mathbf{v}\boldsymbol{\lambda} + \mathbf{r}\boldsymbol{\lambda}_r &= \mathbf{K}(A_1, A_2, C_2), \\
m &= m_0 - \beta t,
\end{aligned}
\tag{5.2}
$$

where C, A_1, A_2, C_2, m_0 are the integration constants.

Analytical Solutions for Extremal Space Trajectories
DOI: http://dx.doi.org/10.1016/B978-0-12-814058-1.00005-9

5.2 CIRCULAR TRAJECTORIES

Similarly to the case with maximum power and variable specific impulse, consider the constraint $r = r_0$ and associated Lagrange multiplier μ_1 in the context of the problem (2.111)–(2.115) for a planar motion. Using a polar coordinate system $O\ r\ \theta$ with the origin at the center of gravitation, and accepting that $v_1 = 0$, one can reduce Eqs. (5.1) to the following form:

$$0 = \frac{c\beta}{m}\frac{\lambda_1}{\lambda} - \frac{\mu}{r^2} + \frac{v_2^2}{r}, \tag{5.3}$$

$$\dot{v}_2 = \frac{c\beta}{m}\frac{\lambda_2}{\lambda}, \tag{5.4}$$

$$\dot{r} = 0, \tag{5.5}$$

$$\dot{\theta} = \frac{v_2}{r}, \tag{5.6}$$

$$\dot{m} = -\beta, \tag{5.7}$$

$$\dot{\lambda}_1 = \lambda_2\frac{v_2}{r} - \lambda_4 + \mu_1, \tag{5.8}$$

$$\dot{\lambda}_2 = -2\lambda_1\frac{v_2}{r} - \frac{\lambda_5}{r}, \tag{5.9}$$

$$\dot{\lambda}_4 = \lambda_1\left(\frac{v_2^2}{r} - 2\frac{\mu}{r^2}\right) + \lambda_5\frac{v_2}{r^2}, \tag{5.10}$$

$$\dot{\lambda}_5 = 0, \tag{5.11}$$

$$\dot{\lambda}_7 = \frac{c\beta}{m^2}\lambda, \tag{5.12}$$

with the Hamiltonian

$$H = -\lambda_1\frac{\mu}{r^2} + \lambda_1\frac{v_2^2}{r} + \lambda_5\frac{v_2}{r} + \left(\frac{c}{m}\lambda - \lambda_7\right)\beta = C. \tag{5.13}$$

Here

$$\lambda = |\boldsymbol{\lambda}|, \quad \mathbf{r}(r, 0), \quad \mathbf{v}(v_1, v_2), \quad \boldsymbol{\lambda}(\lambda_1, \lambda_2),$$
$$\boldsymbol{\lambda}_r\left(\lambda_4 - \mu_1, \lambda_1\frac{v_2}{r} + \frac{\lambda_5}{r}\right).$$

Eqs. (5.3), (5.4), (5.5), (5.9), (5.11), as well as $\lambda_1 = \lambda\sin\varphi$, $\lambda_2 = \lambda\cos\varphi$ allow us to obtain

$$v_2 = \pm\sqrt{\frac{\mu}{r} - \frac{c\beta}{m}r\sin\varphi},$$

$$\dot{v}_2 = \frac{c\beta}{m}\cos\varphi, \tag{5.14}$$

$$\frac{\dot{\lambda}}{\lambda} = \dot{\varphi}\tan\varphi - 2\dot{\theta}\tan\varphi - \frac{\lambda_5}{\lambda r\cos\varphi},$$

By taking the derivative of the first expression of Eqs. (5.14) and equating to the second equation of Eqs. (5.14), one can obtain

$$\dot{\varphi} = -\frac{\beta}{m}\tan\varphi - 2\dot{\theta}. \tag{5.15}$$

Consider now the last equation of Eqs. (5.14) and (5.15). Substitution of the expression

$$\dot{\theta} = \dot{\varphi} + \dot{\psi}$$

into these equations yields

$$\frac{\dot{\lambda}}{\lambda} = -\dot{\varphi}\tan\varphi - 2\dot{\psi}\tan\varphi - \frac{\lambda_5}{\lambda r\cos\varphi}, \tag{5.16}$$

$$3\dot{\varphi} = -\frac{\beta}{m}\tan\varphi - 2\dot{\psi}. \tag{5.17}$$

These two equations allow us to determine φ and $\dot{\varphi}$.

Before to proceed with the solution process, note that as $r = r_0$, the gravity vector is defined as $\mathbf{g} = -\mu/r_0^3\mathbf{r}$ or $\mathbf{g} = -k^2\mathbf{r}$, where $k^2 = \mu/r_0^3$. This means that the gravitational acceleration of spacecraft in a circular trajectory in the central Newtonian field is the same as that of a spacecraft flying in the linear central field. Consequently, the integrals valid for the linear central field can be used in the analysis of circular trajectories. Indeed, as is shown in the next chapter, for this field of gravitation, the equations for the primer vector in terms of coordinates, λ and ψ, given by (see next chapter)

$$\ddot{\lambda} - \lambda\dot{\psi}^2 = -k^2\lambda \tag{5.18}$$

$$2\dot{\lambda}\dot{\psi} + \lambda\ddot{\psi} = 0, \tag{5.19}$$

admit the following integrals:

$$\psi = \arctan\left(\sqrt{\frac{1+e}{1-e}}\tan(kt+\gamma)\right) + \zeta,$$

$$\lambda^2 = \frac{q}{1 + e\cos 2(\psi - \zeta)},$$

$$\lambda^2\dot{\psi} = \sigma, \tag{5.20}$$

$$\dot{\lambda}^2 = \nu^2 - k^2\lambda^2 - \frac{\sigma^2}{\lambda^2}, \tag{5.21}$$

where σ, ν, e, ζ, γ, q are the integration constants, which satisfy the following two expressions:

$$(1-e^2)\sigma^2 = k^2 q^2, \qquad (1-e^2)\nu^2 = 2k^2 q. \tag{5.22}$$

From Eqs. (5.20) it follows that

$$\frac{\dot{\lambda}}{\lambda} = \sqrt{\frac{\nu^2}{\lambda^2} - k^2 - \frac{\sigma^2}{\lambda^4}}, \qquad \dot{\psi} = \frac{\sigma}{\lambda^2}, \tag{5.23}$$

where λ is determined by the second expression of Eqs. (5.20).

By substituting the expression for $\dot{\psi}$ of Eqs. (5.23) into (5.16) and (5.17), one can show that

$$\dot{\varphi}\tan\varphi = -\frac{\dot{\lambda}}{\lambda} - 2\frac{\sigma}{\lambda^2}\tan\varphi - \frac{\lambda_5}{\lambda r\cos\varphi}, \tag{5.24}$$

$$3\dot{\varphi} = -\frac{\beta}{m}\tan\varphi - 2\frac{\sigma}{\lambda^2}, \tag{5.25}$$

where $\frac{\dot{\lambda}}{\lambda}$ is determined using the first expression of Eqs. (5.23). Eqs. (5.24) and (5.25) allow us to determine φ and $\dot{\varphi}$. Indeed, elimination of $\dot{\varphi}$ from these equations leads to the fourth-order algebraic equation with respect to φ:

$$\cos^4\varphi + a_1\cos^3\varphi + b_1\cos^2\varphi + c_1\cos\varphi + d_1 = 0, \tag{5.26}$$

where

$$\begin{aligned}
a_1 &= \frac{6}{a}\left(\frac{\beta}{m} + 3\frac{\dot{\lambda}}{\lambda}\right),\\
b_1 &= \frac{1}{a}\left(-2\left(\frac{\beta}{m} + 3\frac{\dot{\lambda}}{\lambda}\right)\frac{\beta}{m} + 9\frac{\lambda_5^2}{\lambda^2 r^2} - 16\frac{\sigma^2}{\lambda^4}\right),\\
c_1 &= -\frac{6\lambda_5}{a\lambda r}\frac{\beta}{m},\\
d_1 &= \frac{\beta^2}{am^2},
\end{aligned} \tag{5.27}$$

and

$$a = \left(\frac{\beta}{m} + 3\frac{\dot{\lambda}}{\lambda}\right)^2 + 16\frac{\sigma^2}{\lambda^4}.$$

Eq. (5.26) allows us to determine the behavior of the thrust angle, and consequently, the thrust steering law, $\varphi = \varphi(t)$, and it can be solved using Ferrari's method or the trigonometric method developed for solving a fourth-order algebraic equations [146].

Case When Final Polar Angle Is Free and Functional Does Not Depend on This Variable

In the case with free final polar angle, θ_1 and the functional J of the optimal control problem does not depend on this variable, one can use the transversality condition to obtain

$$\lambda_5(t_1) = \lambda_5 = \frac{\partial J}{\partial \theta_1} = 0.$$

Consequently, Eq. (5.26) can be rewritten as a bi-quadratic equation:

$$a_2 \cos^4 \varphi + b_2 \cos^2 \varphi + c_2 = 0, \tag{5.28}$$

where

$$\begin{aligned} a_2 &= \left(\frac{\beta}{m} + 3\frac{\dot{\lambda}}{\lambda}\right)^2 + 16\frac{\sigma^2}{\lambda^4}, \\ b_2 &= -2\left(\frac{\beta}{m} + 3\frac{\dot{\lambda}}{\lambda}\right)\frac{\beta}{m} - 16\frac{\sigma^2}{\lambda^4}, \\ c_2 &= \frac{\beta^2}{m^2}. \end{aligned} \tag{5.29}$$

Solutions of Eq. (5.28) are given as

$$(\cos^2)_{1,2} = \frac{-b_2 \pm \sqrt{b_2^2 - 4a_2c_2}}{2a_2}. \tag{5.30}$$

Of course, these solutions should satisfy the condition

$$\left| \frac{-b_2 \pm \sqrt{b_2^2 - 4a_2c_2}}{2a_2} \right| \leq 1.$$

Note that $\frac{\dot{\lambda}}{\lambda}$ and λ are determined according to Eqs. (5.20).

Determination of Integration Constants

To use the solutions of Eq. (5.26) or Eq. (5.30), it is very important to determine the integration constants depending on the initial conditions.

Assume that the following initial conditions are given:

$$t = t_0 = 0, \quad r = r_0, \quad \theta(t_0) = \theta_0 = 0, \quad m(t_0) = m_0,$$

and $\beta = const$, which can be set according to the propulsion system constraints. Then from the expression

$$v_2 = r\dot{\theta} = \sqrt{\frac{\mu}{r} - \frac{c\beta}{m} r \sin\varphi}$$

of Eqs. (5.14), one can obtain that

$$\dot{\theta}_0 = \sqrt{\frac{\mu}{r_0^3} - \frac{c\beta}{m_0 r_0} \sin\varphi_0} \tag{5.31}$$

where $\varphi_0 = \varphi(t_0)$ and the signs $\pm$ have been omitted for simplicity. Then Eqs. (5.15) and $\dot{\theta} = \dot{\varphi} + \dot{\psi}$ yield

$$\dot{\varphi}_0 = -\frac{\beta}{m_0} \tan\varphi_0 - 2\sqrt{\frac{\mu}{r_0^3} - \frac{c\beta}{m_0 r_0} \sin\varphi_0} \tag{5.32}$$

and

$$\dot{\psi}_0 = \dot{\theta}_0 - \dot{\varphi}_0 = \frac{\beta}{m_0} \tan\varphi_0 + 3\sqrt{\frac{\mu}{r_0^3} - \frac{c\beta}{m_0 r_0} \sin\varphi_0}. \tag{5.33}$$

From the second integral of Eqs. (5.23) and (5.16), it follows that

$$\left(\frac{\sigma^2}{\lambda^4}\right)_0 = \dot{\psi}_0^2 = \left(\frac{\beta}{m_0} \tan\varphi_0 + 3\sqrt{\frac{\mu}{r_0^3} - \frac{c\beta}{m_0 r_0} \sin\varphi_0}\right)^2 \tag{5.34}$$

and

$$\left(\frac{\dot{\lambda}}{\lambda}\right)_0 = (\dot{\varphi}_0 - 2\dot{\theta}_0) \tan\varphi_0 = \left(-\frac{\beta}{m_0} \tan\varphi_0 - 4\sqrt{\frac{\mu}{r_0^3} - \frac{c\beta}{m_0 r_0} \sin\varphi_0}\right) \tan\varphi_0. \tag{5.35}$$

Substituting Eqs. (5.34) and (5.35) into Eq. (5.29), and rewriting Eq. (5.28) for the initial time yield

$$a_2(\varphi_0)\cos^4\varphi_0 + b_2(\varphi_0)\cos^2\varphi_0 + \frac{\beta^2}{m_0^2} = 0. \tag{5.36}$$

This is an algebraic equation with respect to φ_0, and it can be solved using any numerical method of solving nonlinear equations. Once a numerical value of φ_0^* is found, then Eqs. (5.34) and (5.35) allow us to find $\left(\frac{\sigma^2}{\lambda^4}\right)$ and $\left(\frac{\dot\lambda}{\lambda}\right)_0$. Other remaining constants can be determined as follows.

First, as $\theta_0 = 0$, then from $\theta = \psi + \varphi - \pi/2$ it follows that

$$\psi_0 = \frac{\pi}{2} - \varphi_0. \tag{5.37}$$

Also, from Eqs. (5.20) at $t = t_0 = 0$ we obtain

$$\begin{aligned}
\psi_0 &= \arctan\left(\sqrt{\frac{1+e}{1-e}}\tan(\gamma)\right) + \zeta,\\
\lambda_0^2 &= \frac{q}{1 + e\cos 2(\psi_0 - \zeta)},\\
\lambda_0^2\dot\psi_0 &= \sigma, && (5.38)\\
\dot\lambda_0^2 &= \nu^2 - k^2\lambda_0^2 - \frac{\sigma^2}{\lambda_0^2}. && (5.39)
\end{aligned}$$

Corresponding integration constants can be expressed by means of Eqs. (5.23) in the form:

$$\sigma^2 = \frac{(1-e^2)\nu^2}{4k^2}, \qquad q^2 = \frac{(1-e^2)^2\nu^4}{4k^4}. \tag{5.40}$$

Differentiation of the expression for ψ in Eqs. (5.20) yields

$$\dot\psi = \frac{kh}{\cos^2(kt+\gamma) + h^2\sin^2(kt+\gamma)}, \tag{5.41}$$

where $h = \sqrt{(1+e)/(1-e)}$. So, by using Eqs. (5.33), one can obtain

$$\dot\psi_0 = \frac{\beta}{m_0}\tan\varphi_0^* + 3\sqrt{\frac{\mu}{r_0^3} - \frac{c\beta}{m_0 r_0}\sin\varphi_0^*} = \frac{kh}{\cos^2\gamma + h^2\sin^2\gamma},$$

which can be used to determine γ in the form:

$$\cos^2\gamma(e) = \left(\frac{kh}{3Q + \frac{\beta}{m_0}\tan\varphi_0^*} - h^2\right)\frac{1}{1-h^2}, \tag{5.42}$$

where

$$Q = \sqrt{\frac{\mu}{r_0^3} - \frac{c\beta}{m_0 r_0} \sin\varphi_0^*}$$

Expression for ψ_0 of Eqs. (5.38) provides

$$\zeta(e) = \psi_0 - \arctan(h\tan(\gamma(e)))$$

or

$$\zeta(e) = \frac{\pi}{2} - \varphi_0^* - \arctan(h\tan(\gamma(e))). \tag{5.43}$$

Besides that, by equating Eqs. (5.35) and the expression

$$\left(\frac{\dot{\lambda}}{\lambda}\right)_0 = \frac{\nu^2}{\lambda_0^2} - k^2 - \dot{\psi}_0^2,$$

one can show that

$$\frac{\nu^2}{\lambda_0^2} = k^2 + \dot{\psi}_0^2 - (\frac{\beta}{m_0}\tan\varphi_0^* + 4Q). \tag{5.44}$$

Finally, using the expression for λ_0 of Eqs. (5.38) and the second expression of Eqs. (5.40), and substituting Eq. (5.44) into the resulting expression, we obtain that

$$\frac{q^2}{\lambda_0^4} = [1 + e\cos 2(\psi_0 - \zeta(e))]^2 = \frac{(1-e^2)^2}{4k^2}\frac{\nu^4}{\lambda_0^4}$$

or

$$1 + e\cos 2(\frac{\pi}{2} - \varphi_0^* - \zeta(e)) = \frac{(1-e^2)}{2k}\left([\frac{\beta}{m_0}\tan\varphi_0^* + 4Q]^2\tan^2\varphi_0^* + k^2 + \dot{\psi}_0^2\right). \tag{5.45}$$

Eq. (5.45) is a transcendental equation with respect to e, and can be solved using any iterative method to obtain the numerical values of e in terms of the initial conditions:

$$e^* = e(\varphi_0^*, m_0, \beta, c, r_0). \tag{5.46}$$

Substitution of Eq. (5.45) into Eqs. (5.42) and (5.43) yields

$$\gamma^* \;=\; \gamma(e^*) = \gamma(\varphi_0^*, m_0, \beta, c, r_0), \tag{5.47}$$

$$\zeta^* = \zeta(e^*) = \zeta(\varphi_0^*, m_0, \beta, c, r_0). \tag{5.48}$$

The remaining unknown are λ_0, ν, and σ. The first expression of Eqs. (5.40) and (5.44) lead to the expression

$$\frac{\sigma^2}{\lambda_0^4} = \frac{(1-e^{2*})}{4k^2}\frac{\nu^2}{\lambda_0^2}\frac{1}{\lambda_0^2}.$$

By substituting

$$\lambda_0^2\dot{\psi}_0 = \sigma$$

of Eqs. (5.38) into this expression, we show that

$$\dot{\psi}_0^2 = \frac{(1-e^{2*})}{4k^2}\left(k^2 + \dot{\psi}_0^2 - (\frac{\beta}{m_0}\tan\varphi_0^* + 4Q)\right)\frac{1}{\lambda_0^2}$$

or

$$\lambda_0^{2*} = \frac{(1-e^{2*})}{4k^2\dot{\psi}_0^2}\left(k^2 + \dot{\psi}_0^2 - (\frac{\beta}{m_0}\tan\varphi_0^* + 4Q)\right), \tag{5.49}$$

where

$$\dot{\psi}_0 = \frac{\beta}{m_0}\tan\varphi_0 + 3Q.$$

Consequently, substitutions of Eqs. (5.49) into Eqs. (5.38) and (5.40) yield the remaining unknowns to be in the form:

$$q^* = \lambda_0^{2*}(1 + e^*\cos 2(\frac{\pi}{2} - \varphi_0^* - \zeta^*)), \tag{5.50}$$

$$\nu^{2*} = \frac{2k^2 q^*}{1-e^{2*}}, \tag{5.51}$$

$$\sigma^* = \lambda_0^{2*}(\frac{\beta}{m_0}\tan\varphi_0^* + 3Q). \tag{5.52}$$

This completes the determination of the integration constants.

5.3 SYSTEM OF EQUATIONS FOR ARBITRARY THRUST ARCS

Using $\lambda_1 = \lambda\sin\varphi$, $\lambda_2 = \lambda\cos\varphi$, and the equation for λ_1 of Eqs. (2.135),

$$\dot{\lambda}_1 = \lambda_2\frac{v_2}{r} - \lambda_4,$$

one can obtain [137]

$$\lambda_4 = -\dot{\lambda}\sin\varphi + \lambda\dot{\psi}\cos\varphi \tag{5.53}$$

and

$$\dot{\lambda}_4 = (-\ddot{\lambda} - \lambda\dot{\psi}\dot{\varphi})\sin\varphi + (-\dot{\lambda}\dot{\varphi} + \dot{\lambda}\dot{\psi} + \lambda\ddot{\psi})\cos\varphi. \tag{5.54}$$

By equating the right side of this equation to the right side of the equation for λ_4 of Eqs. (2.135)

$$\dot{\lambda}_4 = -\frac{\partial H}{\partial r} = (\frac{v_2^2}{r^2} - 2\frac{\mu}{r^3})\lambda\sin\varphi - \lambda\frac{v_1 v_2}{r^2}\cos\varphi + \lambda_5\frac{v_2}{r^2\cos\varphi}\cos\varphi,$$

where $\cos\varphi \neq 0$, we obtain that

$$\begin{aligned} \ddot{\lambda} - \lambda\dot{\psi}^2 &= 2\lambda\frac{\mu}{r^3} - \lambda\dot{\psi}\dot{\theta} - \lambda\dot{\theta}^2, \\ \lambda\ddot{\psi} + 2\dot{\lambda}\dot{\psi} &= \dot{\lambda}\dot{\theta} - \lambda\frac{v_1}{r}\dot{\theta} + \lambda_5\frac{1}{r\cos\varphi}\dot{\theta}. \end{aligned} \tag{5.55}$$

The projections of the equation

$$\ddot{\boldsymbol{\lambda}} = -\frac{\mu}{r^3}\boldsymbol{\lambda} + 3\frac{\mu}{r^5}(\boldsymbol{\lambda}\mathbf{r})\mathbf{r}$$

onto the thrust direction and its perpendicular can be written in the form:

$$\begin{aligned} \ddot{\psi} &= -3\frac{\mu}{r^3}\sin\varphi\cos\varphi - 2\frac{\dot{\lambda}}{\lambda}\dot{\psi}, \\ \frac{\ddot{\lambda}}{\lambda} &= \dot{\psi}^2 + \frac{\mu}{r^3}(3\sin^2\varphi - 1). \end{aligned} \tag{5.56}$$

From Eqs. (5.55) and (5.56) it follows that

$$\begin{aligned} -(\dot{\theta} + \dot{\psi})\dot{\theta} &= -3\frac{\mu}{r^3}\cos^2\varphi, \\ \left(\frac{\dot{\lambda}}{\lambda} - \frac{v_1}{r} + \lambda_5\frac{1}{r\cos\varphi}\right)\dot{\theta} &= -3\frac{\mu}{r^3}\sin\varphi\cos\varphi. \end{aligned} \tag{5.57}$$

From Hamiltonian of Eqs. (2.135) one can obtain

$$\lambda(-2\frac{\mu}{r^3} + \frac{v_2^2}{r^2})\sin\varphi + (\lambda\frac{v_1 v_2}{r^2} + \lambda_5\frac{v_2}{r^2\cos\varphi})\cos\varphi =$$

$$(-\frac{\mu}{r^3}+v_1\frac{\dot{\lambda}}{r\lambda})\lambda\sin\varphi-(\frac{v_1\dot{\psi}}{r}+\frac{æ\beta-C}{r\lambda\cos\varphi})\lambda\cos\varphi. \tag{5.58}$$

By replacing the right side of Eq. (5.54) by Eq. (5.58), we have

$$\dot{\lambda}_4=(-\frac{\mu}{r^3}+\frac{v_1}{r}\frac{\dot{\lambda}}{\lambda})\lambda\sin\varphi-(\frac{v_1\dot{\psi}}{r}+\frac{æ\beta-C}{r\lambda\cos\varphi})\lambda\cos\varphi. \tag{5.59}$$

Eqs. (5.59) and (5.56) lead to the equations:

$$\begin{aligned}\frac{v_1}{r}\frac{\dot{\lambda}}{\lambda}+\frac{v_2\dot{\psi}}{r} &= \frac{\mu}{r^3}(-3\sin^2\varphi+2),\\ \frac{v_1}{r}\dot{\psi}-\frac{v_2}{r}\frac{\dot{\lambda}}{\lambda} &= 3\frac{\mu}{r^3}\sin\varphi\cos\varphi-\frac{æ\beta-C}{r\lambda\cos\varphi}.\end{aligned} \tag{5.60}$$

Note that Eqs. (5.59), (5.56) and Hamiltonian are equivalent to Eqs. (5.57) and (5.60). By introducing new variables in the form

$$\begin{aligned}&\frac{v_1}{r}=q_1,\quad \frac{v_2}{r}=q_2,\quad \frac{\dot{\lambda}}{\lambda}=p_1,\quad \dot{\psi}=p_2,\\ &\frac{\mu}{r^3}=a,\quad \frac{æ\beta-C}{r\lambda\cos\varphi}=b,\quad \frac{\lambda_5}{r\lambda\cos\varphi}=d,\end{aligned} \tag{5.61}$$

Eqs. (5.57) and (5.60) can be rewritten as

$$q_1p_1+q_2p_2 = a(-3\sin^2\varphi+2), \tag{5.62}$$
$$q_1p_2-q_2p_1 = 3a\sin\varphi\cos\varphi-b, \tag{5.63}$$
$$q_2(q_2+p_2) = 3a\cos^2\varphi, \tag{5.64}$$
$$q_2(p_1-q_1) = -3a\sin\varphi\cos\varphi-dq_2. \tag{5.65}$$

5.4 ANALYTICAL SOLUTIONS FOR CONSTANT THRUST ARCS

Let us prove the following theorem:

Theorem 5.1. *In the case of a flight with a constant power, ($P=const$), and constant specific impulse, ($I_{sp}=const$), Eqs. (2.134) of the problem in Eqs. (2.111)–(2.115) admits at least one family of extremals expressed in quadratures.*

Proof of Theorem 5.1. Taking into account Eqs. (2.135) for v_1, v_2, r, θ and Eq. (5.56), the derivatives in the new variables can be given in the form:

$$\dot{q}_1 = q_3\sin\varphi-a+q_2^2-q_1^2,$$

$$\dot{q}_2 = q_3 \sin\varphi + 2q_1q_2, \quad q_3 = \frac{c\beta}{mr}. \tag{5.66}$$

By rewriting Eqs. (5.64) and (5.65) in the form

$$q_2 + p_2 = F\cos\varphi, \tag{5.67}$$
$$-q_1 + p_1 = -F\sin\varphi \tag{5.68}$$

and by differentiating them, one can obtain

$$\begin{aligned} \dot{q}_2 &= 3a\sin\varphi\cos\varphi + 2p_1p_2 + \dot{F}\cos\varphi - F\sin\varphi\dot{\varphi}, \\ \dot{q}_1 &= 3a\sin^2\varphi - a + p_2^2 - p_1^2 + \dot{F}\sin\varphi + F\cos\varphi\dot{\varphi}. \end{aligned} \tag{5.69}$$

After equating the right sides of Eqs. (5.66) and (5.69), it follows that

$$q_3 - 2p_1F - F^2\sin\varphi = 3a\sin\varphi + \dot{F}. \tag{5.70}$$

Substitution of $\dot{F}$ into this equation yields to the expression:

$$\begin{aligned} &\left(\frac{q_3}{\cos\varphi} - 6a\frac{p_1}{q_2} + 3a\frac{q_1}{q_2} + 3a\frac{q_3}{q_2^2}\right)\cos\varphi + \\ &\left(-9a^2\frac{1}{q_2^2}\cos^2\varphi - 3a\frac{p_2}{q_2}\right)\sin\varphi = 0, \end{aligned}$$

which must be identically satisfied at any point of the thrust arc, and therefore, one can consider two equations of the form:

$$\begin{aligned} \frac{q_3}{\cos\varphi} + \frac{3a}{q_2}\left(q_1 - 2p_1 + \frac{q_3}{q_2}\right) &= 0, \\ \frac{3a}{q_2}\left(\frac{3a\cos^2\varphi}{q_2} - p_2\right) &= 0. \end{aligned} \tag{5.71}$$

Consequently, Eqs. (5.62)–(5.66) and (5.71) allow us to find p_1, p_2, q_1, q_2, b in terms of φ and a. First, from Eqs. (5.62), (5.64), (5.66) and the second equation of Eqs. (5.71) one can find p_1, p_2, q_1, q_2. From the second equation of Eqs. (5.71) it follows that

$$q_2p_2 = -3a\cos^2\varphi. \tag{5.72}$$

Substituting into Eqs. (5.62) and (5.64), we obtain

$$p_1q_1 = 3a\cos 2\varphi + 2a, \tag{5.73}$$

$$q_2 = \pm\sqrt{6a}\cos\varphi. \tag{5.74}$$

Substituting Eq. (5.74) into Eq. (5.72), one can find the expression

$$p_2 = \pm\sqrt{\frac{3}{2}}\sqrt{a}\cos\varphi. \tag{5.75}$$

By rewriting Eq. (5.73) in the form

$$p_1 = \frac{3a\cos 2\varphi + 2a}{q_1}, \tag{5.76}$$

and substituting it and q_2 from Eqs. (5.74) into Eq. (5.65), it follows that

$$q_1 = \frac{1}{2}(\sqrt{\frac{3}{2}}\sqrt{a}\sin\varphi + d \pm \sqrt{(\sqrt{\frac{3}{2}}\sqrt{a}\sin\varphi + d)^2 + 4a(3\cos 2\varphi + 2)}. \tag{5.77}$$

Then Eqs. (5.77) and (5.76) yield

$$p_1 = \frac{2a(3\cos 2\varphi + 2)}{(\sqrt{\frac{3}{2}}\sqrt{a}\sin\varphi + d) \pm \sqrt{(\sqrt{\frac{3}{2}}\sqrt{a}\sin\varphi + d)^2 + 4a(3\cos 2\varphi + 2)}}. \tag{5.78}$$

Consequently, b and q_3 can be expressed in terms of φ and a:

$$b = q_1(\frac{q_2}{2} - d), \quad q_3 = \frac{c\beta}{mr} = \sqrt{6a}\cos\varphi(2p_1(\varphi, a) - q_1(\varphi, a)). \tag{5.79}$$

Using the expression

$$b = \frac{æ\beta - C}{r\lambda\cos\varphi} = \frac{q_3}{\cos\varphi} - \frac{\lambda_7\beta}{\lambda\cos\varphi} + \frac{C}{\lambda\cos\varphi},$$

from Eqs. (5.79) we obtain

$$\frac{q_3}{\cos\varphi} - \frac{\lambda_7\beta}{\lambda\cos\varphi} + \frac{C}{\lambda\cos\varphi} = q_2(\frac{q_1}{2} - d)$$

or

$$\frac{\lambda_7}{\lambda} = \frac{r}{\beta}\left[\sqrt{6a}\cos\varphi 2p_1 - q_1 - \cos\varphi(\frac{q_1}{2} - d) - \frac{C}{\lambda r}\right]. \tag{5.80}$$

Using the equation $\dot{\lambda_7} = \frac{c\beta}{m^2}\lambda$ and its derivative, Eq. (5.80), one can obtain the function $a = a(\varphi)$. But this expression can not be expressed in elementary functions. It will be shown below in the particular case that when

$d = 0$, e.g. when the functional of the problem is not an explicit function of the final polar angle, the function $a = a(\varphi)$ can be expressed in elementary functions.

Particular case. Indeed, from Eqs. (5.77)–(5.80) it follows that

$$q_1 = \frac{v_1}{r} = \sqrt{a}\Phi_1, \tag{5.81}$$

$$p_1 = \frac{\dot{\lambda}}{\lambda} = \sqrt{a}\Phi_2, \tag{5.82}$$

$$q_2 = \frac{v_2}{r} = \sqrt{6}\sqrt{a}\cos\varphi, \tag{5.83}$$

$$p_2 = \dot{\psi} = \sqrt{\frac{3}{2}}a\cos\varphi, \tag{5.84}$$

$$q_3 = \sqrt{6}a\cos\varphi(2\Phi_2 - \Phi_1), \tag{5.85}$$

$$p_3 = \frac{\lambda_7}{\lambda}\frac{\beta}{r} = a\cos\varphi\sqrt{6}(2\Phi_2 - \Phi_1) - \sqrt{\frac{3}{2}}\cos\varphi\Phi_1, \tag{5.86}$$

$$b = \frac{1}{2}q_1 q_2 = \sqrt{\frac{3}{2}}a\Phi_1\cos\varphi, \tag{5.87}$$

where

$$\Phi_1 = \frac{1}{2}\left[\sqrt{\frac{3}{2}}\sin\varphi \pm \sqrt{\frac{3}{2}\sin^2\varphi + 4(3\cos 2\varphi + 2)}\right], \qquad \Phi_2 = \frac{3\cos 2\varphi + 2}{\Phi_1}.$$

Then using $\dot{\lambda_7} = \frac{c\beta}{m^2}\lambda$, the derivative of λ_7 from Eqs. (5.85), and by equating these expressions we can have the solutions for a in the following form:

$$a = \frac{\beta^2}{m^2}\Phi_4^2, \tag{5.88}$$

where

$$\Phi_3 = \sqrt{6}(2\Phi_2 - \Phi_1) - \frac{3}{2}\cos\varphi\Phi_1,$$

$$\Phi_4 = \left[\frac{\sqrt{6}(2\Phi_2 - \Phi_1) + \frac{\dot{\beta}m}{\beta^2}\Phi_3}{\Phi_3(\Phi_2 - 2\Phi_1 - \frac{3}{2}6\sin\varphi) - \frac{3}{2}6\cos\varphi\frac{d\Phi_3}{d\varphi}}\right].$$

Eqs. (5.81) and (5.88) yield

$$r = \frac{\mu}{c^2}\Phi_5, \qquad \frac{m}{\beta} = \frac{\mu}{c^3}\Phi_6, \tag{5.89}$$

where

$$\Phi_5 = \left[\sqrt{6}\cos\varphi\Phi_4(2\Phi_2 - \Phi_1)\right]^2, \quad \Psi_6 = \frac{\Phi_5^2}{\sqrt{6}\cos\varphi\Phi_4(2\Phi_2 - \Phi_1)}.$$

From Eqs. (5.83), (5.84) and the expression $\theta = \psi + \varphi - \frac{\pi}{2}$, one can find that

$$\psi = -\frac{1}{3}\varphi - \frac{1}{3}\theta_0, \quad \theta = \frac{2}{3}\varphi + \frac{2}{3}\theta_0, \tag{5.90}$$

where θ_0 is integration constant. Also, it can be shown that

$$\frac{d\varphi}{dt} = \frac{3}{2}\sqrt{a}\cos\varphi = \frac{3}{2}\sqrt{6}\frac{c^3}{\mu}\frac{\Phi_4}{\Phi_6}. \tag{5.91}$$

Utilization of Eqs. (5.81), (5.82) and (5.91) yields the following quadratures:

$$r = r_0 e^{\int \frac{\Phi_1}{\frac{3}{2}\sqrt{6}\cos\varphi}}, \quad t = \frac{2}{3\sqrt{6}}\frac{\mu}{c^3}\int\frac{\Phi_6}{\Phi_4} + t_0, \quad \lambda = \lambda_0 e^{\int \frac{2}{27}\frac{\Phi_2}{\cos\varphi}d\varphi} \tag{5.92}$$

with integration constants r_0, t_0, λ_0. The components of the velocity can easily be found from Eqs. (5.81) and (5.83) in the form:

$$v_1 = \frac{\mu}{c^2}\sqrt{a}\Phi_2\Phi_5, \qquad v_2 = \sqrt{6}\frac{\mu}{c^2}\sqrt{a}\Phi_5. \tag{5.93}$$

Note that the solutions obtained above are valid in the case when $c = const$, $\beta = const$, and $\lambda_5 = 0$. Consequently, these solutions describe a motion on a maximum thrust (MT) arc and contain the following integration constants: $C, m_0, \lambda_5, r_0, t_0, \theta_0, \lambda_0$.

Theorem 5.1 has been proved. □

In the case when $c = const$, $\lambda_5 = 0$, and β are arbitrary, the same solutions as above can be obtained, and the following solutions for mass and mass-flow rate are valid:

$$m = m_0 e^{\frac{4}{c}\sqrt{\frac{\mu}{r_0}}\int \frac{(\Phi_1 - 2\Phi_2)}{\Phi_1}\cos\varphi d\varphi}, \qquad \beta = \frac{dm}{d\varphi}\frac{d\varphi}{dt}. \tag{5.94}$$

These solutions, in general, correspond to IT arcs, and perhaps, to low-thrust arcs with constant specific impulse depending on the value of $c = const$. In the case when $c \neq const$ and $\beta = const$, one can obtain solutions obtained above and using the equation $c = \frac{2\ constant}{\lambda m}$, as a result of the transversality condition, it can be shown that

$$c = I_{sp}g_0 = \frac{m(\varphi)r(\varphi)}{\beta}\sqrt{6}a(\varphi)\cos\varphi(2\Phi_2 - \Phi_1), \tag{5.95}$$

where the functions a, m, and r are found from the equations presented above. These solutions represent mainly a low-thrust motion with variable specific impulse and maximum power, computed according to the formulae $P_{max} = \frac{1}{2}\beta c^2$.

It can be shown that Eqs. (5.62)–(5.65) can be used to find other classes for low- or high-thrust arcs, or IT and MT arcs with constant or variable specific impulse.

Below it will be shown that MT arcs do not contain conjugate points. The solutions for the auxiliary equations for r and θ have the form:

$$\eta_1 = \frac{\partial r}{\partial r_0} = \int_{\varphi_0}^{\varphi_1} \frac{\Phi_1}{\frac{3}{2}\sqrt{6}\cos\varphi} d\varphi, \qquad \eta_4 = \frac{\partial \theta}{\partial \theta_0} = \frac{3}{2}$$

It can be easily seen that the integral is not zero except for the case when $\varphi = \varphi_0$. This means that the condition $\eta(t_0) = \eta(t') = 0$ is not satisfied, that is the MT arcs do not contain conjugate points.

In this chapter, the thrust arcs with maximum power and constant specific impulse have been considered. The system of equations, which can be used to analyze thrust arcs, has been obtained. On the basis of integration of this system using the first integrals, the new solutions for maximum (or minimum) thrust arcs have been obtained. The solutions obtained can be used to describe a motion with low-thrust or high-thrust propulsion system in the Newtonian field.

CHAPTER 6

Extremal Trajectories in a Linear Central Field

6.1 APPROXIMATION OF THE NEWTONIAN FIELD BY LINEAR CENTRAL FIELD

It is shown that the linear central field assumption is a highly accurate approximation if the thrust vector is nearly tangential to the orbit and the vehicle's trajectory is in a sufficiently thin spherical layer during the maneuver. Utilizing this approximation, the equations of optimal motion admit analytic closed-form solutions describing optimal constant (both low- and high-) thrust space trajectories.

It is known that spacecraft trajectories may consist of combinations of null thrust (NT), intermediate thrust (IT) and maximum thrust (MT) arcs. Previous investigations searching for analytic methods applicable to MT arcs include those of Marec [7], Ehricke [159]. It was shown by Azizov and Korshunova [59] that solving the problem in quadratures requires four first integrals or two integrals in involution for planar motion. For motion in a central Newtonian field, these integrals remain unknown. However, MT arcs can still be studied using (i) numerical techniques, or (ii) analytical methods based on using either known impulsive solutions or utilizing appropriate approximations. This chapter takes the latter approach utilizing a linear central field approximation.

Investigations of optimal motion in a linear central field have been reported by Azizov and Korshunova [59] and Jezewski [118]. In these studies, it is shown that one difficulty in solving the problem using a linear field approximation is determining a valid initial control vector. To address this difficulty, some authors recommended using basic impulsive solutions as starting points and employing an iterative process to satisfy the terminal conditions and corner conditions. Another approach to this problem is to utilize primer vector theory [3]. In this chapter the primer vector methods will be applied to the problem of optimal transfer of a rocket moving with maximum thrust in a thin spherical layer within a central Newtonian field. In this chapter, the new solutions for IT and MT (minimum) arcs for a linear central field will be obtained [137], [138], [158], [160], [161], [162], [163].

Analytical Solutions for Extremal Space Trajectories
DOI: http://dx.doi.org/10.1016/B978-0-12-814058-1.00006-0

The studies of trajectories with high-thrust chemical propulsion system show that it is more convenient to consider the exhaust velocity, c and mass-flow rate, β instead of constant (maximum) power and specific impulse. The relationships between these parameters are given by the formulas: defined by the formulas

$$c = I_{sp}g, \qquad \beta = \frac{2P}{c^2}.$$

The power and specific impulse are constant, and therefore, the solutions obtained in this chapter will be valid for minimum thrust arcs too.

The equation of motion for a spacecraft moving in a central Newtonian field is

$$\ddot{\mathbf{r}} = -\frac{\mu}{r^3}\mathbf{r} + \mathbf{a}, \tag{6.1}$$

where r is the magnitude of the radius vector $\mathbf{r}$ of the spacecraft (represented in a convenient inertial reference frame), and $\mathbf{a}$ is the thrust acceleration vector. The approximation of the equations of motion involve series expansions of the gravitational term, such that

$$\begin{aligned} g &= \frac{\mu}{r^3} = \frac{\mu}{r_0^3} - 3\frac{\mu}{r_0^4}\Delta r + \frac{1}{2!}\frac{\mu}{r_0^5}(\Delta r)^2 + \cdots + \frac{g^{(n)}(r_0)}{n!}(\Delta_r)^{(n)} + O[(\Delta r)^n] \\ &= \frac{\mu}{r_0^3} + \Delta g, \end{aligned} \tag{6.2}$$

where

$$\Delta r = r - r_0, \quad \text{and} \quad \Delta g = \frac{\mu}{2r_0^3}\sum_{k=1}^{\infty}(-1)^k(k+1)(k+2)(\frac{\Delta r}{r_0})^k.$$

Substituting Eq. (6.2) into (6.1) and re-arranging terms yields

$$\ddot{\mathbf{r}} + \frac{\mu}{r_0^3}\mathbf{r} = \mathbf{a} - \Delta\mathbf{R} \tag{6.3}$$

where $\Delta\mathbf{R} = \Delta g\mathbf{r}$.

For problems in which the spacecraft trajectory remains within a thin spherical layer (that is, when Δr remains small), the quantity Δg may be neglected. It follows that when Δg is neglected, Eq. (6.1) reduces to the case of the linear central field, wherein

$$\ddot{\mathbf{r}} + \frac{\mu}{r_0^3}\mathbf{r} = \mathbf{a}. \tag{6.4}$$

We can now evaluate the relative error in the dynamic model associated with neglecting $\Delta\mathbf{R}$ in Eq. (6.3) by first writing

$$\epsilon = \left|\frac{1}{r^3} - \frac{1}{r_0^3}\right| = \frac{\Delta r}{r_0^4}\left|\frac{3 + 3(\frac{\Delta r}{r_0}) + (\frac{\Delta r}{r_0})^2}{1 + 3(\frac{\Delta r}{r_0}) + 3(\frac{\Delta r}{r_0})^2 + (\frac{\Delta r}{r_0})^3}\right|.$$

Thus, when $\Delta r/r_0 \ll 1$, the relative error associated with Eq. (6.4) is of order $O(\Delta r/r_0^4)$.

According to Lawden, the primer vector is governed by [3]

$$\ddot{\boldsymbol{\lambda}} = \lambda\frac{\partial \mathbf{g}}{\partial p}, \tag{6.5}$$

where $\partial\mathbf{g}/\partial p$ denotes differentiation with respect to a displacement in the direction $\boldsymbol{\lambda}$. For the case of a central Newtonian field, we have $\mathbf{g} = -(\mu/r^3)\mathbf{r}$, and we can use Eq. (6.5) to obtain the primer vector governing equations as

$$\ddot{\boldsymbol{\lambda}} = \frac{3\mu}{r^5}(\boldsymbol{\lambda}\cdot\mathbf{r})\mathbf{r} - \frac{\mu}{r^3}\boldsymbol{\lambda}. \tag{6.6}$$

Similarly, in a linear central field, we have $\mathbf{g} = -(\mu/r_0^3)\mathbf{r}$, and we find that the primer vector is governed by

$$\ddot{\boldsymbol{\lambda}} = -\frac{\mu}{r_0^3}\boldsymbol{\lambda}. \tag{6.7}$$

Using the series expansion for g given in Eq. (6.2), we rewrite Eq. (6.6) as

$$\ddot{\boldsymbol{\lambda}} = -\frac{\mu}{r_0^3}\boldsymbol{\lambda} + \frac{3\mu}{r^4}\lambda\sin\varphi\mathbf{r} - \Delta g\boldsymbol{\lambda}, \tag{6.8}$$

where φ is the angle between $\boldsymbol{\lambda}$ and a line perpendicular to $\mathbf{r}$. The coordinate system, defining the angle φ. Comparing Eq. (6.8) with Eq. (6.7), and taking into account Δg, one can conclude that a linear approximation of central Newtonian field is admissible if

$$\sin\varphi \approx 0 \qquad \text{and} \qquad \frac{\Delta r}{r_0} \ll 1. \tag{6.9}$$

When $\sin\varphi \approx 0$ the second term on the right hand-side of Eq. (6.8) is negligible. In this situation, the thrust vector is nearly tangential to the orbit. When $\Delta r/r_0 \ll 1$ and Δg is negligible, the third term on the right hand-side of Eq. (6.8) is negligible.

When solving problems connected with the minimization of fuel expenditure, it is important to evaluate the errors in the minimization functional due to the approximations made in assuming a linear central field. For such problems, the performance functional can be represented in the form

$$\mathcal{J} = c \ln \frac{m_0}{m_1} = \int_0^T |\mathbf{a}(t)|\, dt, \tag{6.10}$$

where m_0 and m_1 are the initial and final rocket masses respectively. Utilizing Eqs. (6.3) and (6.4) in conjunction with Eq. (6.10) to compute the difference in the functionals corresponding to using the linear central field and the central Newtonian field yields the inequality

$$\Delta\mathcal{J} \leq \int_0^T |\Delta\mathbf{R}|\, dt.$$

Define the quantities a_n and S as

$$a_n = (-1)^n (n+1)(n+2)(\frac{\Delta r}{r_0})^n$$

and

$$S = \sum_{n=1}^{\infty} a_n = \sum_{n=1}^{\infty} (-1)^n (n+1)(n+2)(\frac{\Delta r}{r_0})^{n-1}.$$

It then follows that

$$|\Delta g| = \frac{\mu}{2r_0^3} \frac{\Delta r}{r_0} S.$$

It can be shown that

$$\lim_{n\to\infty} \left| \frac{a_{n+1}}{a_n} \right| = \frac{\Delta r}{r_0} \lim_{n\to\infty} (1 + \frac{2}{n+2}) = \frac{\Delta r}{r_0}$$

and if $\Delta r \ll r_0$, we have

$$\lim_{n\to\infty} \left| \frac{a_{n+1}}{a_n} \right| < 1.$$

Then, according to D'Alambert's convergence test, the series $\sum_{n=1}^{\infty} a_n$ converges [135], and

$$|a_1| > |a_2| > |a_3| > \cdots |a_n| > \ldots$$

with $a_n \to 0$ as $n \to \infty$. Consequently, the sum S is bounded above by a finite value, which is denoted here by L. Since $\Delta\mathbf{R} = \Delta g\mathbf{r}$, we have

$$|\Delta\mathbf{R}| \leq \tilde{L}\frac{\Delta r}{r_0}\left(1 + \frac{\Delta r}{r_0}\right), \tag{6.11}$$

where

$$\tilde{L} = \frac{\mu}{2r_0^2}L.$$

Therefore, the error in the performance function due to the linear central field approximation is bounded above by

$$\Delta\mathcal{J} \leq \int_0^T |\Delta\mathbf{R}|\, dt \leq \tilde{L}\frac{\Delta r}{r_0}\left(1 + \frac{\Delta r}{r_0}\right)\int_0^T dt = \tilde{L}\frac{\Delta r}{r_0}\left(1 + \frac{\Delta r}{r_0}\right)T = O(\frac{\Delta r}{r_0}). \tag{6.12}$$

We conclude that the relative error resulting from ignoring $\Delta\mathbf{R}$ is of order $O(\Delta r/r_0)$. The above described approximation is valid in the case of the chemical propulsions when the performance function is given as

$$\mathcal{J} = c\ln\frac{m_0}{m_1} = \int_{t_0}^{t_1} a_t(t)dt \tag{6.13}$$

In the case of the electrical or thermonuclear propulsion the performance function may be represented in the form

$$\mathcal{J} = \int_{t_0}^{t_1} a_t^2(t)dt, \tag{6.14}$$

where t_0 and t_1 are the initial and final times. It was shown that in the first case the error associated with approximation of the Newtonian field by linear central is of order $O(\Delta r/r_0)$ [158]. In order to evaluate the impact of utilizing the linear gravity approximation, the difference of the value of the performance functional in Eq. (6.14) for the two cases (Newtonian field versus linear field) needs to be evaluated. Using Eq. (6.4), this difference may be written by

$$\Delta\mathcal{J} = \left|\frac{1}{2}\int\left[\ddot{\mathbf{r}} + \frac{\mu}{r_0^3}\mathbf{r} + \Delta\mathbf{R}\right]\left[\ddot{\mathbf{r}} + \frac{\mu}{r_0^3}\mathbf{r} + \Delta\mathbf{R}\right]^T dt \right.$$
$$\left. -\frac{1}{2}\int\left[\ddot{\mathbf{r}} + \frac{\mu}{r_0^3}\mathbf{r}\right]\left[\ddot{\mathbf{r}} + \frac{\mu}{r_0^3}\mathbf{r}\right]^T dt\right| =$$

$$\left|\int \mathbf{a}_t \Delta\mathbf{R}^T dt - \frac{1}{2}\int \Delta\mathbf{R}\Delta\mathbf{R}^T dt\right| = \left|\int \frac{c\beta}{m}\sin\varphi\Delta g r dt - \frac{1}{2}\int \Delta\mathbf{R}\Delta\mathbf{R}^T dt\right|,$$

where $a_t = c\beta/m$. If the conditions in Eq. (6.9) are satisfied, then

$$\Delta\mathcal{J} \leq \frac{1}{2}\left|\int \Delta\mathbf{R}\Delta\mathbf{R}^T dt\right| \leq \frac{1}{2}M\left(\frac{\mu}{2r_0^2}\right)^2\left|\int \frac{\Delta r}{r_0}\left(1+\frac{\Delta r}{r_0}\right)dt\right| = O((\Delta r/r_0)^2),$$

where M satisfies condition $\frac{3}{2}|\Delta g| \leq M\frac{\Delta r}{r_0^2}$. Consequently, the relative error in the functional and associated with neglecting the second term on the right hand side of the first equation of Eq. (6.4) is of order $O((\Delta r/r_0)^2)$, and in the dynamical model this error is of order $O(\Delta r/r_0)$.

The results obtained above prove the statement formulated by the following theorem:

Theorem 6.1. *In the context of the problem (2.111)–(2.115), a linear approximation of the central Newtonian field in the vicinity of a reference orbit of radius r_0 is admissible if the trajectory remains within the boundaries of a thin spherical layer the width, Δr, of which is small compared to the magnitude of the radius r_0, and $\sin\varphi \approx 0$. The error associated with this approximation is of order $O(\Delta r/r_0)$ in the dynamic model, and of order $O(\Delta r/r_0)$ in performance functional for chemical propulsion systems, and of order $O((\Delta r/r_0)^2)$ for power limited propulsion systems.*

If the initial point of a MT arc lies in the initial orbit, and the final point of the MT arc lies in a thin spherical layer, then error ϵ_g in the gravitational acceleration associated with the linear approximation can be found from expression:

$$\epsilon_g = \left|\frac{1}{r^3(f_2)} - \frac{1}{r_0^3}\right|, \tag{6.15}$$

where r_0 is the initial radius vector's magnitude, f_2 is true anomaly of the final point. The relative errors at the final point of MT arc may be computed as

$$O\left(\frac{\Delta r(f_2)}{r_0}\right)^2 = \left(\frac{r(f_2) - r_0}{r_0^2}\right)^2,$$
$$O\left(\frac{\Delta r(f_2)}{r_0^4}\right) = \frac{r(f_2) - r_0}{r_0^4}$$

in the performance functional and dynamic model, respectively.

In summary, a linear approximation of the central Newtonian field is admissible if the rocket trajectory remains within the boundaries of a thin spherical layer the width, Δr, of which is small compared to the magnitude of the radius r_0, and $\sin\varphi \approx 0$. The linear central field approximation leads to a relative error on the order $O(\Delta r/r_0)$ in the dynamic model, and relative error on the order $O(\Delta r/r_0)$ in computation of the performance function in the case of chemical systems and of order $O((\Delta r/r_0)^2)$ in the case of electrical or thermonuclear propulsion.

6.2 CANONICAL EQUATIONS AND FIRST INTEGRALS

Let us prove the following lemma.

Lemma 6.1. *In the context of the problem (2.111)–(2.115), the linear approximation of the central Newtonian field in the vicinity of a reference orbit allows us to obtain four new integrals, and the hodograph of the primer vector ($\boldsymbol{\lambda}$ trajectory) is a central ellipse ($e < 1$), circle ($e = 0$) or a straight line ($e = 1$), where e is an integration constant.*

Proof of Lemma 6.1. Assume that the linear approximation of the central Newtonian field is carried out. The canonical system of equations for a rocket in a linear central field are given by [158]:

$$\begin{aligned}
\dot{\mathbf{r}} &= \mathbf{v}, \\
\dot{\mathbf{v}} &= \frac{c\beta}{m}\frac{\boldsymbol{\lambda}}{\lambda} - k^2\mathbf{r}, \\
\dot{m} &= -\beta, \\
\dot{\boldsymbol{\lambda}} &= -\boldsymbol{\lambda}_r, \\
\dot{\boldsymbol{\lambda}}_r &= k^2\boldsymbol{\lambda}, \\
\dot{\lambda}_7 &= \frac{c\beta}{m^2}\lambda.
\end{aligned} \tag{6.16}$$

The Hamiltonian is

$$H = -k^2(\boldsymbol{\lambda}\cdot\mathbf{r}) + (\boldsymbol{\lambda}_r\cdot\mathbf{v}) + \beta\chi, \tag{6.17}$$

where the switching function is defined to be

$$\chi = \frac{c}{m}\lambda - \lambda_7.$$

In a polar reference frame (for the planar case), we use the coordinates (r, θ). Let λ_i, $i = 1, 2$ denote the components of the primer vector, $\boldsymbol{\lambda}$, where

$\lambda = ||\boldsymbol{\lambda}|| \neq 0$ for MT arcs, except for cases when a reversal of thrust direction occurs. Define the vector $\boldsymbol{\lambda}_r$, conjugated to the radius vector, as

$$\boldsymbol{\lambda}_r = \begin{bmatrix} \lambda_4 & \lambda_1 \frac{v_2}{r} - \lambda_2 \frac{v_1}{r} + \frac{\lambda_5}{r} \end{bmatrix}^T. \tag{6.18}$$

Denote the multipliers conjugated to r, θ as λ_4, λ_5, respectively, and λ_7 as the multiplier conjugated to the mass, m. In planar polar coordinates, the canonical system in Eq. (6.16) can be expanded yielding

$$\begin{aligned}
\dot{r} &= v_1, \\
\dot{\theta} &= \frac{v_2}{r} \\
\dot{v}_1 &= \frac{c\beta}{m}\frac{\lambda_1}{\lambda} - k^2 r + \frac{v_2^2}{r}, \\
\dot{v}_2 &= \frac{c\beta}{m}\frac{\lambda_2}{\lambda} - \frac{v_1 v_2}{r}, \\
\dot{m} &= -\beta, \\
\dot{\lambda}_1 &= \lambda_2 \frac{v_2}{r} - \lambda_4, \\
\dot{\lambda}_2 &= -2\lambda_1 \frac{v_2}{r} + \lambda_2 \frac{v_1}{r} - \frac{\lambda_5}{r}, \\
\dot{\lambda}_4 &= \lambda_1 \left(\frac{v_2^2}{r^2} + k^2\right) - \lambda_2 \left(\frac{v_1 v_2}{r^2}\right) + \lambda_5 \frac{v_2}{r^2}, \\
\dot{\lambda}_5 &= 0, \\
\dot{\lambda}_7 &= \frac{c\beta}{m^2}\lambda,
\end{aligned} \tag{6.19}$$

and the Hamiltonian in Eq. (6.17) reduces to

$$H = \lambda_1 \left(\frac{v_2^2}{r} - k^2 r\right) - \lambda_2 \frac{v_1 v_2}{r} + \lambda_4 v_1 + \lambda_5 \frac{v_2}{r} + \beta\chi.$$

From the canonical system in Eq. (6.16), we determine that the primer vector is governed by

$$\ddot{\boldsymbol{\lambda}} = -k^2 \boldsymbol{\lambda}$$

and, in (λ, ψ) coordinates, has projections

$$\begin{aligned}
\ddot{\lambda} - \lambda\dot{\psi}^2 &= -k^2\lambda, \\
2\dot{\lambda}\dot{\psi} + \lambda\ddot{\psi} &= 0.
\end{aligned} \tag{6.20}$$

There exists four first integrals associated with the primer vector projections in Eq. (6.20). Two of the first integrals are

$$\lambda^2\dot{\lambda}^2 = \lambda^2\nu^2 - k^2\lambda^4 - \sigma^2 \qquad \text{and} \qquad \lambda^2\dot{\psi} = \sigma. \tag{6.21}$$

When $\dot{\psi} \neq 0$, we have the additional first integrals

$$\lambda^2 = \frac{q}{1 + e\cos 2(\psi - \zeta)} \qquad \text{and} \qquad kt = \arctan\left(\sqrt{\frac{1-e}{1+e}}\tan(\psi - \zeta)\right) - \gamma. \tag{6.22}$$

The variables ν, σ, γ, and ζ are independent integration constants, and q and e are related to ν and σ via

$$(1 - e^2)\sigma^2 = k^2q^2 \qquad \text{and} \qquad (1 - e^2)\nu^2 = 2k^2q. \tag{6.23}$$

From Eq. (6.21), it follows that when $\dot{\psi} = 0$, we have $\sigma = 0$ and γ, ζ, q and e play no role in the solution.

End of proof of Lemma 6.1. □

The first integrals in Eqs. (6.21) and (6.22) define hodographs of the primer vector for the MT arcs. Depending on the quantity e, the hodograph of the primer vector is a central ellipse ($e < 1$), a straight line ($e = 1$), or a circle ($e = 0$). It can be shown that choosing the central ellipse leads to analytical closed form solutions representing motion along circular trajectories. In this paper, the primer vector trajectories are taken to be straight lines. Three other first integrals associated with the canonical system are found to be

$$\begin{aligned} \lambda_1\left(\frac{v_2^2}{r} - k^2 r\right) - \lambda_2\frac{v_1 v_2}{r} + \lambda_4 v_1 + \lambda_5\frac{v_2}{r} + \chi\beta &= C, \\ m &= m_0 - \beta t, \\ \lambda_5 &= \lambda_{50}, \end{aligned} \tag{6.24}$$

where C, m_0, and λ_{50} are independent integration constants.

Consider a fixed Cartesian coordinate system $OXYZ$ and a coordinate system $Oxyz$ which is rotating with angular velocity, $\dot{\theta}$, with respect to the $OXYZ$. If $\dot{\boldsymbol{\lambda}}$ is a derivative of $\boldsymbol{\lambda}$ in the $OXYZ$ system, and $\breve{\boldsymbol{\lambda}}$ is a derivative of $\boldsymbol{\lambda}$ in the $Oxyz$ system, then

$$\dot{\boldsymbol{\lambda}} = \breve{\boldsymbol{\lambda}} + \dot{\theta}\mathbf{k}_1 \times \boldsymbol{\lambda}. \tag{6.25}$$

In $(r,\ \theta)$ coordinates, the equations of motion are

$$\ddot{r}-\dot{\theta}^2 r=\frac{c\beta\lambda_1}{m\lambda}-k^2 r \qquad \text{and} \qquad r\ddot{\theta}+2\dot{r}\dot{\theta}=\frac{c\beta\lambda_2}{m\lambda}. \tag{6.26}$$

Introducing the transformation $\boldsymbol{\lambda}=r\mathbf{q}$, and with the definition

$$\mathbf{q}'=\frac{d\mathbf{q}}{d\theta}$$

we compute the derivatives

$$\breve{\mathbf{q}}=\dot{\theta}\mathbf{q}' \quad \text{and} \quad \frac{d\breve{\mathbf{q}}}{dt}=\dot{\theta}^2\mathbf{q}''+\ddot{\theta}\mathbf{q}'.$$

Using Eqs. (6.25) and (6.26), and the derivatives defined above, we obtain

$$\begin{aligned}\ddot{\boldsymbol{\lambda}} &= r\dot{\theta}^2\mathbf{q}''+\frac{c\beta}{m}\frac{\lambda_2}{\lambda}\mathbf{q}'+\left(\frac{c\beta\lambda_1}{m\lambda}-k^2 r\right)\mathbf{q}+\frac{c\beta\lambda_2}{m\lambda}\mathbf{k}_1\times\mathbf{q}+2r\dot{\theta}^2\mathbf{k}_1\times\mathbf{q}' \\ &\quad +r\dot{\theta}^2\mathbf{k}_1\left(\mathbf{q}\cdot\mathbf{k}_1\right).\end{aligned} \tag{6.27}$$

With $\mathbf{q}=\mathbf{q}(u_1,u_2)$, it follows that $\lambda_1=ru_1$ and $\lambda_2=ru_2$, and using Eq. (6.27), the projections of $\ddot{\boldsymbol{\lambda}}$ on the polar coordinate system axes are

$$\begin{aligned}(\ddot{\boldsymbol{\lambda}})_r &= r\dot{\theta}^2u_1''+\frac{c\beta}{m\lambda}ru_2u_1'+\frac{c\beta}{M\lambda}ru_1^2-k^2ru_1-\frac{c\beta}{m\lambda}ru_2^2-2r\dot{\theta}^2u_2'=-k^2\lambda_1, \\ (\ddot{\boldsymbol{\lambda}})_\theta &= r\dot{\theta}^2u_2''+\frac{c\beta}{m\lambda}ru_2u_2'+\frac{c\beta}{m\lambda}ru_1u_2-k^2ru_2+\frac{c\beta}{m\lambda}ru_1u_2+2r\dot{\theta}^2u_1'=-k^2\lambda_2.\end{aligned} \tag{6.28}$$

Utilizing the relationships given in Eq. (6.28), together with the integral for the Hamiltonian given by the first expression in Eq. (6.24), it can be shown that

$$\begin{aligned}\dot{\theta}^2(u_1''-2u_2')+\frac{c\beta}{m\lambda}(u_2u_1'+u_1^2-u_2^2) &= 0, \\ \dot{\theta}^2(u_2''+2u_1')+\frac{c\beta}{m\lambda}(u_2u_2'+2u_1u_2) &= 0.\end{aligned} \tag{6.29}$$

The first integral associated with Eq. (6.29) is

$$u_1^2+u_2^2+2C_1u_2=0, \tag{6.30}$$

where $C_1\neq 0$ is an integration constant. Using the fact that

$$\lambda_1=\lambda\sin\varphi, \qquad \lambda_2=\lambda\cos\varphi, \tag{6.31}$$

substituting $\lambda_1 = ru_1$ and $\lambda_2 = ru_2$ into Eq. (6.30), and simplifying yields

$$r = \frac{\lambda C_2}{\cos\varphi}, \tag{6.32}$$

where $C_2 = -1/2C_1$. It can be seen that a singularity occurs in Eq. (6.32) when $\cos\varphi = 0$. However, if $\cos\varphi = 0$, then the linear field assumption (that is, $\sin\varphi \approx 0$) is violated. Therefore, it is reasonable in our case to assume that $\cos\varphi \neq 0$. Also, $\cos\varphi = 0$ implies that the rocket is thrusting radially from the central attracting body. This particular scenario is not considered here.

Using the facts that $\dot{r} = v_1$, $r\dot{\theta} = v_2$, and $\lambda_2 = \lambda\cos\varphi$ and

$$\varphi = \frac{\pi}{2} + \theta - \psi \quad \text{(which implies} \quad \dot{\varphi} = \dot{\theta} - \dot{\psi}),$$

we take the time derivative of Eq. (6.32), and substitute appropriately into the equation for $\dot{\lambda}_2$ from the canonical system given in Eq. (6.19) yielding the important relationship between λ_5 and $\dot{\psi}$:

$$\lambda_5 = -2r\lambda\sin\varphi\dot{\psi}. \tag{6.33}$$

Utilizing Eqs. (6.21) and (6.31), we find that Eq. (6.33) can also be written as

$$\lambda_5 = -\frac{2\sigma r\sin\varphi}{\lambda}. \tag{6.34}$$

6.3 ANALYTIC SOLUTIONS FOR MAXIMUM THRUST ARCS

In this section, we present the analytic solutions to the problem of optimal motion in a linear central field. The solutions are general in the sense that the particular performance functional, $\mathcal{J}$, is not specified.

Using the expressions in Eqs. (6.19), (6.21), (6.31), (6.34), and the fact that

$$\dot{\varphi} = \dot{\theta} - \dot{\psi} = \frac{v_2}{r} - \frac{\sigma}{\lambda^2},$$

we obtain the relationship

$$\lambda_2 v_1 - \lambda_1 v_2 - \lambda_5 - \frac{\lambda_2}{\lambda^2} rz - \sigma\frac{\lambda_1}{\lambda^2} r = 0, \tag{6.35}$$

where

$$z = \sqrt{\lambda^2 v^2 - k^2\lambda^4 - \sigma^2}.$$

Similarly, starting with the expression for $\dot{\lambda}_1$ from the canonical system given in Eq. (6.19), namely

$$\dot{\lambda}_1 = \lambda_2 \frac{v_2}{r} - \lambda_4,$$

we make appropriate substitutions to obtain the relationship

$$\lambda_4 = \sigma \frac{\lambda_2}{\lambda^2} - \frac{\lambda_1}{\lambda^2} z. \tag{6.36}$$

Using Eqs. (6.32), (6.35), and (6.36), we rewrite the integral for the Hamiltonian (the first expression in Eq. (6.24)) as

$$\frac{d}{dt}(C_2 \lambda \tan\varphi) + \left[\frac{\lambda^4 k^2 + \sigma^2}{\lambda^2 z}\right](C_2 \lambda \tan\varphi) - \left[\frac{\lambda(\chi\beta - C)}{z}\right] = 0,$$

which can be integrated in quadratures yielding $\tan\varphi$:

$$\tan\varphi = \frac{h_1 + Q}{C_2 h_2 \lambda},$$

where

$$\begin{aligned}
h_1 &= -\int_0^t \frac{\lambda(\chi\beta - C)h_2}{z} dt, \\
h_2 &= e^{h_3}, \\
h_3 &= \int_0^t \frac{\lambda^4 k^2 + \sigma^2}{\lambda^2 z} dt, \\
\chi &= \int_0^t \frac{c}{m\lambda} z dt + \chi_0,
\end{aligned} \tag{6.37}$$

and σ, C, C_2, χ_0, and Q are integration constants. Collecting all the relevant equations, the general solution is found to be

$$\begin{aligned}
v_1 &= r\left[\frac{z}{\lambda^2} + d\tan\varphi\right], \\
v_2 &= r\left[\frac{\sigma}{\lambda^2} + d\right], \\
r &= \frac{\lambda C_2}{\cos\varphi}, \\
\theta &= \varphi + \psi - \frac{\pi}{2}, \\
m &= m_0 - \beta t,
\end{aligned} \tag{6.38}$$

$$\begin{aligned}
\lambda_1 &= \lambda \sin\varphi, \\
\lambda_2 &= \lambda \cos\varphi, \\
\lambda_4 &= \sigma\frac{\lambda_2}{\lambda^2} - \frac{\lambda_1}{\lambda^2}z, \\
\lambda_5 &= \lambda_{50}, \\
\lambda_7 &= c\beta\int_0^t \frac{\lambda}{m^2}d\tau + \lambda_{70},
\end{aligned}$$

where

$$d = \left(\frac{\chi\beta - C}{C_2}\cos^2\varphi - \frac{\nu^2}{2}\sin 2\varphi\right)\frac{1}{z}$$

and m_0, λ_{50}, ν, and λ_{70} are integration constants. The values of the various constants are problem specific and depend on the boundary conditions and on the selected performance index. Note that the variables λ and ψ are explicit functions of time [see Eqs. (6.21) and (6.22)]. The expressions (6.38) are the solutions of the canonical system of Eqs. (6.27) and describes MT arcs in the linear central field regardless the functional of the problem and the fact that whether time is predetermined or free.

Theorem 6.2. *Assume that in the context of the problem (2.111)–(2.115), a linear approximation of the central Newtonian field in the vicinity of a reference orbit is considered. If the flight time is not fixed and the fuel expenditure is minimized, then the hodograph of the primer vector on MT arcs is a straight line, the thrust direction is inertially fixed and the corresponding analytical solutions for MT arcs are expressed in terms of integral sine and integral cosine.*

Proof of Theorem 6.2. In a particular case of free final time (when $C = 0$) and minimum-fuel (that is, $\mathcal{J} = m_0 - m_1$), it follows from the transversality condition that

$$\lambda_{51} = -\frac{\partial \mathcal{J}}{\partial \theta_1} = 0,$$

where the subscript "1" denotes the value at the final time. According to the first integrals of the canonical system of equations [see Eq. (6.24)], we know that λ_5 is a constant, and according to the Weierstrass condition has to be continuous over the entire trajectory. Therefore, we have

$$\lambda_5 = \lambda_{50} = \lambda_{51} = 0.$$

Recalling Eq. (6.34), we obtain the important result that

$$\sigma \sin\varphi \equiv 0.$$

Thus, we can consider two cases: (i) $\sigma \equiv 0$ and $\sin\varphi \neq 0$, or (ii) $\sin\varphi \equiv 0$ and $\sigma \neq 0$. If $\sigma \neq 0$ and $\sin\varphi \equiv 0$, we have the case of tangential thrust which, while representing a problem of practical interest, is not considered in this paper. Accepting that $\sin\varphi \neq 0$ and $\sigma \equiv 0$, we find from Eq. (6.21) that $\sigma = 0$ implies $\dot{\psi} = 0$ (since $\lambda \neq 0$), hence

$$\psi = \psi_0,$$

where ψ_0 is a constant. With $\dot{\psi} = 0$, Eq. (6.20) can be solved for λ:

$$\lambda = a\sin(kt + \alpha),$$

where a and α are integration constants, and in terms of ν in Eq. (6.21), we have the relationship $\nu = ak$. The fact that $\dot{\psi} = 0$ implies that the hodograph of the primer vector is a straight line and the direction of thrust is inertially fixed. We make the following definitions:

$$x = \frac{km_0}{\beta} - kt, \qquad x_0 = \frac{km_0}{\beta}, \qquad \alpha_0 = \alpha + \frac{km_0}{\beta}.$$

Then, with $\dot{\psi} = 0$ and $\lambda = a\sin(kt + \alpha)$, the switching function is

$$\begin{aligned}
\chi &= akc\int_0^t \frac{\cos(kt+\alpha)}{m_0 - \beta t}dt = \frac{akc}{\beta}\left[\int_{x_0}^{x} \frac{\sin(\alpha_0)\sin(x) - \cos(\alpha_0)\cos(x)}{x}dx\right], \\
&= \frac{akc}{\beta}[\sin(\alpha_0)(Si(x) - Si(x_0)) - \cos(\alpha_0)(Ci(x) - Ci(x_0))], \qquad (6.39) \\
&= -\frac{akc}{\beta}[F_1(x_0, x)\sin(\alpha_0) + F_2(x_0, x)\cos(\alpha_0)],
\end{aligned}$$

where

$$\begin{aligned}
F_1 &= F_1(x_0, x) = Si(x) - Si(x_0), \\
F_2 &= F_2(x_0, x) = Ci(x) - Ci(x_0)
\end{aligned}$$

and the functions $Si(x)$ and $Ci(x)$ are integral sine and integral cosine functions:

$$\begin{aligned}
Si(x) &= \sum_{i=1}^{\infty} \frac{(-1)^{i+1}x^{2i-1}}{(2i-1)(2i-1)!}, \\
Ci(x) &= C_0 + \ln(x) + \sum_{i=1}^{\infty} \frac{(-1)^i x^{2i}}{(2i)(2i)!},
\end{aligned}$$

where $C_0 = 0.577216$ is the Euler-Mascheroni constant.

The complete solution to the free final time, minimum fuel problem, obtained from the general solution described in Eq. (6.38), is given by

$$
\begin{aligned}
v_1 &= r\left[k\cot(kt+\alpha)+\dot{\varphi}\tan\varphi\right],\\
v_2 &= -aC_2k\frac{\sin\varphi}{\cos(kt+\alpha)}+\frac{\chi\beta}{ak}\frac{2\cos^2\varphi}{aC_2\sin 2(kt+\alpha)},\\
r &= aC_2\frac{\sin(kt+\alpha)}{\cos\varphi},\\
\theta &= \varphi+\psi_0-\frac{\pi}{2},\\
m &= m_0-\beta t,\\
\lambda_1 &= a\sin(kt+\alpha)\sin\varphi,\\
\lambda_2 &= a\sin(kt+\alpha)\cos\varphi,\\
\lambda_4 &= -ak\cos(kt+\alpha)\sin\varphi,\\
\lambda_7 &= acm_0^2\left[\frac{\sin(kt+\alpha)}{m_0-\beta t}\right]-\chi+\lambda_{70},
\end{aligned}
\tag{6.40}
$$

where

$$
\begin{aligned}
\tan\varphi &= \frac{\tan\alpha\tan\varphi_0}{\tan(kt+\alpha)}+\frac{\chi\beta}{ak}\frac{1}{aC_2k}+\frac{cs}{aC_2k\tan(kt+\alpha)},\\
\dot{\varphi} &= -\frac{k\sin 2\varphi}{\sin^2(kt+\alpha)}+\frac{\chi\beta}{ak}\frac{2\cos^2\varphi}{aC_2\sin 2(kt+\alpha)},\\
s &= F_2\sin\alpha_0-F_1\cos\alpha_0
\end{aligned}
$$

and φ_0 is a new integration constant. The expressions (6.40) are the closed form analytical solutions of the canonical equations (6.27) for the free final time problem and describe motion along the MT arcs.

End of proof of Theorem 6.2. □

We note that in our problem statement the mass-flow rate is limited. Therefore our solution is not equivalent to the case of instantaneous velocity change which, as pointed out by Battin, is generally an inadequate assumption if one is interested in developing rocket guidance solutions [98]. The last expression of Eq. (6.40) can be modified to represent an approximate optimal rocket engine steering law in a realistic gravitational field. This is an extension of the idea of employing a linear-tangent steering law which can be obtained by using constant gravity vector.

Existence of Conjugate Points

Consider the existence of conjugate points on MT arcs obtained in this section. The integration constants are $a, \alpha, \psi_0, \varphi_0, m_0, \lambda_{70}, C_2$. Derivatives $\eta_2 = \eta_{23} = \frac{\partial\theta}{\partial\psi_0} = 1,\ \eta_5 = \eta_{55} = \frac{\partial m}{\partial m_0} = 1$ show that the solutions for MT arcs do not satisfy the conditions $\eta_2(t_0) = \eta_2(t') = 0,\ \eta_5(t_0) = \eta_5(t') = 0$. Therefore, the MT arcs do not contain the conjugate points.

6.4 FIRST INTEGRALS FOR INTERMEDIATE THRUST ARCS

In this section, the optimal control problem on determination of optimal trajectories of a point in a linear central field is considered in the presence of a constraint on mass-flow rate [3]. As mentioned above, the trajectory may contain the arcs of zero thrust (ZT), intermediate thrust (IT) and maximum thrust (MT). The MT arcs in the linear field in the planar case have been considered in the previous sections. It is known that one first integral is necessary to reduce the system of canonical equations to quadratures in the case of three dimensional IT arcs. Below it will be shown that a linearity of the gravitational field allows to obtain the solutions for the three dimensional IT arcs in quadratures. It will also be shown an impossibility of transfer between ZT and IT arcs.

The differential equations of motion with IT arc can be written in the form [59]:

$$\begin{aligned}
\dot{\mathbf{v}} &= \frac{c\beta}{m}\frac{\boldsymbol{\lambda}}{\lambda} - k^2\mathbf{r},\\
\dot{\mathbf{r}} &= \mathbf{v},\\
\dot{M} &= -\beta,\\
\dot{\boldsymbol{\lambda}} &= -\boldsymbol{\lambda}_{\mathbf{r}},\\
\dot{\boldsymbol{\lambda}}_{\mathbf{r}} &= k^2\boldsymbol{\lambda},\\
\dot{\lambda}_7 &= \frac{c\beta}{m^2}\lambda
\end{aligned} \tag{6.41}$$

with

$$H = -k^2\boldsymbol{\lambda}\mathbf{r} + \boldsymbol{\lambda}_{\mathbf{r}}\mathbf{v} + \chi\beta \tag{6.42}$$

and

$$0 \le \beta \le \bar{\beta},$$

and $k^2 = \frac{\mu}{r_0^3}$ is the frequency of Shuler, $\mathbf{r} = (r, 0, 0)$ is the radius-vector originated from the center of attraction placed at the origin of the spherical coordinate system with (r, θ, δ), $\mathbf{v} = (v_1, v_2, v_3)$ is the velocity vector, M is the mass, m is mass-flow rate, $c = const$ is the exhaust velocity, $\lambda = (\lambda_1, \lambda_2, \lambda_3)$ is the primer-vector,

$$\lambda_{\mathbf{r}} = (\lambda_4; \quad \lambda_1 \frac{v_2}{r} + \lambda_2 \frac{v_3 tg\delta - v_1}{r} - \lambda_3 \frac{v_2}{r} tg\delta + \frac{\lambda_5}{r\cos\delta}; \quad \frac{\lambda_1 v_3 - \lambda_3 v_1 + \lambda_6}{r})$$

is the vector conjugate to the radius vector, $\lambda_4, \lambda_5, \lambda_6$ are the multipliers associated with (r, θ, δ) respectively, λ_7 is the multiplier conjugate to the mass, $\chi = \frac{c}{M}\lambda - \lambda_7$ is the switching function. In the spherical coordinate system Eq. (6.41) can be rewritten as [59], [65]:

$$\begin{aligned}
\dot{v}_1 &= \frac{c\beta}{m}\frac{\lambda_1}{\lambda} - k^2 r + \frac{v_2^2}{r} + \frac{v_3^2}{r}, \\
\dot{v}_2 &= \frac{c\beta}{m}\frac{\lambda_2}{\lambda} - \frac{v_1 v_2}{r} + \frac{v_2 v_3}{r} tg\delta, \\
\dot{v}_3 &= \frac{c\beta}{m}\frac{\lambda_3}{\lambda} - \frac{v_2^2}{r} tg\delta - \frac{v_1 v_3}{r}, \\
\dot{r} &= v_1, \\
\dot{\theta} &= \frac{v_2}{r\cos\delta}, \\
\dot{\delta} &= \frac{v_3}{r}, \\
\dot{M} &= -\beta. \\
\dot{\lambda}_1 &= \lambda_2 \frac{v_2}{r} + \lambda_3 \frac{v_3}{r} - \lambda_4, \\
\dot{\lambda}_2 &= -2\lambda_1 \frac{v_2}{r} - \lambda_2 \frac{v_3}{r} tg\delta + \lambda_2 \frac{v_1}{r} + 2\lambda_3 \frac{v_2}{r} tg\delta - \frac{\lambda_5}{r\cos\delta}, \\
\dot{\lambda}_3 &= -2\lambda_1 \frac{v_3}{r} - \lambda_2 \frac{v_2}{r} tg\delta + \lambda_3 \frac{v_1}{r} - \frac{\lambda_6}{r}, \\
\dot{\lambda}_4 &= \lambda_1 (\frac{v_2^2}{r^2} + \frac{v_3^2}{r^2} + k^2) + \lambda_2 v_2 (\frac{v_3}{r^2} tg\delta - \frac{v_1}{r^2}) - \\
&\quad \lambda_3 v_2 (\frac{v_1}{r^2} + \frac{v_2}{r^2} tg\delta) - \lambda_5 \frac{v_2}{r^2\cos\delta} + \lambda_6 \frac{v_3}{r^2}, \\
\dot{\lambda}_5 &= 0, \\
\dot{\lambda}_6 &= \frac{\lambda_3 v_2^2 - \lambda_2 v_2 v_3 - \lambda_5 v_2 \sin\delta}{r\cos^2\delta}, \\
\dot{\lambda}_7 &= \frac{c\beta}{m^2}\lambda.
\end{aligned} \tag{6.43}$$

The Hamiltonian (6.42) of Eqs. (6.43) is expressed as

$$H=\lambda_1(\frac{v_2^2}{r}+\frac{v_3^2}{r}-k^2r)+\lambda_2(\frac{v_2v_3}{r}tg\delta-\frac{v_1v_2}{r})-$$
$$\lambda_3(\frac{v_2^2}{r}tg\delta+\frac{v_1v_3}{r})+\lambda_4v_1+\lambda_5\frac{v_2}{r\cos\delta}+\lambda_6\frac{v_3}{r}. \tag{6.44}$$

In the case of IT arcs, for the system Eq. (6.43) with the Hamiltonian Eq. (6.44) there exist following first integrals:

$$H=C,\quad \lambda=const,\quad \lambda_5=a_3,\quad \lambda_6=a_1\sin\theta-a_2\cos\theta,\quad \lambda_7M=a_4,$$
$$\lambda_3v_2-\lambda_2v_3=a_1\cos\theta\cos\delta+a_2\sin\theta\sin\delta+a_3\sin\delta, \tag{6.45}$$
$$\lambda_1v_1+\lambda_2v_2+\lambda_3v_3-2\lambda_4r+c\lambda\ln\frac{m_0}{m}-3Ct=a_5,$$

where $a_i(i=1,...,5)$, C are integration constants. Then $\ddot{\lambda}=-k^2\lambda$ has following first integrals:

$$\Lambda_1=b_1\cos kt+c_1\sin kt,$$
$$\Lambda_2=b_2\cos kt+c_2\sin kt,$$
$$\Lambda_3=b_3\cos kt+c_3\sin kt,$$

where $\Lambda_1,\Lambda_2,\Lambda_3$ Cartesian components of the primer vector; $b_1,b_2,b_3,c_1,$ c_2,c_3 are integration constants. Then using the relationships between spherical and Cartesian components, we obtain:

$$\begin{aligned}\lambda_1 &= \Lambda_1\cos\theta\cos\delta+\Lambda_2\sin\theta\cos\delta+\Lambda_3\sin\delta,\\ \lambda_2 &= -\Lambda_1\sin\theta+\Lambda_2\cos\theta,\\ \lambda_3 &= -\Lambda_1\cos\theta\sin\delta-\Lambda_2\sin\theta\sin\delta+\Lambda_3\cos\delta.\end{aligned} \tag{6.46}$$

6.5 ANALYTICAL SOLUTIONS FOR INTERMEDIATE THRUST ARCS

Below we prove the following theorem:

Theorem 6.3. *In the context of the problem Eqs. (2.111)–(2.115), consider a linear approximation of the central Newtonian field in the vicinity of a reference orbit. The canonical system for extremals of the problem is given by Eqs. (6.41) with Hamiltonian in Eq. (6.42). Then, for intermediate thrust arcs, Eqs. (6.41) admit at least one family of extremals expressed in quadratures.*

Proof of Theorem 6.3. By computing derivatives and equating them to the first three equations of Eq. (6.43), we find following:

$$\lambda_4 = -\dot{\lambda}_1 \cos\theta \cos\delta + \dot{\lambda}_2 \sin\theta \cos\delta + \dot{\lambda}_3 \sin\delta, \tag{6.47}$$

$$\lambda_1 v_2 - \lambda_2 v_1 = x_4 - x_2 r, \tag{6.48}$$

$$\lambda_1 v_3 - \lambda_3 v_1 = -\lambda_6 - x_1 r, \tag{6.49}$$

where

$$\begin{aligned} x_1 &= \dot{\Lambda}_1 \cos\theta \sin\delta + \dot{\Lambda}_2 \sin\theta \sin\delta - \dot{\Lambda}_3 \cos\delta, \\ x_2 &= -\dot{\Lambda}_1 \sin\theta + \dot{\Lambda}_2 \cos\theta, \\ x_3 &= a_1 \cos\theta \sin\delta + a_2 \sin\theta \sin\delta - a_3 \cos\delta, \\ x_4 &= a_1 \cos\theta \cos\delta + a_2 \sin\theta \cos\delta + a_3 \sin\delta. \end{aligned}$$

By comparing Eq. (6.49) and fourth relationship of Eq. (6.45), we will have

$$\lambda_3(\lambda_1 v_2 - \lambda_2 v_1 = -\lambda_2 \lambda_6 + \lambda_2 r x_3. \tag{6.50}$$

Besides that, from Eq. (6.48) and Eq. (6.50) it follows that

$$r = \frac{\lambda_3 x_4 + \lambda_2 \lambda_6 - \lambda_1 x_3}{\lambda_2 x_1 + \lambda_3 x_2}. \tag{6.51}$$

The last integral Eq. (6.45) taking into account Eq. (6.48) and Eq. (6.49) gives

$$r = \lambda_2 \frac{\lambda_2 \lambda_6 - \lambda_1 x_3 + \lambda_3 x_4}{\lambda_1 \lambda_3 \lambda_4 + x_2(\lambda_2^2 + \lambda_3^2)}. \tag{6.52}$$

By equating Eq. (6.51) and (6.52), we obtain following relationship:

$$\lambda_2 x_2 = \lambda_1 \lambda_4 + \lambda_3 x_1. \tag{6.53}$$

Using Eq. (6.46) and (6.47), one can easily show that Eq. (6.53) is a direct consequence of the integral $\lambda = const.$ Taking into account Eq. (6.48), (6.49), and (6.53), from Eq. (6.44) we can find that

$$r = \frac{x_2 x_4 + x_1 \lambda_6 + \lambda_1 C}{x_1^2 + x_2^2 - \lambda_1^2 k^2}. \tag{6.54}$$

From Eq. (6.51), (6.54), and Eq. (6.53) it follows that:

$$-x_1x_4\lambda_4 - x_1^2x_3 + x_2\lambda_2\lambda_6 - x_2^2x_3 -$$
$$\lambda_1 k^2(\lambda_3x_4 + \lambda_2\lambda_6 - \lambda_1x_3) - C(\lambda_2x_1 + \lambda_3x_2) = 0. \tag{6.55}$$

Substituting $x_i(i = 1, ..., 4)$, $\lambda_j(j = 1, ..., 6)$, into the latter equation and after some manipulations we obtain

$$n_1 tg^2kt + n_2 tgkt + n_3 = 0, \tag{6.56}$$

where

$$n_1 = x_4B_1B_6 - \lambda_3B_2B_6 - B_2^2x_3 - k^2x_4B_3B_5 -$$
$$k^2\lambda_6b_3B_4 - k^2x_3B_3^2 - x_3B_1^2 - CB_1B_4,$$
$$n_2 = x_4(A_1B_6 + A_6B_1) - 2x_3A_1B_1 - \lambda_6(A_2B_6 + A_6B_2) - 2x_3A_2B_2 -$$
$$k^2x_4(A_3B_5 + A_5B_3) - k^2\lambda_6(A_3B_4 + A_4B_3) - 2k^2x_3A_3B_3 - C(A_1B_4 + A_4B_1),$$
$$n_3 = x_4a_1\lambda_6 - A_1^2x_3 - \lambda_6A_2A_6 - x_3A_2^2 - k^2x_4A_3A_5 -$$
$$k^2\lambda_6a_3A_4 - k^2x_3A_3^2 - CA_1A_4,$$
$$A_1 = k(c_1g_1h_2 - c_2g_1h_1 - c_3g_2), \quad B_1 = -k(b_1g_1h_2 + b_2g_1h_1 - b_3g_2),$$
$$A_2 = k(-c_1h_1 + c_2h_2), \quad B_2 = -k(-b_1h_1 + b_2h_2),$$
$$A_3 = b_1h_2g_2 + b_2g_2h_1 + b_3g_1, \quad B_3 = c_1h_2g_2 + c_2g_2h_1 + c_3g_1,$$
$$A_4 = -b_1h_1 + b_2h_2, \quad B_4 = -c_1h_1 + c_2h_2,$$
$$A_5 = -b_1h_2g_2 + b_2g_1h_1 + b_3g_2, \quad B_5 = -c_1h_2g_1 - c_2g_1h_1 + c_3g_2,$$
$$A_6 = k(c_1h_2g_2 + c_2g_2h_1 + c_3g_1, \quad B_6 = -k(b_1h_2g_2 + b_2g_2h_1 + b_3g_1,$$
$$h_1 = \sin\theta, \quad h_2 = \cos\theta, \quad g_1 = \sin\delta, \quad g_2 = \cos\delta.$$

From Eq. (6.56) we can find t in terms of φ and θ:

$$t = \frac{1}{k} arctg(\frac{-n_1 + \sqrt{n_2^2 - 4n_1n_3}}{2n_1}) + \pi\frac{n}{k}, \qquad n = 0, 1, 2, ... \tag{6.57}$$

Sign of the square root is determined such that $t > 0$. Furthermore, after differentiation of Eq. (6.57) and utilizing Eq. (6.48) and Eq. (6.49), we obtain formulas for the velocity components:

$$v_1 = \frac{(x_4 - x_2r)\Lambda_1 + \Lambda_2\cos\delta(-\lambda_6 + x_1r)}{k(1 + \Lambda_1^2)\lambda_1 r\cos\delta - \lambda_2\Lambda_1 - \lambda_3\Lambda_2},$$
$$v_2 = \frac{\lambda_2v_1 + x_4 - x_2r}{\lambda_1}, \tag{6.58}$$

$$v_3 = \frac{\lambda_3 v_1 + \lambda_6 - x_1 r}{\lambda_1}.$$

From the fifth integral Eq. (6.45) it follows that

$$m = m_0 \exp\left[-\frac{\lambda_1 v_1 + \lambda_2 v_2 + \lambda_3 v_3 - r\lambda_4 - 3C - t - a_5}{c\lambda}\right],$$

where r, v_1, v_2, v_3 are found from Eq. (6.51), and (6.58). The multipliers $\lambda_4, \lambda_6, \lambda_7$ can be determined from Eq. (6.47), fourth and fifth integrals of Eq. (6.45). So, variables r, v_1, v_2, v_3, m, t and multipliers $\lambda_1, \lambda_2, \lambda_3, \lambda_4, \lambda_6, \lambda_7$ are expressed through spherical coordinates θ and δ. The angle δ can be expressed in terms of θ utilizing Eq. (6.43) in the form:

$$\frac{d\delta}{d\theta} = \frac{v_3(\theta, \delta)\cos\delta}{v_2(\theta, \delta)}. \tag{6.59}$$

If $t = 0$ is initial moment, then while θ and δ are changed from 0 to such values, at which the condition

$$tg\delta = \frac{(c_2 b_3 - b_3 c_3)\cos\theta + (c_3 b_1 - b_3 c_1)\sin\theta}{c_1 b_2 - b_1 c_2},$$

is satisfied, the magnitude r is changed from

$$r = \frac{a_1 b_1 + a_2 b_2 + a_3 b_3}{b_2 c_1 - b_3 c_2}$$

to ∞. Analysis of Eq. (6.54), (6.58), and (6.59) show that the trajectories are the smooth curves around the center of attraction.

Now it will shown below that the optimal transfer between ZT and IT arcs in the linear central field does not exist. It is known that for the IT arcs $\chi \leq 0$ [3]. If a ZT arc is changed to an IT arc, then latter starts at $\chi = 0$. From $\frac{d\chi}{df} = \frac{c}{M}\frac{d\lambda}{df}$, where f is the true anomaly, it follows that on IT arcs the functions λ and χ take their maximum values at the same f. Studies of the primer vector for circular and elliptical orbits shown that on ZT arcs λ takes its maximum value at $f = 0$ or $f = \frac{\pi}{2}$ [3]. This means that if the circular or elliptical orbit is a part of the optimal trajectory, then the thrust is switched at $f = 0$ or $f = \frac{\pi}{2}$. But at all such points, a sit was shown in Ref. [65], $\lambda_1 = 0$, $\lambda_3 = 0$, that is the thrust is aligned along a perpendicular to the radius vector. From the other side, on IT arcs at such values of λ_1 and λ_3, as it follows from Eq. (6.54) and (6.55), the magnitude of the radius vector is $r = 0$ at $x_2 \neq 0$ or $r = \infty$ at $x_2 = 0$, that is the thrust can not be aligned in a manner above mentioned. Consequently, the three dimensional optimal transfer between ZT and IT arcs (or vice versa) in the linear central filed is impossible. End of proof of Theorem 6.3. □

In this chapter, the new analytical solutions for IT and MT (minimum) arcs have been obtained for the linear central field. It has been shown that a linear field approximation of the central Newtonian field is admissible if the ratio of the width of the layer to the radius of a reference orbit and thrust angle are negligibly small. On MT arcs the thrust vector is directed under a constant angle to horizon. In the solutions obtained for MT arcs, the power and specific impulse are constant, and therefore, these solutions are also valid for minimum thrust arcs. The thrust arcs considered in this chapter represent the trajectories of motion with a high-thrust propulsion.

CHAPTER 7

Extremal Trajectories in a Uniform Gravity Field

7.1 OPTIMAL CONTROL PROBLEM FOR POWERED DESCENT

The studies presented below describe the first attempt to demonstrate that the integration constants play an important role in the design of an envelope of the descent trajectories and in the real-time targeting and guidance design for precision landing. In particular, the proposed solutions can be used to determine a manifold of the initial conditions from which the lander can be guided to prescribed LS or to its vicinity determined by a terminal manifold in the state space. One of the new utilities of the analytical solutions is that the LS can be re-designated on-board as many times as needed by redefining the integration constants, thereby solving the real-time retargeting problem to allow for a hazard detection and avoidance and to provide a safe pin-point landing.

The problem of a planetary powered descent and landing at a specified landing site (LS) is considered [125], [122], [126]. A series of simulations have been conducted to demonstrate the utility of the proposed trajectory control solutions to achieve safe landing on Mars [123]. The algorithms used in the simulations do not have iterative procedure or approximations except for the numerical computation of some constants. These algorithms were designed to incorporate the solution of the minimum landing error problem by appropriate selection of the integration constants, maneuver time and the control parameters. The simulation results show that the analytical solutions obtained in this work can be implemented to generate feasible and extremal or optimal powered descent and landing trajectories with minimum landing error [164]. Feasibility of the trajectories is understood in the sense of connecting the initial and final conditions with some nonzero landing errors in the final position and velocity vectors, and satisfying the mass, control and time constraints. The results were comparable to those of the studies that use convex optimization and other numerical-analytical methods [121]. Due to their explicitness, the proposed solutions can be used to determine manifolds of the initial and final conditions for atmospheric entry and powered descent, to generate the trajectory envelopes,

Analytical Solutions for Extremal Space Trajectories
DOI: http://dx.doi.org/10.1016/B978-0-12-814058-1.00007-2

to provide re-targeting design aimed to achieve the landing and to incorporate attitude guidance. Integration of the trajectory control, targeting and guidance solutions with attitude guidance is subject of further studies.

7.1.1 Problem Statement

The minimum-fuel optimal control problem is considered below as part of the optimal trajectory design problem for Mars powered descent and landing at a desired LS [126]. Consider a spacecraft as a point with variable mass moving in a drag-free uniform gravity field with the gravity and thrust accelerations. The equations of motion of the point are

$$\ddot{\mathbf{r}} = \mathbf{u} + \mathbf{g}, \qquad \dot{m} = -\alpha T, \tag{7.1}$$

where $\mathbf{r} \in R^3$ is the state vector, $\mathbf{u} \in R^3$ is the thrust acceleration control vector, and $\mathbf{g} = \mathbf{g}_0$ is the constant gravitational acceleration vector. Let us introduce an inertial coordinate system, $OXYZ$ with the origin, O at the Mars center of mass, m is the spacecraft mass, $\alpha = 1/(I_{sp}g_e)$ is the given positive constant, I_{sp} is the specific impulse, g_e is the sea level gravitational acceleration on Earth, T is the thrust. The X-axis of this system is directed towards the point of interest, the Z-axis is directed parallel to the velocity at initial time of the powered descent maneuvers. The Y-axis completes a right-handed triad. The magnitude of the control vector, $||\mathbf{u}||$ is limited by the following constraint [123]:

$$0 \leq a \leq ||\mathbf{u}|| \leq b,$$

where a and b are the given constants. As the constant a can be zero, the descent trajectory may consist of zero thrust or ballistic arcs and thrusting arcs. The components of the control vector can be given in the form: $\mathbf{u}_i = ue_i$, $i = 1, 2, 3$, where $u = ||\mathbf{u}||$, and e_i are the components of $\mathbf{e} = \mathbf{u}/u$. All these components are given in the system $OXYZ$. Consequently, the control vector is to satisfy the following constraints [164]:

$$\Phi_1 = e_1^2 + e_2^2 + e_3^2 - 1 = 0, \qquad \Phi_2 = (u - a)(b - u) - \eta^2 = 0, \tag{7.2}$$

where η_2 is the unknown slack variable. In this case, the control vector-function is extended to have the following components: u, e_i $(i = 1, 2, 3)$, η.

The initial and final conditions are given as follows:

$$\mathbf{r}(t_i) \;=\; \mathbf{r}_0,$$

$$\begin{aligned}\dot{\mathbf{r}}(t_i) &= \dot{\mathbf{r}}_0,\\ \mathbf{r}(t_f) &= \mathbf{r}_1 = \mathbf{0},\\ \dot{\mathbf{r}}(t_f) &= \dot{\mathbf{r}}_1 = \mathbf{0},\\ m(t_i) &= m_0.\end{aligned} \tag{7.3}$$

Here t_i and t_f are the initial and final instants of time. t_i is assumed known, but t_f is to be determined in the solution process.

The performance index of the problem is given as

$$J = \int_{t_i}^{t_f} ||\mathbf{u}||dt. \tag{7.4}$$

Now the problem under consideration can be stated as follows: it is required to find the state vector, $\mathbf{x}(\mathbf{r}, \dot{\mathbf{r}}, m)$, and $\mathbf{u}$ that can satisfy Eqs. (7.1)–(7.2) and minimize J, given in Eq. (7.4).

7.1.2 First-Order Optimality Conditions

By defining $\mathbf{r} = \mathbf{r}(x_1, x_2, x_3)$, $\mathbf{v} = \mathbf{v}(v_1, v_2, v_3)$, and $\mathbf{g} = \mathbf{g}(g_1, g_2, g_3)$, where the components of the vectors are given in the $OXYZ$ coordinate system, and accepting that $T = mu$, Eqs. (7.1) can be rewritten in the following first-order form ($j = 1, 2, 3$):

$$\begin{aligned}\dot{x}_j &= v_j,\\ \dot{v}_j &= ue_j + g_j,\\ \dot{m} &= -\alpha mu.\end{aligned} \tag{7.5}$$

The corresponding boundary conditions, Eqs. (7.3) are

$$\begin{aligned}\mathbf{r}(t_i) &= \mathbf{r}_0,\\ \mathbf{v}(t_i) &= \mathbf{v}_0,\\ \mathbf{r}(t_f) &= \mathbf{r}_1 = \mathbf{0},\\ \mathbf{v}(t_f) &= \mathbf{v}_1 = \mathbf{0},\\ m(t_i) &= m_0.\end{aligned} \tag{7.6}$$

If

$$\mathbf{x} = \mathbf{x}(\mathbf{r}, \mathbf{v}, m), \qquad \boldsymbol{\lambda} = \boldsymbol{\lambda}(\boldsymbol{\lambda}_r, \boldsymbol{\lambda}_v, \lambda_m),$$

then the extremality conditions for the problem Eqs. (7.1)–(7.4) can be given as [121]:

$$\dot{\mathbf{x}} = \left[\frac{\partial H}{\partial \boldsymbol{\lambda}}\right]^T, \qquad \dot{\boldsymbol{\lambda}} = -\left[\frac{\partial H}{\partial \mathbf{x}}\right]^T \tag{7.7}$$

where the Hamiltonian function, H is given as

$$\begin{aligned} H &= \sum \lambda_i(ue_i + g_i) - \lambda_m \alpha m u + \mu_1[e_1^2 + e_2^2 + e_3^2 - 1] \\ &\quad + \mu_2[(u-a)(b-u) - \eta^2] + \lambda_0 u. \end{aligned} \tag{7.8}$$

Here μ_1 and μ_2 are additional slack variables, and λ_0 is assumed to be a non-zero Lagrange multiplier associated with the integrand function of Eq. (7.4). The second half of Eq. (7.7) can be re-written in the form:

$$\begin{aligned} \dot{\lambda}_i &= -\frac{\partial H}{\partial v_i} = -\lambda_{i+3}, \\ \dot{\lambda}_{i+3} &= -\frac{\partial H}{\partial x_i} = -\sum_j \lambda_j \frac{\partial g_j}{\partial x_i}, \\ \dot{\lambda}_m &= \lambda_m \alpha u. \end{aligned} \tag{7.9}$$

In addition to Eqs. (7.9), the extremality of H with respect to the control variables can be expressed with the following conditions:

$$\begin{aligned} \frac{\partial H}{\partial e_i} &= \lambda_i u + 2\mu_1 e_i = 0, \\ \frac{\partial H}{\partial u} &= \sum \lambda_i e_i - \lambda_m \alpha m + \mu_2(a + b - 2u) + \lambda_0 = 0, \\ \frac{\partial H}{\partial \eta} &= -2\mu_2 \eta = 0. \end{aligned} \tag{7.10}$$

Eqs. (7.5) and (7.9) can be used to find all 14 components of $\mathbf{x}$ and $\boldsymbol{\lambda}$, and Eqs. (7.10) can serve to determine the control variables, e_i, u, and η by employing the Weierstrass condition. Note that without loss of generality, one can choose λ_0 to be $\lambda_0 = 1$.

7.2 LAGRANGE MULTIPLIERS AND OPTIMAL CONTROL REGIMES

Note that $\mathbf{g} = \mathbf{g}(g_1, g_2, g_3)$ is a constant vector, which can be determined as $\mathbf{g} = \mathbf{g}(-g_0,\ 0,\ 0)$, where $g_0 = ||\mathbf{g}|| = const$. Consequently, taking into account that α is also a constant parameter, Eqs. (7.9) can be integrated as

$$\begin{aligned}\lambda_i &= a_i t + b_i, \quad i=1,2,3;\\ \lambda_{i+3} &= a_i, \quad i=1,2,3;\\ \lambda_m &= \lambda_{m0}\exp[\alpha\int u dt],\end{aligned} \tag{7.11}$$

where $a_i,\ b_i\ (i=1,2,3)$ and λ_{m0} are the integration constants. The first group of formulas of Eqs. (7.11) show that

$$\boldsymbol{\lambda}_v = \mathbf{a}t + \mathbf{b}, \tag{7.12}$$

the $\mathbf{p}$-trajectory or the hodograph of the primer vector, $\boldsymbol{\lambda}_v(\lambda_1, \lambda_2, \lambda_3)$ is a straight line, which passes through zero if all constants $b_i\ (i=1,2,3)$ are zeros. λ is the distance from the origin of the coordinate system with axes $\lambda_1, \lambda_2, \lambda_3$ to the hodograph. At the same time, not all a_i and $b_i\ (i=1,2,3)$ can be zeros. In general, the magnitude

$$\lambda = \sqrt{\lambda_1^2 + \lambda_2^2 + \lambda_3^2} = ||\boldsymbol{\lambda}_v||$$

is a monotonic, and either increasing or decreasing function of time.

As is known, the Weierstrass condition can be given in the form [3]:

$$\lambda_i \dot{x}_i \geq \lambda_i \dot{x}_i^*,$$

where $\dot{x}_i$ are computed on the optimal trajectory, and $\dot{x}_i^*$ are computed on the admissible trajectory. For the problem Eqs. (7.1)–(2.115) this condition has the form:

$$u(\sum \lambda_i e_i - \lambda_m \alpha m) \geq u^*(\sum \lambda_i e_i^* - \lambda_m \alpha m). \tag{7.13}$$

If $\eta = 0$ and $u = a$ (see Eqs. (7.2)), then Eq. (7.13) can only be satisfied if $(\sum \lambda_i e_i - \lambda_m \alpha m) \leq 0$. In the same manner, if $\eta = 0$ and $u = b$, then Eq. (7.13) can be satisfied if $(\sum \lambda_i e_i - \lambda_m \alpha m) \geq 0$. If $\eta \neq 0$ and $a < u < b$, then Eq. (7.13) can be satisfied if $(\sum \lambda_i e_i - \lambda_m \alpha m) = 0$ over a non-zero interval of time, on which u takes intermediate values: $a < u < b$. Based on these analyses, one can determine

$$\chi = \sum \lambda_i e_i - \lambda_m \alpha m$$

as a switching function. Consequently, there may exist three options for the control regimes:

$$(1)\ u=a \quad if \quad \chi \le 0; \quad (2)\ u=b \quad if \quad \chi \ge 0; \quad (3)\ a<u<b \quad if \quad \chi \equiv 0. \tag{7.14}$$

From the last equation of Eqs. (7.10), one can find that there may exist three cases:

1. If $\mu_2=0,\ \eta \neq 0$, then there may exist a variable u, such that $a<u<b$. In this case, from the second equation of Eqs. (7.10) and from Eqs. (7.14) it can be seen that $\lambda_0=0$ which contradicts to the assumption made above that $\lambda_0 \neq 0$.

2. If $\mu_2 \neq 0,\ \eta=0$, then from Eqs. (7.2) it can be seen that u takes boundary values: $u=a$ or $u=b$.

3. If $\mu_2=0,\ \eta=0$, then one can obtain the same contradiction as described in the case 1.

Consequently, an optimal trajectory of the problem in Eqs. (7.1)–(2.115) includes the thrust arcs with boundary values of the control function: $u=a$ or $u=b$. The intermediate values of the control are not optimal.

From the first group of equations of Eqs. (7.10), it can be seen that μ_1 can be chosen to show that $\boldsymbol{\lambda}||\mathbf{e}$, or simply,

$$e_i=\lambda_i/\lambda, \quad i=1,2,3. \tag{7.15}$$

In this case one can find that $\chi=\lambda-\lambda_m\alpha m$ and $\dot{\chi}=\dot{\lambda}$. Consequently, from Eq. (7.12) it follows that, in general, there may exist the following cases (see Fig. 7.1):

(a) λ decreases $\forall t \in [t_i, t_f]$, and it may or may not reach its minimum value. This case means that λ decreases from t_i until t_f, and $t_f \le t^*$, where t^* is the time instant at which λ reaches its minimum value. In this case, $\dot{\lambda} \le 0$ and $\dot{\chi} \le 0\ \forall t \in [t_i, t_f]$ and correspondingly, χ is decreasing function of time and may cross zero only at once. So, either $\chi>0$ or $\chi<0$ or $\chi>0,\ \chi<0\ \forall t \in [t_i, t_f]$. Correspondingly, the control regime is either $u=b$ or $u=a$ or $u=b,\ u=a\ \forall t \in [t_i, t_f]$ respectively. If $t_f=t^*$, then $\dot{\lambda}(t_f)=\dot{\chi}(t_f)=0$.

(b) λ increases $\forall t \in [t_i, t_f]$, and it may or may not start from its minimum value. This case means that λ increases from $t_i \ge t^*$ until t_f. In this case, $\dot{\lambda} \ge 0$ and $\dot{\chi} \ge 0\ \forall t \in [t_i, t_f]$, and so χ is increasing function of time and may cross zero only at once. As in the previous case, either $\chi>0$ or $\chi<0$ or $\chi<0,\ \chi>0\ \forall t \in [t_i, t_f]$. Correspondingly, the control regime is either $u=b$ or $u=a$ or $u=a,\ u=b\ \forall t\in[t_i, t_f]$ respectively. If $t_i=t^*$, then $\dot{\lambda}(t_i)=\dot{\chi}(t_i)=0$.

(c) λ decreases from t_i to t^*, reaches its minimum value at t^*, then increases from t^* to t_f. So, $\dot{\lambda}=\dot{\chi}<0\ \forall t\in[t_i, t^*]$, then $\dot{\lambda}(t^*)=\dot{\chi}(t^*)=0$, and then $\dot{\lambda}=\dot{\chi}>0\ \forall t\in[t^*, t_f]$. So, χ can cross zero once (cases c1 and c2 below) or twice (case c3) with following sequences:

(c1) $\chi(t)>0\ \forall t\in[t_i, t_1^*]$; $\chi(t_1^*)=0$; $\chi(t)<0\ \forall t\in[t_1^*, t^*]$; $\dot{\chi}(t^*)=0$; $\chi(t)<0\ \forall t\in[t^*, t_f]$. The corresponding control regime has the following sequence: $u=b,\ u=a$.

(c2) $\chi(t)<0\ \forall t\in[t_i, t^*]$; $\dot{\chi}(t^*)=0$; $\chi(t)<0\ \forall t\in[t^*, t_1^*]$; $\chi(t_1^*)=0$; $\chi(t)>0\ \forall t\in[t_1^*, t_f]$. The corresponding control regime has the following sequence: $u=a,\ u=b$.

(c3) $\chi(t)>0\ \forall t\in[t_i, t_1^*]$; $\chi(t_1^*)=0$; $\chi(t)<0\ \forall t\in[t_1^*, t^*]$; $\dot{\chi}(t^*)=0$; $\chi(t)<0\ \forall t\in[t^*, t_2^*]$; $\chi(t_2^*)=0$; $\chi(t)>0\ \forall t\in[t_2^*, t_f]$. Corresponding control regime has the following sequence: $u=b,\ u=a,\ u=b$.

To summarize, an optimal trajectory in the problem Eqs. (7.1)–(2.115) may have one of the following regimes for control:

Case 1: $u=a\ \forall t\in[t_i, t_f]$.
Case 2: $u=b\ \forall t\in[t_i, t_f]$.
Case 3: $u=b\ \forall t\in[t_i, t_1^*]$; $u=a\ \forall t\in[t_1^*, t_f]$.
Case 4: $u=a\ \forall t\in[t_i, t_1^*]$; $u=b\ \forall t\in[t_1^*, t_f]$.
Case 5: $u=b\ \forall t\in[t_i, t_1^*]$, $u=a\ \forall t\in[t_1^*, t_2^*]$, and $u=b\ \forall t\in[t_2^*, t_f]$.

7.3 OPTIMAL TRAJECTORY ARCS

First, let us re-write the boundary conditions, Eqs. (7.6) in the form:

$$\begin{aligned}\mathbf{r}_0 &= \mathbf{r}_0(x_{10}, x_{20}, x_{30}),\\ \mathbf{v}_0 &= \mathbf{v}_0(v_{10}, v_{20}, v_{30}),\\ \mathbf{r}_1 &= \mathbf{r}_1(x_{11}, x_{21}, x_{31}),\\ \mathbf{v}_1 &= \mathbf{v}_1(v_{11}, v_{21}, v_{31}),\\ m(t_i) &= m_0.\end{aligned} \tag{7.16}$$

Using Eqs. (7.15) and (7.11), one can rewrite the second group of Eqs. (7.5) as

$$\dot{v}_1 = \frac{(a_1 t + b_1)u}{\lambda} - g_0,$$

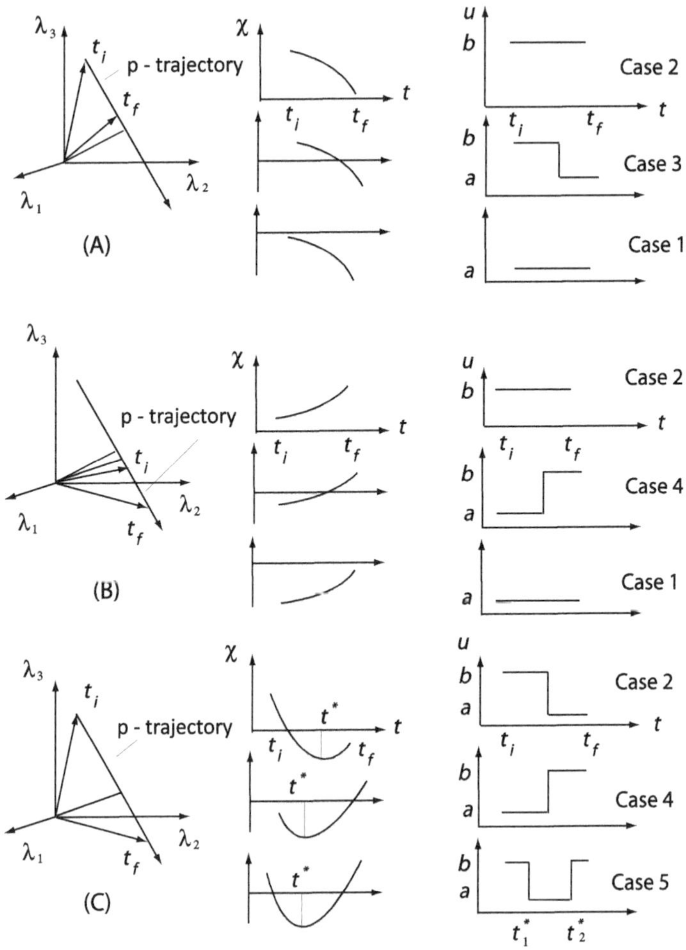

Figure 7.1 p-Trajectory, switching function and optimal control regimes.

$$\begin{aligned} \dot{v}_2 &= \frac{(a_2 t + b_2)u}{\lambda}, \\ \dot{v}_3 &= \frac{(a_3 t + b_3)u}{\lambda}, \end{aligned} \tag{7.17}$$

where

$$\lambda = \sqrt{k_1 t^2 + k_2 t + k_3},$$
$$k_1 = a_1^2 + a_2^2 + a_3^2, \quad k_2 = 2(a_1 b_1 + a_2 b_2 + a_3 b_3), \quad k_3 = b_1^2 + b_2^2 + b_3^2.$$

Taking into account that the control variable takes the boundary values, that is either $u = a$ or $u = b$ or $u = b$, $u = a$ and $u = b$, one can integrate Eqs. (7.17) in the following form:

$$\begin{aligned} v_1 &= A_1 u - g_0 t + v_{10}, \\ v_2 &= A_2 u + v_{20}, \\ v_3 &= A_3 u + v_{30}, \end{aligned} \tag{7.18}$$

where

$$A_i = \bar{A}_i + A_{i0}, \qquad i = 1, 2, 3, \tag{7.19}$$

with

$$\bar{A}_i = \frac{a_i}{k_1}\lambda + \left(\frac{b_i}{\sqrt{k_1}} - \frac{a_i k_2}{2\sqrt{k_1^3}} \right) \ln(\tau),$$

$$A_{i0} = -\bar{A}_i(t_0), \quad \tau = 2\lambda\sqrt{k_1} + 2k_1 t + k_2.$$

By substituting Eqs. (7.18) into the first group of Eqs. (7.5), the latter can be integrated in the form:

$$\begin{aligned} x_1 &= B_1 u - \frac{g_0}{2} t^2 + v_{10} t + x_{10}, \\ x_2 &= B_2 u + v_{20} t + x_{20}, \\ x_3 &= B_3 u + v_{30} t + x_{30}, \end{aligned} \tag{7.20}$$

where

$$B_i = \bar{B}_i + B_{i0}, \qquad i = 1, 2, 3,$$

with

$$\bar{B}_i = \frac{a_i}{k_1}\left[\frac{2k_1 t + k_2}{4k_1}\lambda + \frac{4k_1 k_3 - k_2^2}{8\sqrt{k_1^3}} \ln(\tau) \right] +$$

$$\left(\frac{b_i}{\sqrt{k_1}} - \frac{a_i k_2}{2\sqrt{k_1^3}} \right) \left[\frac{\tau(\ln\tau - 1)}{4k_1} + \frac{k_2^2 - 4k_1 k_3}{4k_1} \frac{\ln\tau + 1}{\tau} \right], \tag{7.21}$$

$$B_{i0} = -\bar{B}_i(t_0).$$

As u is constant ($u=a$ or $u=b$), the last equations of Eqs. (7.5) and (7.11) can be integrated to obtain the mass and corresponding Lagrange multiplier:

$$m = m_0 \exp[-\alpha u t], \tag{7.22}$$

$$\lambda_m = \lambda_{m0} \exp[\alpha u t], \tag{7.23}$$

where λ_{m0} is the integration constant. The performance index in Eq. (2.115) of the problem can now be computed in a form depending on which case among the cases (1–5) is under consideration (see Fig. 7.1):
Case 1: $u=a\ \forall t \in [t_i, t_f]$:

$$J = a(t_f - t_i). \tag{7.24}$$

Case 2: $u=b\ \forall t \in [t_i, t_f]$:

$$J = b(t_f - t_i). \tag{7.25}$$

Case 3: $u=b\ \forall t \in [t_i, t_1^*]$; $u=a\ \forall t \in [t_1^*, t_f]$

$$J = b(t_1^* - t_i) + a(t_f - t_1^*). \tag{7.26}$$

Case 4: $u=a\ \forall t \in [t_i, t_1^*]$; $u=b\ \forall t \in [t_1^*, t_f]$

$$J = a(t_1^* - t_i) + b(t_f - t_1^*). \tag{7.27}$$

Case 5: $u=b\ \forall t \in [t_i, t_1^*]$, $u=a\ \forall t \in [t_1^*, t_2^*]$, and $u=b\ \forall t \in [t_2^*, t_f]$:

$$J = b(t_1^* - t_i) + a(t_2^* - t_1^*) + b(t_f - t_2^*). \tag{7.28}$$

The constants a_i and b_i, $(i=1,2,3)$ can be determined from Eq. (7.18) and (7.20) using Eqs. (7.16). Note that due to presence of λ_i, $i=1,2,3$, Eq. (7.8) explicitly depends on time, and therefore Eqs. (7.7) do not possess an integral for Pontryagin function:

$$H = \sum \lambda_i(ue_i + g_i) - \lambda_m \alpha m u + \lambda_0 u. \tag{7.29}$$

It can be seen that the procedure of determining the constants and t_f described above is valid for each thrust arc with corresponding value of u. In the cases (3–5), the time instants, t_1^* and/or t_2^* can be found from the continuity conditions for the position, velocity and Lagrange multipliers at the junction points. The analytical solutions obtained above allow us to make further progress in the complete design of powered descent. In particular, these solutions can be used to determine a manifold of the initial conditions from which the lander can be guided to prescribed LS or to its vicinity in the case of landing errors. Qualitative analysis can be conducted to generate an envelope of descent and landing trajectories which can be formulated explicitly. Another important aspect of this topic is the design of attitude guidance, and as the Lagrange multipliers and corresponding trajectory solutions are determined explicitly, the trajectory guidance design can easily facilitate or incorporate the attitude guidance.

CHAPTER 8

Number of Thrust Arcs for Extremal Orbital Transfers

8.1 METHOD OF APPLICATION OF ANALYTICAL SOLUTIONS FOR THRUST ARCS

The functional dependencies of the number of TAs of one type, two different types and the total number of TAs on the number of integration constants of the analytical solutions for these arcs have been determined. They allow for determination of the structure of the maneuver trajectory and the number of junction points. The fundamental importance of existence of the analytical solutions for TA in the synthesis of the trajectories. The results are given in the form of theorems, and primarily applicable to the minimum-fuel problems.

It will be shown that, in essence, the solutions of the optimal control problem can be reduced to the solution of a system of continuity equations obtained from the continuity conditions at the junction points [165]. This is one of the fundamental results of the monograph. The solution of the continuity equations and synthesis of the maneuver trajectory are, in general, the last stages of the methodology of determining extremal trajectories developed in this work. Following this methodology, it is concluded that the solution of the problem can be reduced to an analytical closed-form which is considered as the general result of this monograph.

8.1.1 Continuity Equations

It is known that the construction of extremals or, in a general case, a synthesis of various TAs is associated with the question of determination of the number of TA on the maneuver trajectory. Consider the variational problem of a transfer between given orbits which lie in the same plane that contain the center of gravitational attraction. Assume that the analytical solutions for TAs have been obtained as a result of solving the canonical system of equations in the polar coordinates [127], [165]:

$$r = r(\varphi, C_1, \ldots, C_{10}),$$
$$v_1 = v_1(\varphi, C_1, \ldots, C_{10}),$$

Analytical Solutions for Extremal Space Trajectories
DOI: http://dx.doi.org/10.1016/B978-0-12-814058-1.00008-4

$$
\begin{aligned}
v_2 &= v_2(\varphi, C_1, \ldots, C_{10}),\\
\theta &= \theta(\varphi, C_1, \ldots, C_{10}),\\
m &= m(\varphi, C_1, \ldots, C_{10}).
\end{aligned}
$$

The solutions for the Lagrange multipliers have a form:

$$\lambda_i = \lambda_i(\varphi, C_1, \ldots, C_{10}), \quad i = 1, \ldots, 5$$

where φ is independent variable, $C_1, \ldots, C_{10}$ are integration constants. Here we accept that $\lambda_3, \lambda_4, \lambda_5$ are conjugated with r, θ, m respectively. As is known, a motion in an arbitrary orbit, which represents a conic section, is described by formulas:

$$
\begin{aligned}
r &= r(p, e, f),\\
v_1 &= v_1(p, e, f),\\
v_2 &= v_2(p, e, f),\\
\theta &= f + \omega,
\end{aligned}
$$

where f is true anomaly, p is semi-latus rectum, e is eccentricity of the orbit, ω is the longitude of perigee. Besides that, the general solutions for the primer vector and its derivative for zero thrust (ZT) in arcs can be written in the form:

$$
\begin{aligned}
\boldsymbol{\lambda} &= \boldsymbol{\lambda}(f, e, A, B, C, D),\\
\dot{\boldsymbol{\lambda}} &= \dot{\boldsymbol{\lambda}}(f, e, A, B, C, D),
\end{aligned}
$$

where A, B, C, D are integration constants, e is the eccentricity of the orbit. The eccentricity can take any value, including zero, which corresponds to a circular orbit. In the chapter we consider arbitrary extremal solutions for TAs. The tests of these solutions with various optimality conditions can be considered as part of the methodology mentioned above, but they are not considered in this chapter. The next stage in determining the solutions of the variational problem is the synthesis of various TAs into the extremal trajectory solutions.

Assume that it is required to determine a maneuver trajectory between two arbitrary elliptical orbits with parameters p_1, e_1, ω_1 and p_2, e_2, ω_2 respectively. The initial and final positions of the spacecraft in the orbits and the maneuver time are not fixed. The initial mass is given, and the functional of the problem is the difference between the initial and final masses. The number of junction points which allows us to determine the number of TAs is of critical importance in determining the structure of the trajectory. The simplest structure of the maneuver trajectory is ZTA1-FTA-ZTA2 that contain two junction points, where ZTA means zero thrust arc, that is a part of the orbit, and the number next to ZTA indicates an order of appearance of this arc on the trajectory, FTA means an arbitrary finite thrust arc. Different and more complicated structures with various TAs are possible. Below, the questions of constructing the continuity equations at the junction points of the trajectories containing one, two, three, and n number of TAs are considered in detail. The possibilities of existence of various types of TAs on the maneuver trajectories are discussed.

8.1.2 Case of Trajectory With One Thrust Arc

In this case, the maneuver trajectory will contain two junction points. At these points, the continuity conditions for radius vector, velocity vector, mass, primer vector and its derivative, and the conditions for switching function have to be satisfied. Also, the transversality conditions at initial and final points of the trajectory must be satisfied. As the phases of motion in the boundary orbits are not fixed, then the one can have the following expressions for position and velocity components:

$$r_1 = R(f_1), \qquad v_{11} = V_1(f_1), \qquad v_{21} = V_2(f_1) \tag{8.1}$$

$$r_2 = R(f_2), \qquad v_{12} = V_1(f_2), \qquad v_{22} = V_2(f_2) \tag{8.2}$$

The functional of the problem is given by

$$J' = J + \nu_s F_s(x_0, x_1, t_0, t_1),$$

where $J = m_0 - m_f$, ν is the constant Lagrange multiplier, s is the number of initial and final conditions, and F is the functions that express these conditions. The functional J' can be determined as [130]

$$\begin{aligned} J' = J &+ \nu_{r1}[r_1 - R(f_1)] + \nu_{v11}[v_{11} - V_1(f_1)] + \nu_{v21}[v_{21} - V_2(f_1)] \\ &+ \nu_{r2}[r_2 - R(f_2)] + \nu_{v12}[v_{12} - V_1(f_2)] + \nu_{v22}[v_{22} - V_2(f_2)]. \end{aligned}$$

Consequently, the transversality conditions will have the form:
at the initial point:

$$\lambda_{11} = \frac{\partial J'}{\partial v_{11}} = \nu_{v11}, \quad \lambda_{21} = \frac{\partial J'}{\partial v_{21}} = \nu_{v21}, \quad \lambda_{31} = \frac{\partial J'}{\partial r_1} = \nu_{r1},$$

$$\lambda_{41} = \frac{\partial J'}{\partial f_1} = -\nu_{r1}\frac{\partial R(f_1)}{\partial f_1} - \nu_{v11}\frac{\partial V_1(f_1)}{\partial f_1} - \nu_{v21}\frac{\partial V-2(f_1)}{\partial f_1},$$

and at the final point:

$$\lambda_{12} = \frac{\partial J'}{\partial v_{12}} = \nu_{v12}, \quad \lambda_{22} = \frac{\partial J'}{\partial v_{22}} = \nu_{v22}, \quad \lambda_{32} = \frac{\partial J'}{\partial r_2} = \nu_{r2},$$

$$\lambda_{42} = \frac{\partial J'}{\partial f_2} = -\nu_{r2}\frac{\partial R(f_2)}{\partial f_2} - \nu_{v12}\frac{\partial V_1(f_2)}{\partial f_2} - \nu_{v22}\frac{\partial V-2(f_2)}{\partial f_2},$$

$$\lambda_{52} = \frac{\partial J'}{m_f} = -1.$$

The equations obtained allow us to find ν_r, ν_v if λ_i, $i = 1, 2, 3, 4$ are given. The values of ν_r and ν_v do not affect the procedure of computation of other unknowns of the problem, and therefore, their computation will be omitted. The continuity conditions of the variables have the following form:

For the first junction point:

$$\begin{aligned} r(p_1, e_1, f_1) &= r(\varphi_1, C_1, \ldots, C_{10}), \\ v_1(p_1, e_1, f_1) &= v_1(\varphi_1, C_1, \ldots, C_{10}), \\ v_2(p_1, e_1, f_1) &= v_2(\varphi_1, C_1, \ldots, C_{10}), \\ \theta(f_1) &= \theta(\varphi_1, C_1, \ldots, C_{10}). \end{aligned} \tag{8.3}$$

For the second junction point:

$$\begin{aligned} r(p_2, e_2, f_2) &= r(\varphi_2, C_1, \ldots, C_{10}), \\ v_1(p_2, e_2, f_2) &= v_1(\varphi_2, C_1, \ldots, C_{10}), \\ v_2(p_2, e_2, f_2) &= v_2(\varphi_2, C_1, \ldots, C_{10}), \\ \theta(f_2) &= \theta(\varphi_2, C_1, \ldots, C_{10}). \end{aligned} \tag{8.4}$$

Besides that, the switching function conditions must be satisfied at the junction points:

$$\begin{aligned} \chi_1 &= \frac{c\lambda(\varphi_1, C_1, \ldots, C_{10})}{m_0} - \lambda_{51} = 0, \\ \chi_2 &= \frac{c\lambda(\varphi_2, C_1, \ldots, C_{10})}{m(\varphi_2, C_1, \ldots, C_{10})} - \lambda_{52}(\varphi_2, C_1, \ldots, C_{10}) = 0. \end{aligned} \tag{8.5}$$

The continuity conditions for the primer vector and its derivative can be given in the following form:

For the first junction point:

$$\begin{aligned}
\lambda_{11}(f_1, e_1, A_1, B_1, C_1, D_1) &= \lambda_1(\varphi_1, C_1, \ldots, C_{10}), \\
\lambda_{21}(f_1, e_1, A_1, B_1, C_1, D_1) &= \lambda_2(\varphi_1, C_1, \ldots, C_{10}), \\
\lambda_{31}(f_1, e_1, A_1, B_1, C_1, D_1) &= \lambda_3(\varphi_1, C_1, \ldots, C_{10}), \\
\lambda_{41}(f_1, e_1, A_1, B_1, C_1, D_1) &= \lambda_4(\varphi_1, C_1, \ldots, C_{10}).
\end{aligned} \tag{8.6}$$

For the second junction point:

$$\begin{aligned}
\lambda_{12}(f_2, e_2, A_2, B_2, C_2, D_2) &= \lambda_1(\varphi_2, C_1, \ldots, C_{10}), \\
\lambda_{22}(f_2, e_2, A_2, B_2, C_2, D_2) &= \lambda_2(\varphi_2, C_1, \ldots, C_{10}), \\
\lambda_{32}(f_2, e_2, A_2, B_2, C_2, D_2) &= \lambda_3(\varphi_2, C_1, \ldots, C_{10}), \\
\lambda_{42}(f_2, e_2, A_2, B_2, C_2, D_2) &= \lambda_4(\varphi_2, C_1, \ldots, C_{10}).
\end{aligned} \tag{8.7}$$

Eqs. (8.3)–(8.7) provide a continuous transfer from one TA to another, the main aspect of the synthesis of TAs. These equations, in general, allow us to find unknowns variables, $\varphi_k, f_k,\ (k = 1, 2)$ and unknown constants, $C_i,\ (i = 1, \ldots, 10),\ A_k, B_k, C_k, D_k,\ (k = 1, 2),\ \lambda_{51}$. It can be easily seen that the first equation of Eqs. (8.5), Eqs. (8.6) and (8.7) allow us to determine constants $\lambda_{51},\ A_k, B_k, C_k, D_k,\ (k = 1, 2)$. Eqs. (8.3) and (8.4), obtained as the continuity conditions for r, v_1, v_2, θ at both junction points. Also, the second equation of Eqs. (8.5), expressing the equality of the switching function to zero at the second junction point, serve to find 14 unknown variables, $\varphi_k, f_k,\ (k = 1, 2)$ and unknown constants, $C_i,\ (i = 1, \ldots, 10)$. Assuming that, in general, the thrust direction, φ and the spacecraft position, that is the true anomaly, f in the boundary conditions, are not determined, one can come to the following conclusion: in order to determine the unknowns mentioned above uniquely, it is necessary to have in the analytical solutions for TAs for the radius vector, the velocity components and the polar angle, 5 integration constants instead of 10. In this case, the 9 unknowns are $\varphi_1, \varphi_2, f_1, f_2$, and $C_j,\ (j = 1, \ldots, 5)$. Note that 10 constants allow us to find the families of the maneuver trajectories that can be found by changing the additional constants, $C_6, \ldots, C_{10}$. This yields the following important result: *given the conditions of the problem being considered, only particular solutions of the canonical equations for TAs can be used to construct the extremal maneuver trajectory.* Base on what has been described above, one can conclude that in the case of one TA on the trajectory with two junction points, the problem of synthesizing the trajectory is reduced to the solution of 9 algebraic equations with 9 unknowns. As mentioned above, these equations are the consequence of the continuity conditions at the junction points, and below they will be referred as *continuity equations.*

Particular cases of the problem conditions.

1. Let $\varphi_1, \varphi_2, f_1, f_2$ be unknowns and the number of integration constants is less than 5, for example, $C_1, \ldots, C_p < 5$. Then the unknowns of the problem are found from $4 + p$ equations. Besides that, additional $5 - p$ equations must be satisfied and this imposes a constraint on application of the analytical solutions.

2. If any one angle among $\varphi_1, \varphi_2, f_1, f_2$ is given, for example, f_1, then to construct the maneuver trajectory uniquely, it is necessary that the analytical solutions contain 6 constants. This conclusion follows from the necessity of increasing the number of unknowns to equate to the number of equations above mentioned, that is 9 continuity equations are enough in order to find 9 unknowns, $\varphi_1, \varphi_2, f_2, C_j, \ (j = 1, \ldots, 6)$.

3. If any two angles among $\varphi_1, \varphi_2, f_1, f_2$ are given, for example, φ_1, f_1, then to construct the maneuver trajectory uniquely, it is necessary that the analytical solutions contain 7 constants. The conclusion may follow on the same basis as in the previous case.

4. If all angles $\varphi_1, \varphi_2, f_1, f_2$ are given, then to construct the maneuver trajectory uniquely, it is necessary that the analytical solutions contain 9 constants. If it is possible to solve 9 equations, mentioned above, then this means that the given analytical solutions allow us to construct the maneuver trajectory between any given positions in the boundary orbits with given thrust angles at the initial and final times.

5. If p number of continuity equations are not used, then the number of constants, which is enough to find the solutions uniquely, will be equal to $5 - p$.

Analysis of these particular cases considered above leads to conclusion that the more integration constants in the solutions of the canonical system, the more flexibility in the construction of the maneuver trajectory.

8.1.3 Case of Trajectory With Two Thrust Arcs

In this case the maneuver trajectory will contain four junction points. The trajectory structure has a form: ZTA1-FTA1-ZTA2-FTA2-ZTA3. Assume that FTA1 and FTA2 are described by different analytical solutions with constants C_k and K_k, $(k = 1, \ldots, 10)$. Note that at the junction points, besides the conditions mentioned in the previous section, the condition, which shows the equality of the magnitudes of the primer vector at the final point of FTA1 and the initial point of FTA2, must be satisfied. For simplicity, consider the following notations:

$$r = x_1, \quad v_1 = x_2, \quad v_2 = x_3.$$

Then the continuity conditions will have a form:
At the first junction point:

$$\begin{aligned} x_i(p_1, e_1, f_1) &= x_i(\varphi_1, C_1, \ldots, C_{10}), \qquad i = 1, 2, 3. \\ \theta = f_1 + \omega_1 &= \theta(\varphi_1, C_1, \ldots, C_{10}). \end{aligned}$$

At the second junction point:

$$\begin{aligned} x_i(\varphi_2, C_1, \ldots, C_{10}) &= x_i(p_2, e_2, f_2), \quad i = 1, 2, 3 \\ \theta(\varphi_2, C_1, \ldots, C_{10}) &= \theta = f_2 + \omega_2, \\ \lambda(\varphi_2, C_1, \ldots, C_{10}) &= \lambda(\varphi_3, K_1, \ldots, K_{10}), \\ \frac{c\lambda(\varphi_2, C_1, \ldots, C_{10})}{m(\varphi_2, C_1, \ldots, C_{10})} - \lambda_{52}(\varphi_2, C_1, \ldots, C_{10}) &= 0. \end{aligned}$$

At the third junction point:

$$\begin{aligned} x_i(p_2, e_2, f_3) &= x_i(\varphi_3, K_1, \ldots, K_{10}), \qquad i = 1, 2, 3 \\ \theta = f_3 + \omega_2 &= \theta(\varphi_3, K_1, \ldots, K_{10}). \end{aligned}$$

At the fourth junction point:

$$\begin{aligned} x_i(\varphi_4, K_1, \ldots, K_{10}) &= x_i(p_3, e_3, f_4), \quad i = 1, 2, 3 \\ \theta(\varphi_3, K_1, \ldots, K_{10}) = \theta &= f_4 + \omega_3, \\ \frac{c\lambda(\varphi_4, K_1, \ldots, K_{10})}{m(\varphi_4, K_1, \ldots, K_{10})} - \lambda_{52}(\varphi_4, K_1, \ldots, K_{10}) = \lambda_{54} &= \frac{J}{m_{4f}} = -1. \end{aligned}$$

Here p_2, e_2 are ω_2 are the parameters of the elliptical transfer orbit subject to determination. Note that the eccentricities of the boundary orbits, e_1, e_3 may be arbitrary, including zeros. The number of the equations constructed above is 19, and the number of unknowns is 31 containing 11 unknown variables, $\varphi_j,\ f_j\ (j = 1, \ldots, 4),\ p_2, e_2, \omega_2$, and 20 unknown constants, $C_k,\ K_k,\ (k = 1, \ldots, 10)$. Consequently, to obtain equal number of equations and unknowns, the number of existing unknowns need to be decreased. Assuming that the thrust angles and the true anomalies at the junction points are not determined, the general number of the constants should be decreased from 20 to 8. This may be possible, for example, in the case when the solutions for each arc have 4 constants. This means that to make the analytical solutions useful in solving the maneuver problem being considered, each of these solutions should have 4 constants. In a particular case, these solutions may the same. Note that other combination of the constants in the solutions may lead to difficulties in constructing the continuity equations.

Based on the analysis presented above, one can conclude that in the case of two TAs, the problem of synthesis of the maneuver trajectory is reduced to the solution of 19 algebraic continuity equations with 19 unknowns. Consequently, the result obtained above about particular solutions is also valid in this case:

given the conditions of the problem being considered, only particular solutions of the canonical equations for TAs allow us to construct the maneuver trajectory.

Particular cases of the problem conditions.

1. Assume we have unknowns, φ_j, f_j $(j = 1, ..., 4)$, $p_2, e_2, \omega_2, C_k, K_k$, $(k = 1, ..., p < 4)$. Then $11 + 2$ equations are enough to find these unknowns, and the remaining $2(4 - p)$ equations impose constraints on the behavior of the unknowns and make the range of the applications of the solutions shorter.

2. If any one unknown among φ_j, f_j $(j = 1, ..., 4)$, $p_2, e_2, \omega_2, C_k, K_k$, $(k = 1, ..., 4,)$ is given, then one can have 19 equations with 18 unknowns. One equation in this case will serve as a constraint imposed on all other unknowns. If each of the solutions for TAs have 5 constants instead of 4, then we would have 19 equations with 20 unknowns. In this case, the auxiliary unknown would serve as a parameter to construct a family of extremal maneuver trajectories.

3. If any two unknowns among φ_j, f_j $(j = 1, ..., 4)$, $p_2, e_2, \omega_2, C_k, K_k$, $(k = 1, ..., 5)$ are given and each of the solutions for TAs have 5 constants, then one can have 19 equations with 19 unknowns. In this case, the continuity equations allow us to synthesize a unique extremal trajectory.

4. If any three unknowns among φ_j, f_j $(j = 1, ..., 4)$, $p_2, e_2, \omega_2, C_k, K_k$, $(k = 1, ..., 5)$ are given and each of the solutions for TAs have 4 constants, then one will have 19 equations with 16 unknowns. This means that three additional constraints must be satisfied. If each of the solutions for TAs have 5 constants, then we can have only one constraint equation. If each of the solutions for TAs have 6 constants, then as in the particular case 2, the continuity equations allow us to construct a one-parameter family of extremal maneuver trajectories.

8.1.4 Case of Trajectory With Three Thrust Arcs

In this case, the maneuver trajectory will contain six junction points. One possible structure of the trajectory is ZTA1-FTA1-ZTA2-FTA2-ZTA3-FTA3-ZTA4. Assume that FTA1, FTA2, and FTA3 are described by different analytical solutions with constants $C_k,\ K_k,\ B_k\ (k=1, ..., 10)$. Similar to the previous sections, construct the continuity conditions:
At the first junction point:

$$\begin{aligned} x_i(p_1, e_1, f_1) &= x_i(\varphi_1, C_1, \ldots, C_{10}), \quad i=1,2,3. \\ \theta = f_1 + \omega_1 &= \theta(\varphi_1, C_1, \ldots, C_{10}). \end{aligned}$$

At the second junction point:

$$\begin{aligned} x_i(\varphi_2, C_1, \ldots, C_{10}) &= x_i(p_2, e_2, f_2), \quad i=1,2,3, \\ \theta(\varphi_2, C_1, \ldots, C_{10}) = \theta &= f_2 + \omega_2, \\ \lambda(\varphi_2, C_1, \ldots, C_{10}) &= \lambda(\varphi_3, K_1, ..., K_{10}), \\ \frac{c\lambda(\varphi_2, C_1, \ldots, C_{10})}{m(\varphi_2, C_1, \ldots, C_{10})} - \lambda_{52}(\varphi_2, C_1, \ldots, C_{10}) &= 0. \end{aligned}$$

At the third junction point:

$$\begin{aligned} x_i(p_2, e_2, f_3) &= x_i(\varphi_3, K_1, ..., K_{10}), \quad i=1,2,3 \\ \theta = f_3 + \omega_2 &= \theta(\varphi_3, K_1, ..., K_{10}). \end{aligned}$$

At the fourth junction point:

$$\begin{aligned} x_i(\varphi_4, K_1, ..., K_{10}) &= x_i(p_3, e_3, f_4), \quad i=1,2,3 \\ \theta(\varphi_3, K_1, ..., K_{10}) = \theta &= f_4 + \omega_3, \\ \lambda(\varphi_4, K_1, ..., K_{10}) &= \lambda\varphi_5, B_1, ..., B_{10}), \\ \frac{c\lambda(\varphi_4, K_1, ..., K_{10})}{m(\varphi_4, K_1, ..., K_{10})} - \lambda_{52}(\varphi_4, K_1, ..., K_{10}) &= 0. \end{aligned}$$

At the fifth junction point:

$$\begin{aligned} x_i(p_3, e_3, f_5) &= x_i(\varphi_5, B_1, ..., B_{10}), \quad i=1,2,3 \\ \theta = f_5 + \omega_3 &= \theta(\varphi_5, B_1, ..., B_{10}). \end{aligned}$$

At the sixth junction point:

$$\begin{aligned} x_i(\varphi_6, B_1, ..., B_{10}) &= x_i(p_4, e_4, f_6), \quad i=1,2,3 \\ \theta(\varphi_6, B_1, ..., B_{10}) = \theta &= f_6 + \omega_4, \\ \frac{c\lambda(\varphi_6, B_1, ..., B_{10})}{m(\varphi_6, B_1, ..., B_{10})} - \lambda_{56}(\varphi_6, B_1, ..., B_{10}) &= \frac{J}{m_{4f}} = -1. \end{aligned}$$

Here p_2, e_2, ω_2 and p_3, e_3, ω_3 are the unknown parameters of two intermediate elliptical transfer orbits, and p_1, e_1, ω_1 and p_4, e_4, ω_4 are the parameters of the initial and final elliptical orbits. Note that the eccentricities of the boundary orbits are arbitrary, including zeros. The number of equations constructed above are 29, whereas the number of unknowns is 48:

$$\varphi_j,\ f_j\ (j=1, ..., 6),\ p_2, e_2, \omega_2,\ p_3, e_3, \omega_3,\ C_k, K_k, B_k\ (k=1, ..., 10).$$

As it follows from this, to obtain equal number of equations and unknowns, the number of unknowns in this case has to be decreased. Assuming that the thrust angles and true anomalies at the junction points are usually not determined, and noting that the parameters of the intermediate elliptical orbits are arbitrary, we can conclude that the total number of constants need to be reduced from 30 to 11, that is the total number of constants of the analytical solutions for all three TAs should be 11. Consequently, the result obtained in the previous sections about the particular solutions is also valid in this case: *given the conditions of the problem being considered, only particular solutions of the canonical equations for TAs can be used to construct the maneuver trajectory.* The solutions for TAs, as mentioned in the previous section, can be arbitrary, and in a particular case, two those solutions may be the same. Consequently, in the case of three TAs, it is required to solve 29 equations with 29 unknowns, including 11 unknown constants. As in the previous sections, some particular cases are possible in the case being considered.

On the basis of the method of application of the analytical solutions in the synthesis of maneuver trajectories, one can conclude that ***in the case of one, two and three TAs, the solution of the problem can be reduced to the solution of a certain number of continuity equations.***

Consequently, *the possibility of application of a general solution for TAs with a complete number of integration constants is limited, and therefore such a solution has only a theoretical value.*

8.1.5 Case of Trajectory With *n* Thrust Arcs

Based on the analyzes of the number of equations in each case considered above, one can show that if there are *n* TAs of the same type and have the same solutions, then the total number of equations is equal to $10n + (n - 1) = 11n - 1$. In the same manner, one can find that the total number of unknowns is equal to $N = 7n + nm - 3$, where n is the number of TAs, m is the number of constants. Note these results are valid if all TAs are of the same type and have the same solutions.

These results generalize the conclusion made in the previous section, that is ***the solution of the problem can be reduced to the solution of a certain number of continuity equations.***

8.2 NUMBER OF THRUST ARCS ON EXTREMAL TRAJECTORY

8.2.1 Number of Equations at Junction Points

The analysis conducted in the previous sections allow us to make the following conclusions about the number of continuity equations depending on the number of thrust arcs:

- Equations for radius vector, velocity vector and polar angle $\mathbf{r}, \mathbf{v}, \theta$:

Number of FTAs	Number of eqs.	Junction point
1	8	At each
2	16	junction
3	24	point

- Equation for primer vector, $\boldsymbol{\lambda}$:

Number of FTAs	Number of eqs.	Junction point
1	0	final point of each FTA
2	1	initial point of 2nd FTA
3	2	initial points of 2nd and 3rd FTAs

- Equation for switching function, χ:

Number of FTAs	Number of eqs.	Junction point
1	1	at final point of each FTA
2	2	at final points of 1st and 2nd FTAs
3	3	at final points of 1st, 2nd, and 3rd FTAs

By computing the number of equations in each case, one can conclude that

- If there is one FTA, then the total number of equations is 9;
- If there are two FTAs, then the total number of equations is 19;

- If there are three FTAs, then the total number of equations is 29;
- If there are n FTAs, then the total number of equations is $10n - 1$.

8.2.2 Number of Unknowns in the Continuity Equations

Here we assume that only one type of TAs with the same analytical solutions with a certain number of constants will be used to construct the maneuver trajectory. Consider a functional dependence of the number of unknowns in the continuity equations and in the transversality conditions on the number of FTAs and their constants. Let the solutions for FTA contain m constants. The analysis of the previous sections show that the following cases may take place:

- Case with one FTA:
  - The number of unknowns is $N = 4 + m$:
    φ_i, f_i, C_k $(i = 1, 2;\ k = 1., ..., m)$
- Case with two FTAs:
  - The number of unknowns is $N = 11 + 2m$:
    $\varphi_i, f_i, p_2, e_2, \omega_2, C_k, K_k$ $(i = 1, ..., 4;\ k = 1., ..., m)$
- Case with three FTAs:
  - The number of unknowns is $N = 18 + 3m$:
    $\varphi_i, f_i, p_2, e_2, \omega_2, p_3, e_3, \omega_3, C_k, K_k, B_k$ $(i = 1, ..., 6;\ k = 1., ..., m)$
- Case with n FTAs:
  - The number of unknowns is $N = 7n + mn - 3$:
    $\varphi_i, f_i, p_r, e_r, \omega_r, C_k$ $(r = 1, ..., n - 1;\ i = 1, ..., 2n;\ k = 1., ..., nm)$.
    Here C_k represents the total number of constants.

The results obtained above are given below in the table.

DETERMINATION OF THE NUMBER OF UNKNOWNS, N

Number of FTAs	1	2	3	4	n
φ	2	4	6	8	$2n$
f	2	4	6	8	$2n$
C_k	m	$2m$	$3m$	$4m$	nm
p_r	0	1	2	3	$n-1$
e_r	0	1	2	3	$n-1$
ω_r	0	1	2	3	$n-1$

Summing the quantities in the last column, one can find the total number of unknowns:

$$N = 7n + nm - 3, \tag{8.8}$$

where n is the number of FTAs, m is the number of constants in the solutions for r, v_1, v_2, θ for the FTA. Note that Eq. (8.8) is valid in the case when all FTAs are described by the same analytical solutions.

8.2.3 Number of Finite Thrust Arcs

Theorem 8.1. *The number of FTAs of the same type on extremal transfer trajectory between two coplanar orbits is*

$$n = \frac{2}{m-3}, \tag{8.9}$$

where m is the total number of constants in all analytical solutions for radius vector, velocity vector and polar angle.

Proof of Theorem 8.1. As mentioned in the previous sections, to obtain solutions of the continuity equations, the number of equations and the number of unknowns should be equal. Consequently, if the number of equations is $10n - 1$, then using Eq. (8.8), one can have

$$10n - 1 = 7n + nm - 3,$$

or

$$n = \frac{2}{m-3}. \tag{8.10}$$

End of proof of Theorem 8.1. □

Eq. (8.10) determines the number of FTAs depending on the number of constants of the solutions for these FTAs. Consequently, there exist only to cases:

$$a. \quad n = 2 \quad m = 4; \qquad b. \quad n = 1 \quad m = 5.$$

This means that no more than two FTAs can be used to accomplish the transfer maneuver. At the same time, this does not mean that other classes of FTAs can not be used in this maneuver. From Eq. (8.10) it also follows that as n is integer, then the general solution for FTA with $m = 10$ integration constants can not be used for the maneuver being considered.

Now, consider the case when two different types of FTAs are used to synthesize the maneuver trajectory. The number of these arcs may be arbitrary. Below we prove the following theorem:

Theorem 8.2. *The number of FTAs of two different types on extremal trajectory of transfer between two arbitrary coplanar elliptical orbits is equal to*

$$n = \frac{q_1(m_1 - m_2) - 2}{3 - m_2}, \tag{8.11}$$

where m_1 *and* m_2 *are total number of constants in the analytical solutions for radius vector, velocity vector and polar angle for both types of FTAs, and* q_1 *is the number of FTAs of the first type, and* q_2 *is the number of FTAs of the second type, where* $q_1 + q_2 = n$.

Proof of Theorem 8.2. Let n be the total number of FTAs, and all solutions of the first and second types of FTAs contain m_1 and m_2 constants respectively. Then the total number of unknowns can be determined according to the formulae:

$$N = 2n + 2n + q_1 m_1 + q_2 m_2 + 3(n-1),$$

where q_1, q_2 are the numbers of the first and second types of FTAs, where $q_1 + q_2 = n$. Using the condition that the number of equations and the number of unknowns are equal, we can have:

$$10n - 1 = N = 2n + 2n + q_1 m_1 + q_2 m_2 + 3(n-1)$$

or

$$n = \frac{1}{3}(q_1 m_1 + q_2 m_2 - 2).$$

Substituting $q_2 = n - q_1$ into the latter formulae yields

$$n = \frac{q_1(m_1 - m_2) - 2}{3 - m_2}. \tag{8.12}$$

If $q_2 = 0$, then $q_1 = n$, $m_1 = m$, and consequently, from Eqs. (8.12) leads to Eq. (8.10). End of proof of Theorem 8.2. □

As a general case, consider now a transfer trajectory with various FTAs.

Theorem 8.3. *The total number of FTAs on extremal trajectory of transfer between two coplanar elliptical orbits is equal to*

$$n = \frac{1}{3}\left(\sum_{j=1}^{e} q_j m_j - 2\right), \tag{8.13}$$

where q_j, $(j = 1, ..., e)$ *is the number of FTAs with* m_j *constants of integration, and* $q_1 + q_2 + ... + q_e = n$, *and* e *is the number of types of FTAs used in the maneuver.*

Proof of Theorem 8.3. It can be shown that the total number of unknowns is determined according to the formulae:

$$N = 2n + 2n + q_1 m_1 + q_2 m_2 + ... + q_e m_e + 3(n - 1),$$

where q_j, $(j = 1, ..., e)$ is the number of FTAs with m_j integration constants, where $q_1 + q_2 + ... + q_e = n$, e is the number of types of FTAs. Then the condition that the number of equations and the number of unknowns are equal, is given by

$$10n - 1 = 2n + 2n + \sum_{j=1}^{e} q_j m_j + 3(n - 1)$$

or

$$n = \frac{1}{3}\left(\sum_{j=1}^{e} q_j m_j - 2\right). \tag{8.14}$$

It can also be shown that Eq. (8.14) can be reduced to Eq. (8.10) or Eq. (8.12), if one or two types of FTAs respectively are used. End of proof of Theorem 8.3. □

Consequently, Eq. (8.14) allows us to obtain the number of FTAs on the maneuver trajectory depending on the integration constants of the solutions for these FTAs. Eq. (8.14) leads to the following particular cases:

$$\begin{aligned}
n &= 1: \quad q_1 = 1,\ q_2 = q_3 = ... = 0, \quad m_1 = 5 \\
n &= 2: \quad q_1 = 1,\ q_2 = 1, \quad m_1 + m_2 = 8 \\
n &= 2: \quad q_1 = 2,\ q_2 = q_3 = ... = 0, \quad m_1 = 4 \\
n &= 3: \quad q_1 = 2,\ q_2 = 1, \quad q_3 = ... = 0, \quad 2m_1 + m_2 = 11 \\
n &= 4: \quad q_1 = 2,\ q_2 = 1, \quad q_3 = 1, \quad 2m_1 + m_2 + m_3 = 14,
\end{aligned}$$

and etc., where m_1, m_2, m_3 satisfy the conditions (or expressions) in the corresponding cases.

Note that if the continuity equations are solvable, then the corresponding solution allows us, in particular, to determine the sequence of ZTAs and

FTAs, that is the structure of the trajectory, and to synthesize the maneuver trajectory of interest.

In this chapter, the determination of the number of FTAs on the maneuver trajectory is described. The system of continuity equations with equal number of unknowns, including the integration constants, has been obtained. The number of equations and the number of unknowns have been determined for the cases of trajectories containing one, two and three FTAs. The results obtained have been generalized for the case of an arbitrary number of FTAs. The formulae of computation of the number of FTAs depending on the number of constants has been derived. The results have been presented in the form of theorems with their proofs. It was shown that the particular solutions with incomplete number of integration constants are crucial in the synthesis of extremal maneuver trajectories. It is concluded that *the possibility of application of a general solution for TAs with a complete number of integration constants is limited, and therefore such a solution has only a theoretical value.*

8.3 MAIN RESULT AND GENERAL CONCLUSION OF THIS STUDY

8.3.1 Main Result of This Study

On the basis of the method of application of the analytical solutions in the synthesis of maneuver trajectories analyzed above, and the functional relationships for the number of thrust arcs, one can conclude that ***the solution of the optimal control problem for trajectory synthesis can be reduced to the solution of a system of algebraic continuity equations formed for the junction points (that is the equations obtained from the continuity conditions at the junction points).***

Note however, that although this system of equations has equal number of unknowns, these equations must be solvable for these unknowns, which allows for a unique synthesis of the trajectory. The solutions of the continuity equations and the synthesis of the maneuver trajectory are the last stages of the methodology of determination of extremal trajectories.

8.3.2 General Conclusion of This Study

Following the methodology of determination of the extremal trajectories, we conclude that, ***in general, the solution of the optimal control problem for***

trajectory synthesis can be reduced to a closed analytical form thereby allowing for an analytical synthesis of the desired trajectory, and the existence of the solutions depends on the solvability of the system of algebraic continuity equations formed for the junction points.

CHAPTER 9

Some Problems of Trajectory Synthesis in the Newtonian Field

9.1 TRANSFER BETWEEN ELLIPTICAL ORBITS VIA AN INTERMEDIATE THRUST ARC WITH CONSTANT SPECIFIC IMPULSE

Consider the problem of transfer between elliptical orbits which are given by (p_1, e_1, ω_1) and (p_2, e_2, ω_2). The maneuver time is considered free [40]. It is also assumed that the spacecraft initial and final positions are not fixed. The analytical solutions for the radius vector and velocity vector obtained for thrust arcs with constant specific impulse and variable power have two integration constants: n and θ_0. Here Eq. (8.10) is not applicable as the total number of unknowns is less than the number of equations. This case corresponds the particular case described in subsection 7.1.2. Consequently, this case with insufficient number of constants imposes a constraint in the application of the solutions. It will be shown below that this constraint may mean that not all parameters of the final orbit are free. This indicates that the total number of thrust arcs should be equal to one. This means that the maneuver trajectory consists of two zero-thrust arcs connected with one finite thrust arc, that is there are two junction points. These points are determined by the continuity conditions for the primer vector and its derivative. According to Eq. (2.102), at these points $\mathbf{r}$, $\mathbf{v}$, and $\boldsymbol{\lambda}$ must be continuous [3].

The values of the variables at the junction points will be denoted by the order number of these points. So, the continuity conditions at the first junction point are [149]

$$\frac{p_1}{1 + e_1 \cos f_1} = \frac{\mu}{n^2} L_1(s_1), \tag{9.1}$$

$$\frac{\mu}{p_1} e_1 \sin f_1 = n L_2(s_1), \tag{9.2}$$

$$\frac{\mu}{p_1}(1 + e_1 \cos f_1) = n L_3(s_1), \tag{9.3}$$

$$\theta_1 = f_1 + \omega_1 = \theta_0 - 3\cot\varphi_1 - 4\varphi_1, \tag{9.4}$$

Analytical Solutions for Extremal Space Trajectories
DOI: http://dx.doi.org/10.1016/B978-0-12-814058-1.00009-6

where

$$s_1 = \sin\varphi_1, \quad k_1 = \cos\varphi_1.$$

At the second junction point these conditions are:

$$\frac{\mu}{n^2}L_1(s_2) = \frac{p_2}{1+e_2\cos f_2}, \tag{9.5}$$

$$nL_2(s_2) = \frac{\mu}{p_2}e_2\sin f_2, \tag{9.6}$$

$$nL_3(s_2) = \frac{\mu}{p_2}(1+e_2\cos f_2), \tag{9.7}$$

$$\theta_2 = \theta_0 - 3\cot\varphi_2 - 4\varphi_2 = f_2 + \omega_2, \tag{9.8}$$

where

$$s_2 = \sin\varphi_2, \qquad k_2 = \cos\varphi_2.$$

8 conditions in Eqs. (9.1)–(9.8) allows us to find 6 unknowns $n, \varphi_1, \varphi_2, f_1, f_2$, and θ_0 depending on the elements of the initial and final orbits. As the number of unknowns is less than the number of equations, in total two elements of the initial and final orbits must satisfy additional conditions which are derived from Eqs. (9.1)–(9.8). It will be shown below that these two elements can be represented by p_2 and ω_2. Excluding $\frac{\mu}{p_2}$ from Eqs. (9.5) and (9.6), we obtain

$$4n^2\frac{k_2^2}{s_2^4}\frac{(1-2s_2^2)^2}{(3-5s_2^2)^2} = n^2\frac{(1-3s_2^2)}{9s_2^6(1+e_2\cos f_2)}e_2^2\sin^2 f_2$$

or

$$\frac{e_2^2\sin^2 f_2}{1+e_2\cos f_2} = \frac{36k_2^2s_2^2(1-2s_2^2)^2}{(1-3s_2^2)(3-5s_2^2)^2}, \tag{9.9}$$

where

$$\frac{\mu}{p_2} = n^2\frac{(1-3s_2^2)}{9s_2^6(1+e_2\cos f_2)}. \tag{9.10}$$

Assuming that $1-3s_2^2 > 0$, from Eqs. (9.6) and (9.7) it follows that

$$\frac{e_2\sin f_2}{1+e_2\cos f_2} = \frac{6k_2s_2(1-2s_2^2)}{3-13s_2^2+12s_2^4}. \tag{9.11}$$

Eqs. (9.9) and (9.11) can serve to determine φ_2 and f_2 depending on eccentricity of the final orbit, e_2. From Eqs. (9.9) and the square of Eq. (9.11), after some simplifications, one can obtain

$$e_2 \cos f_2 = \frac{(3 - 13s_2^2 + 12s_2^4)^2}{(1 - 3s_2^2)(3 - 5s_2^2)^2} - 1. \tag{9.12}$$

Besides that, Eq. (9.9) can be rewritten as

$$e_2 \sin f_2 \frac{e \sin f_2}{1 + e \cos f_2} = \frac{36k_2^2 s_2^2 (1 - 2s_2^2)^2}{(1 - 3s_2^2)(3 - 5s_2^2)^2}. \tag{9.13}$$

After substituting the right hand side of Eq. (9.11) into Eq. (9.13), the latter is represented in the form:

$$e_2 \sin f_2 \frac{6k_2 s_2 (1 - 2s_2^2)}{3 - 13s_2^2 + 12s_2^4} = \frac{36k_2^2 s_2^2 (1 - 2s_2^2)^2}{(1 - 3s_2^2)(3 - 5s_2^2)^2}$$

or

$$e_2 \sin f_2 = \frac{6k_2 s_2 (3 - 13s_2^2 + 12s_2^4)(1 - 2s_2^2)}{(1 - 3s_2^2)(3 - 5s_2^2)^2}. \tag{9.14}$$

Let us introduce the following variables:

$$\begin{aligned} q_1(\varphi_2) &= 3 - 13s_2^2 + 12s_2^4, \\ q_2(\varphi_2) &= 1 - 3s_2^2, \\ q_3(\varphi_2) &= 1 - 2s_2^2, \\ q_4(\varphi_2) &= 3 - 5s_2^2. \end{aligned}$$

Then Eqs. (9.12) and (9.14) can be rewritten as

$$e_2 \cos f_2 = \frac{q_1^2 - q_2 q_4^2}{q_2 q_4^2}, \tag{9.15}$$

$$e \sin f_2 = \frac{6k_2 s_2 q_1 q_3}{q_2 q_4^2}. \tag{9.16}$$

Exclusion of f_2 from these expressions yields the function $\varphi_2 = \varphi_2(e_2)$ that can be found from the equation:

$$(q_1^2 - q_2 q_4^2)^2 + 36k_2^2 s_2^2 q_1^2 q_3^2 - e_2^2 q_2^2 q_4^4 = 0. \tag{9.17}$$

In the same manner, one can find the function $\varphi_1 = \varphi_1(e_1)$. Consequently, initial and final thrust angles, φ_1 and φ_2 are the functions of only eccentricities of the boundary orbits [149]. In other words, the ellipticity (form) of the boundary orbits determines the thrust angle at the junction points located in these orbits. This functional relationship is illustrated in Fig. 4.13. The true anomaly, f_2 is easily found from Eqs. (9.15) and (9.16) in the form:

$$f_2 = f_2(e_2) = \tan^{-1}\left[\frac{k_2 s_2 q_1 q_3}{q_1^2 - q_2 q_4^2}\right]. \tag{9.18}$$

The constant, n can be determined from Eqs. (9.1)–(9.3) as a function of p_1 and e_1 in the form:

$$n = n(p_1, e_1) = \sqrt{\frac{p_1}{9\mu}\frac{q_2^2 q_4^2}{s_2^6 q_1^2}}. \tag{9.19}$$

It follows from this expression that the ellipticity and size of the initial orbit determine the constant n. Now, Eq. (9.5) allows us to determine p_2 in the form:

$$p_2 = p_2(p_1, e_1, e_2) = 9\frac{\mu}{n^2}\frac{s_2^6(1 + e_2\cos f_2)}{1 - 3s_2^2}. \tag{9.20}$$

From this equality it can be easily seen that p_2 of the final orbit is not arbitrary. The initial and final polar angles, θ_1 and θ_2, and the constant θ_0 can be determined from Eqs. (9.4)–(9.8) in the form (see Fig. 4.15):

$$\begin{aligned} \theta_1 &= f_1 + \omega_1, \\ \theta_0 &= \theta_1 + 3\cot\varphi_1 + 4\varphi_1, \\ \theta_2 &= \theta_0 - 3\cot\varphi_2 - 4\varphi_2. \end{aligned} \tag{9.21}$$

As mentioned above, the number of unknowns in the continuity equations (9.1)–(9.8) is less than the number of these equations, and this leads to the limitation of applicability of the analytical solutions for the thrust arc. The satisfaction of the equalities in Eq. (9.8) leads to the limitation of the selection of the final orbit's orientation, that is the angular distance of the orbit's perigee ω_2 on the basis of the equality:

$$\omega_2 = \theta_2 - f_2. \tag{9.22}$$

So, Eqs. (9.20) and (9.22) impose constraints on the selection of the final orbit. Only eccentricity of this orbit can be arbitrary. Note that only two elements among (p_1, e_1, ω_1) and (p_2, e_2, ω_2) must satisfy two certain conditions. These conditions can be obtained by solving Eqs. (9.1)–(9.8). In the problem considered, these elements are p_2 and ω_2, and the two conditions are Eqs. (9.20) and (9.22).

Note that, the usual mission analysis and computation of the mission parameters, including the flight equations are integrated in Cartesian coordinates, in which the center of the coordinate system coincides with the center of attraction. The magnitude of radius vector, components of the velocity vector, primer vector and its derivative can be transformed into a Cartesian coordinate system using the following equalities: (see Fig. 4.15):

$$\begin{aligned} x &= r\cos\theta, \\ y &= r\sin\theta, \\ v_x &= \dot{r}\cos\theta - r\dot{\theta}\sin\theta, \\ v_y &= \dot{r}\sin\theta + r\dot{\theta}\cos\theta, \end{aligned} \tag{9.23}$$

and

$$\begin{aligned} \Lambda_1 &= \lambda_1\cos\theta - \lambda_2\sin\theta, \\ \Lambda_2 &= \lambda_1\sin\theta + \lambda_2\cos\theta, \\ \Lambda_4 &= -\lambda_4\cos\theta + (\lambda_1\frac{v_2}{r} - \lambda_2\frac{v_1}{r} + \frac{\lambda_5}{r})\sin\theta, \\ \Lambda_5 &= -\lambda_4\sin\theta - (\lambda_1\frac{v_2}{r} - \lambda_2\frac{v_1}{r} + \frac{\lambda_5}{r})\cos\theta. \end{aligned} \tag{9.24}$$

Below are the equations of motion and costate equations (canonical equations) of the problem:

$$\begin{aligned} \dot{x} &= v_x, \\ \dot{y} &= v_y, \\ \dot{v}_x &= \frac{c\beta}{m}\frac{\Lambda_1}{\Lambda} - \frac{\mu}{r^3}x, \\ \dot{v}_y &= \frac{c\beta}{m}\frac{\Lambda_2}{\Lambda} - \frac{\mu}{r^3}y, \\ \dot{m} &= -\beta, \end{aligned} \tag{9.25}$$

and

$$\begin{aligned} \dot{\Lambda}_1 &= -\Lambda_4, \\ \dot{\Lambda}_2 &= -\Lambda_5, \\ \dot{\Lambda}_4 &= \frac{\mu}{r^3}\Lambda_1 - 3\frac{\mu}{r^5}x(x\Lambda_1 + y\Lambda_2), \end{aligned} \tag{9.26}$$

$$\dot{\Lambda}_5 = \frac{\mu}{r^3}\Lambda_2 - 3\frac{\mu}{r^5}y(x\Lambda_1 + y\Lambda_2),$$
$$\dot{\Lambda}_7 = \frac{c\beta}{m^2}\Lambda,$$

where

$$v_1 = \dot{r}, \quad v_2 = r\dot{\theta}, \qquad \Lambda = \sqrt{\Lambda_1^2 + \Lambda_2^2}$$
$$\boldsymbol{\lambda}_r = -\dot{\boldsymbol{\lambda}} = \left(\lambda_4,\ \lambda_1\frac{v_2}{r} - \lambda_2\frac{v_1}{r} + \frac{\lambda_5}{r}\right).$$

Note that for the numerical integration, the control parameter is the thrust angle, obtained from the analytical solutions. This angle is used as the independent variable in the analytical solutions considered above. The components of $\boldsymbol{\lambda}$ at the junction points are computed according to formulas [26]:

$$\lambda_{1i} = \lambda \sin\varphi_i,$$
$$\lambda_{2i} = \lambda \cos\varphi_i, \quad i = 1, 2$$

As is known, the primer vector for elliptical orbit with eccentricity e is determined as follows:

$$\lambda_1 = A\cos f + Be\sin f,$$
$$\lambda_2 = -A\sin f + B(1 + e\cos f) + \frac{D - A\sin f}{1 + e\cos f},$$

where A, B, and D are constants, and $C = 0$. It can be shown that if $(C = 0)$, then $A = 0$ [3]. If $\lambda_1 = \lambda\sin\varphi$, $\lambda_2 = \lambda\cos\varphi$, then in the problem considered, the continuity conditions for primer vector allow us to find the constants (Lawden's constants) in the form [3], [26]:

$$B_i = \frac{\lambda_{1i}}{e_i \sin f_i},$$
$$D_i = (\lambda_{2i} - B_i(1 + e_i\cos f_i))(1 + e_i\cos f_i).$$

It follows from this that the solution of the variational problem is reduced to the solution of some number of continuity equations at the junction points.

Note that by adding the errors in the right sides of Eqs. (9.1)–(9.8), one can construct the main equations to analyze the guidance scheme. Based on the above described, the spacecraft can be transferred from the thrust arc extremal into given elliptical orbit or vice versa. This means that the corresponding class of extremals can be used to escape from an elliptical orbit. Besides that, the form of the analytical solutions allows us to separate kinematical analysis of the problem parameters from their dynamical analysis, and also to separately consider the behavior of the primer vector.

9.1.1 Numerical Example

Consider the transfer between elliptical orbits given by the following parameters: $e_1 = 0.1,\ p_1 = 9900$ km, $\omega_1 = 1.0$ rad. For the elliptical orbit, $e_2 = 0.12$, and the other parameters of the final orbit are subject to determination from the continuity conditions constructed at the junction points. This is because the number of the unknowns is less that the number of equations. Also, we have $I_{sp} = const = 1000$ s. Assuming that the transfer is started at $t_1 = 0$ s. The numerical results show that the maneuver ends at $t_2 = 39{,}045.23$ s, (or $t_{2d} = 0.4519$ days), also $p_2 = 2.9547554$ km, $\omega_2 = 1.5848$ rad. The thrust angle, φ is increased from 0.05 to 0.06 (rad) with respect to horizon. The thrust direction is changed in the area bounded by horizon and velocity directions (see Fig. 9.1). From Figs. 9.1 and 9.2 it can be seen that the thrust vector magnitude is decreased, the thrust angle is increased at all points of the thrust arc. The behaviors of the magnitude and direction of the thrust vectors are illustrated in Fig. 9.1, where the proportionality coefficient is 10^8. It can also be shown that the constants of the problem are sensitive to the initial conditions and can be computed after the computations of the primer vector. The initial Lagrange multipliers, including the primer vector, its derivative, the multiplier for mass at the initial and final junction points are given as follows:

Junction points	λ_x	λ_y	λ_{rx}	λ_{ry}	λ_m
First	-1	-8.364×10^{-3}	-5.294×10^{-6}	6.33×10^{-4}	1.02
Second	0.856	0.5168	6.308×10^{-5}	-1.045×10^{-4}	101.94

The analytical solutions of the problem, obtained above have been compared to the results of the numerical integration of the canonical equations using the 4th order Runge-Kutta's method. The equations of the optimal motion and costate equations for the Lagrange multipliers have been integrated in the polar coordinate system with the origin at the center of

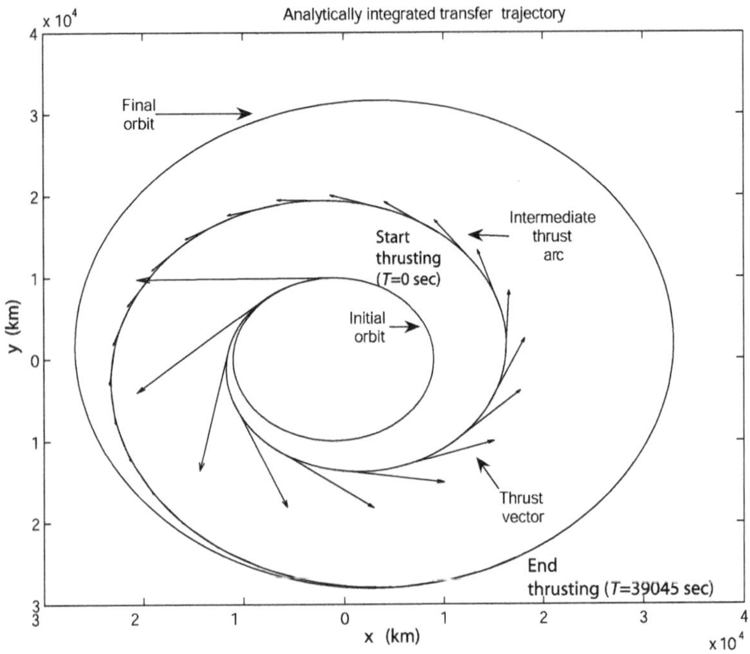

Figure 9.1 Transfer between elliptical orbits via intermediate thrust arc with constant specific impulse.

attraction using the thrust angle behavior obtained from the analytical solutions. Note that this angle is the independent variable of the analytical solutions, and the time is the independent variable of the numerical integration. The computations have shown that the results of the analytical and numerical integrations coincide for any given time during the maneuver. This means that the analytical solutions for the thrust arcs can be used as the nominal trajectory solutions in the guidance problem. In particular, the analytical solutions provide exact values of the initial Lagrange multipliers, which can be used in the numerical integration, and allow us to decrease the number of iterations and the time to produce the guidance commands by an on-board flight software.

It can be concluded that the transfer between the elliptical orbits can be accomplished via intermediate thrust arc. This maneuver is possible when two parameters of the final orbit are specified. From the general theory of optimization it follows that the thrust arc can not start from and can not end at a circular orbit.

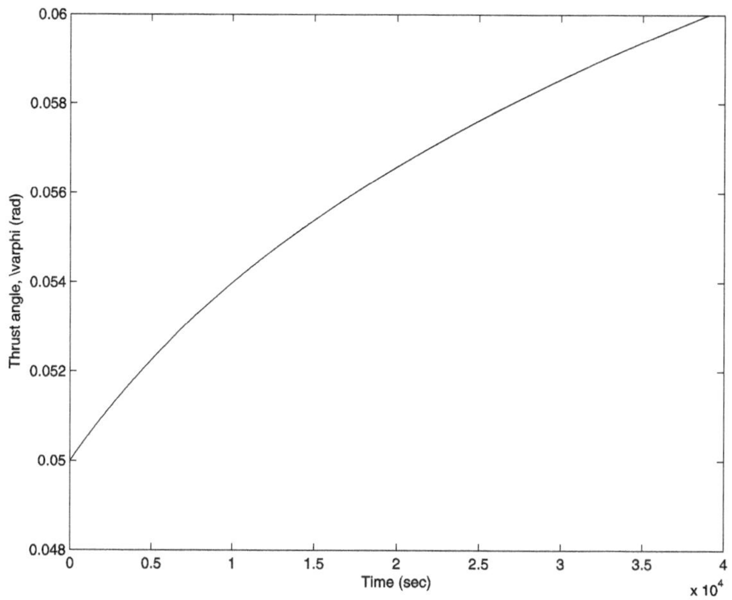

Figure 9.2 Thrust angle vs. time.

9.2 TRANSFER FROM GIVEN POSITION INTO ELLIPTICAL ORBIT VIA AN INTERMEDIATE THRUST ARC WITH VARIABLE SPECIFIC IMPULSE

Consider this problem with assumption that the maneuver is started when the spacecraft enters the Earth sphere of influence (SOI) with radius $r_{si} = 9.24820 \times 10^5$ km and terminated when the spacecraft is in the given elliptical parking orbit [138], [144]. The perigee and apogee of this orbit and the final mass are assumed given. The initial true anomaly in the parking orbit and the orientation of this orbit are subject to determination. Only one thrust arc is assumed to accomplish the maneuver. The initial and final points of the thrust arc are accepted as the junction points with order numbers 1 and 2 respectively. An intermediate thrust (IT) arc with variable specific impulse (VSI) is considered as the thrust arc [166]. The continuity conditions at these junction points are

$$r_1 = \frac{\mu}{d^2} F_1(s_1), \tag{9.27}$$

$$v_{11} = dF_2(s_1), \tag{9.28}$$

$$v_{21} = dF_3(s_1), \tag{9.29}$$

$$\theta_0 = 0 \tag{9.30}$$

and

$$\frac{\mu}{d^2}F_1(s_2) = \frac{p}{1 + e\cos f_2} = r_2, \tag{9.31}$$

$$dF_2(s_2) = \sqrt{\frac{\mu}{p}}e\sin f_2 = v_{12}, \tag{9.32}$$

$$dF_3(s_2) = \sqrt{\frac{\mu}{p}}(1 + e\cos f_2) = v_{22}, \tag{9.33}$$

$$\theta_2 = f_2 + \omega = \frac{1}{4}(\frac{3}{\tan\varphi_1} + \varphi_1) - \frac{1}{4}(\frac{3}{\tan\varphi_2} + \varphi_2) + \theta_1, \tag{9.34}$$

where

$$s_1 = \sin^2\varphi_1, \quad s_2 = \sin^2\varphi_2.$$

The functions F_1, F_2, F_3 are determined in the previous chapters. Below it will be shown that $r_1, v_{11}, v_{21}, s_2, \alpha, f_2$ can be determined as functions of φ_1, p, and e. Then from Eqs. (9.32) and (9.33) one can find that

$$s_2 = 3\frac{6 + q^2 - 2\sqrt{9 - 6q^2}}{36 + q^2}, \tag{9.35}$$

$$\alpha(f_2) = 3\frac{d^4}{\mu} = \frac{\mu}{p}\frac{1 + e^2 + 2e\cos f_2}{(F_1^2(s_2) + F_2^2(s_2))^2}, \quad q = \frac{e\sin f_2}{1 + e\cos f_2}. \tag{9.36}$$

Substitution of these expressions into Eq. (9.31) yields

$$\frac{p}{1 + e\cos f_2} - \frac{3\mu}{\alpha(f_2)}F_1(s_2) = 0,$$

from which one can find the function $f_2 = f_2(p, e)$. This expression shows that the final position of the spacecraft depends only on the final orbit parameters [166]. The unknowns r_1, v_{11}, v_{21} are found from Eqs. (9.27)–(9.32). Substituting $f_2(p, e)$ into Eqs. (9.35) and (9.36), and using $s_2 = \sin^2\varphi_2$, we have

$$\varphi_2(p, e) = \arcsin\sqrt{3\frac{6 + q^2(p, e) - 2\sqrt{9 - 6q^2(p, e)}}{36 + q^2(p, e)}}, \tag{9.37}$$

$$\alpha(p, e) = \frac{3}{p}\frac{1 + e^2 + 2e\cos f_2(p, e)}{F_1^2(\varphi_2(p, e)) + F_2^2(\varphi_2(p, e))}. \tag{9.38}$$

In the case when $r_1 = r_0$ is given, the thrust angle, φ_1 is found from Eq. (9.27) in the form:

$$\varphi_1(p, e) = \arcsin\left[\frac{a^2}{\mu^{\frac{2}{3}}} r_0^{\frac{4}{3}}\right], \quad d = d(p, e) = \left[\frac{\mu\alpha(p, e)}{3}\right]^{\frac{1}{4}}.$$

Substituting φ_1 and d into Eqs. (9.28) and (9.29), it can be shown that

$$v_{11} = v_{11}(p, e, r_0) = d(p, e)F_1(p, e, r_0)), \tag{9.39}$$

$$v_{21} = v_{21}(p, e, r_0) = d(p, e)F_2(p, e, r_0)). \tag{9.40}$$

Then the final polar angle is found from Eqs. (9.34) in the form:

$$\theta_2(p, e, r_0) = \frac{1}{4}\left(\frac{3}{\tan\varphi_1} + \varphi_1\right) - \frac{1}{4}\left(\frac{3}{\tan\varphi_2} + \varphi_2\right) + \theta_1. \tag{9.41}$$

Number of revolutions on the spiral trajectory can be computed according to the formulae:

$$N_{rev} = \frac{\theta_2(p, e, r_0)}{2\pi}.$$

The orientation of the final orbit satisfies the equality $\omega = \theta_2(p, e, r_0) - f_2(p, e)$. Note that the constant d (or α) is considered as a function of p and e, is valid for Eqs. (9.27)–(9.33). If m_2, $P(= P_{max})$, and t_2 are given, then all constants, m_1, the initial and final values of the primer vector and I_{sp} are found as the functions of p, e, r_0, m_2, P, and t_2 in the form (see the related formulas provided in Section 3.3):

$$\frac{\lambda_v}{b} = \frac{2}{5P}\left(\frac{\xi}{t_2} - 3\alpha\right), \quad \frac{C}{\lambda_v} = \frac{P}{2}\frac{\lambda_v}{b} - \alpha, \quad \frac{C_1}{\lambda_v} = \frac{3k_1 - z_1(1 - 5s_1)}{3 - 5z_1},$$

$$m_1 = \left[\frac{1}{m_2} - \frac{P}{2}t_2\left(\frac{\lambda_v}{b}\right)^2\right]^{-1}, \quad I_{sp1} = \frac{2}{g_0 m_1}\frac{b}{\lambda_v}, \quad I_{sp2} = \frac{2}{g_0 m_2}\frac{b}{\lambda_v},$$

$$\beta 1 = \frac{2P}{I_{sp1}^2 g_0^2}, \quad \beta 2 = \frac{2P}{I_{sp2}^2 g_0^2}, \quad a_t = \frac{I_{sp1} g_0 \beta}{m_1} = constant, \quad J = \frac{1}{2}a_t^2 t_2,$$

where

$$\xi = \frac{1}{t_c}\left[\frac{3z_2k_2(1 - 5s_2)}{3 - 5s_2} - \frac{C_1}{\lambda_v}\right], \quad z_1 = \sqrt{\frac{\mu}{r_1}}(1 - 3s_1), \quad z_2 = \sqrt{\frac{\mu}{r_2}}(1 - 3s_2),$$

$$k_1 = \cos\varphi_1 \quad k_2 = \cos\varphi_2, \quad t_c = \frac{5}{2}P\frac{\lambda_v}{b} - 3\alpha, \quad \frac{\bar{C}}{\lambda_v} = -\alpha.$$

Based on the solution process described above, it can be noted that in essence, the solution of the variational problem is reduced to the solution of the continuity equations constructed at the junction points. The numerical results are given below.

9.2.1 Numerical Example

If p, e, and r_0 are given, then after finding α (or d), one can obtain $(r, \theta), (v_1, v_2), \varphi_1, \varphi_2$, and N_{rev}. The solutions show that these values are independent on P, I_{sp}, m, β, a_t. Here it is accepted that $r_{min} = 6870$ $(h_0 = 500)$, $r_{max} = 6880$ km, or $p = 6874$ km, $e = 0.00072$, and also and $\omega = 4.2502$ rad. Then it can be found that $f_2 = -1.5708$ rad, and $\alpha = 1.2165e^{-12}$. Furthermore, if $r_1 = r_0 = r_{si}$, then the initial thrust angle can easily be computed. The results are given in the table below:

Variables	r	θ	v_1	v_2	θ	φ
Initial	924,820.0	1.1735	−0.0125	0.6565	0.0	−0.0095
Final	6874.0	6.2832	−0.0055	7.6143	$2\pi N_{rev}$	−0.00036

Also, $\frac{C_1}{\lambda_v} = 0.65619$, $\frac{\bar{C}}{\lambda_v} = -1.21651 \times 10^{-12}$. By specifying t_2, m_2, P, one can compute the other parameters.

The following intervals have been considered: $30 \leq t_2 \leq 180$ days; $3 \leq P \leq 12$ kw; $200 \leq m_2 \leq 600$ kg. The apogee of the final orbit is changed in the interval: $6880 \leq r_{max} \leq 7370$ km. The values of all parameters for various values of apogee, r_{max} for the case when $t_2 = 120$ days, $m_2 = 200$ kg, $P = 3$ kw, $e_{eff} = 0.68$, $r_{min} = 6870$ km. Here eccentricity is changed from $e = 0.0007272$ to $e = 0.03511$. If the final orbit is more elliptical, then the maneuver becomes more effective. The number of revolutions satisfies the condition: $315.75 \leq N_{rev} \leq 6.54$, and the initial thrust angle is deviated from the tangential direction to the trajectory (see Figs. 9.3 and 9.4). The analysis show that the initial velocity and the number of revolutions are sensitive to the initial thrust angle, which is a function of initial distance (see Fig. 9.5). The thrust angle and the velocity are increased during the maneuver. The perigee and the apogee $(6880 \leq r_{max} \leq 7370$ km) and the initial distance determines the form of the spiral trajectory. 45 times increase in eccentricity yields the decrease of the number of revolutions from 315 to 4.5. The solutions allowed us to obtain the time histories of the variables for various values of the final mass and the power (see Figs. 9.6–9.8 for $P_{max} = P = 3$ kw and $m_1 = 200$ kg). These figures show almost a linear dependence between the flight time and specific impulse, which becomes

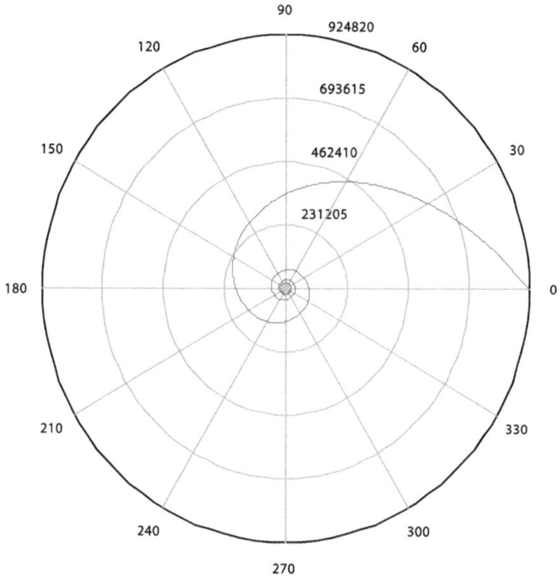

Figure 9.3 Example of maneuver trajectory: Second class of IT arcs.

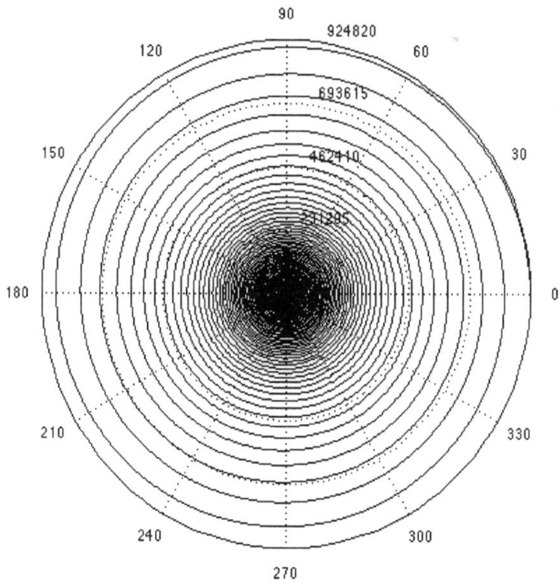

Figure 9.4 Maneuver trajectory.

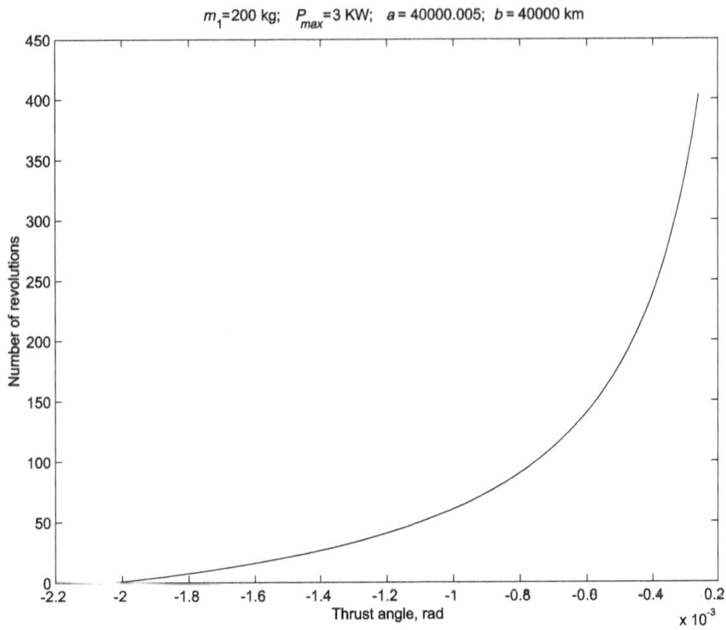

Figure 9.5 Dependence of the number of revolutions on thrust angle.

almost constant when the maneuver time is increased. The other characteristics of the maneuver are presented in Fig. 9.9, Fig. 9.10, and Fig. 9.11. Fig. 9.12 shows the behavior and sensitivity of the functional to changes in eccentricity and in the form of the final orbit which is changed, in particular, when the apogee is increased. The functional can take more lower values for higher eccentricities and greater number of revolutions around the center of attraction. Fig. 9.12 shows applicability of the solutions obtained depending on perigee, apogee and eccentricity, which satisfies the inequality $0 < e \leq 0.043$. Note that these solutions can not be used to transfer from a spiral trajectory to a circular orbit. However, any other orbits with non-zero eccentricity can be used for this purpose.

9.3 TRANSFER BETWEEN ELLIPTICAL ORBITS VIA AN INTERMEDIATE THRUST ARC WITH VARIABLE SPECIFIC IMPULSE

Consider the problem of free time transfer between two arbitrary elliptical orbits (p_1, e_1, ω_1) and (p_2, e_2, ω_2) [26]. The scenario of the transfer is

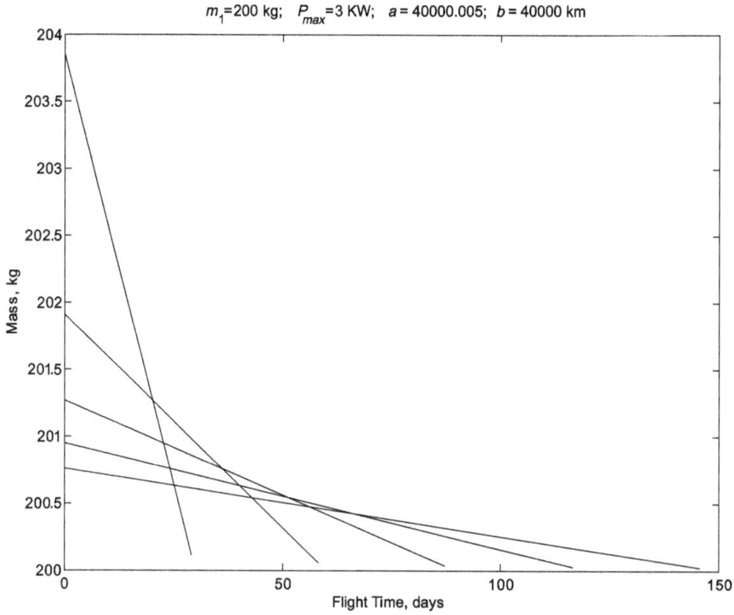

Figure 9.6 Mass vs. time.

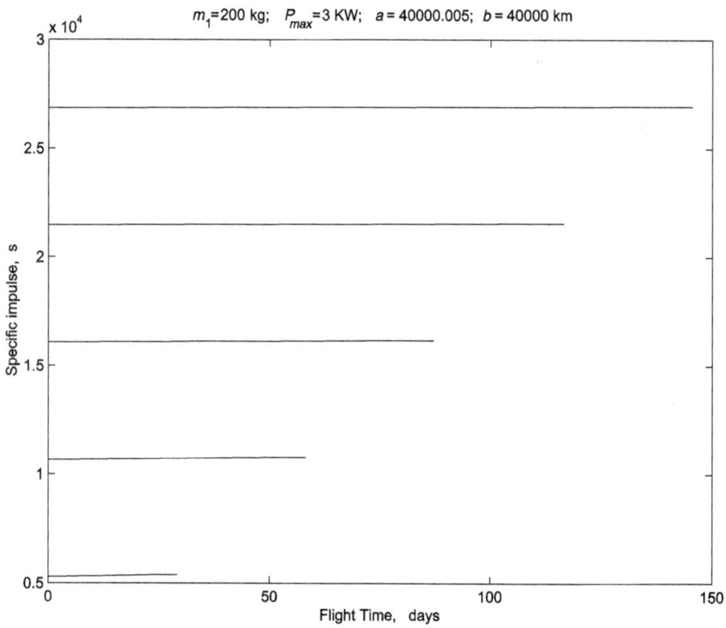

Figure 9.7 Specific impulse vs. time.

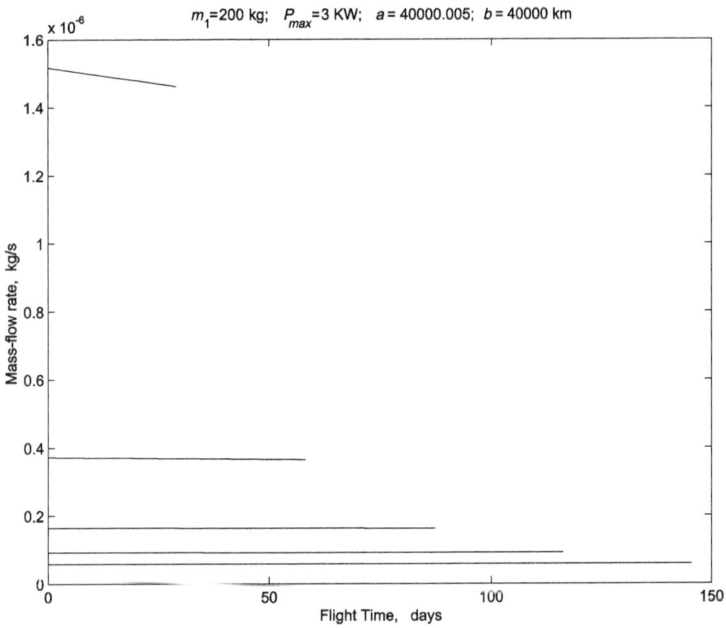

Figure 9.8 Mass-flow rate vs. time.

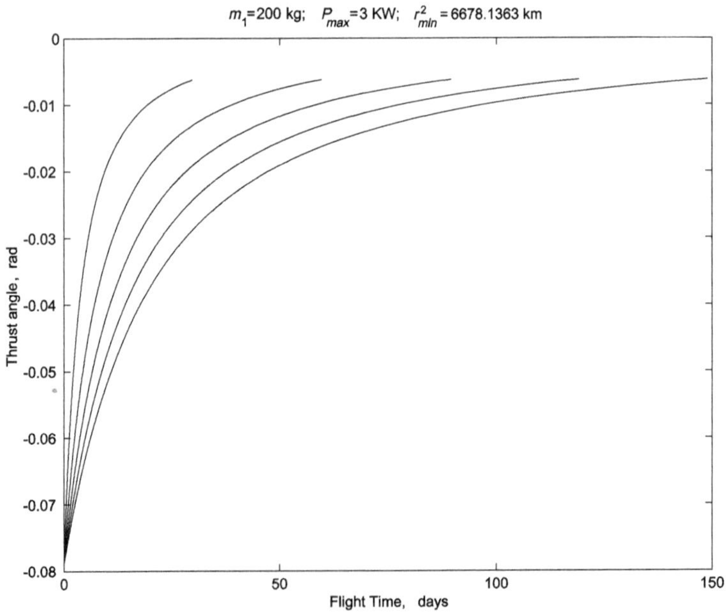

Figure 9.9 Thrust angle vs. time.

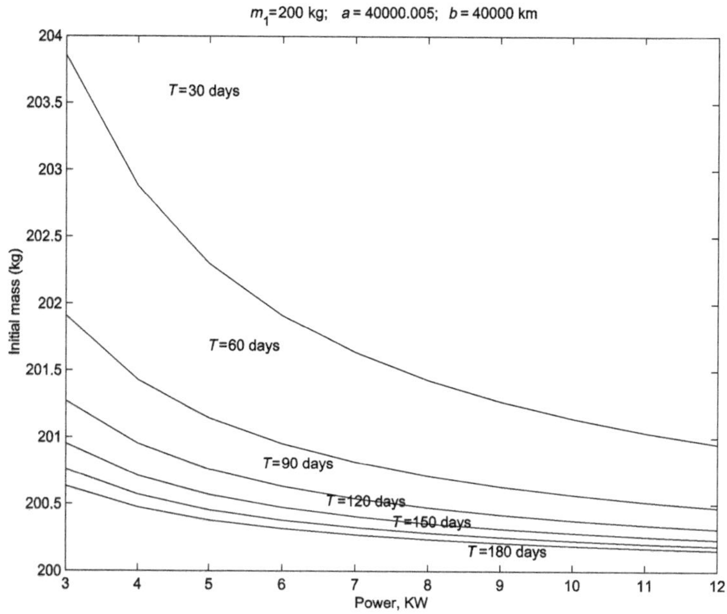

Figure 9.10 Dependence of the initial mass on power.

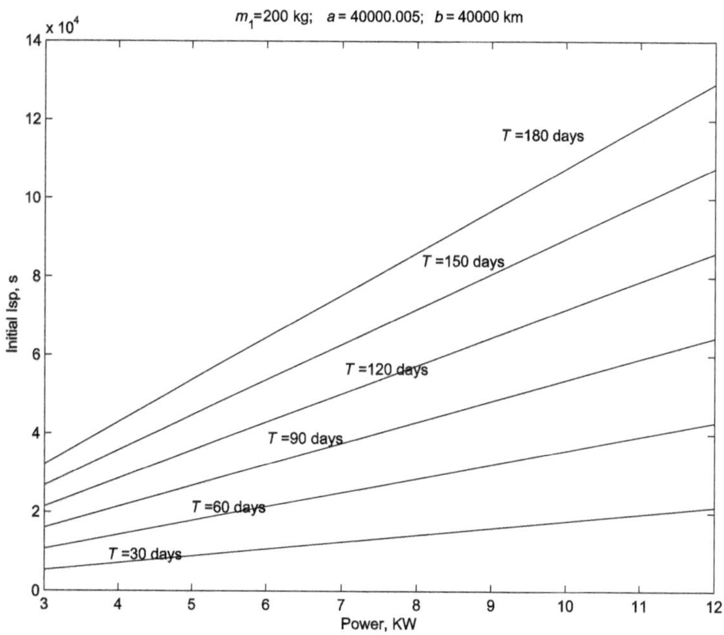

Figure 9.11 Dependence of the initial specific impulse on power.

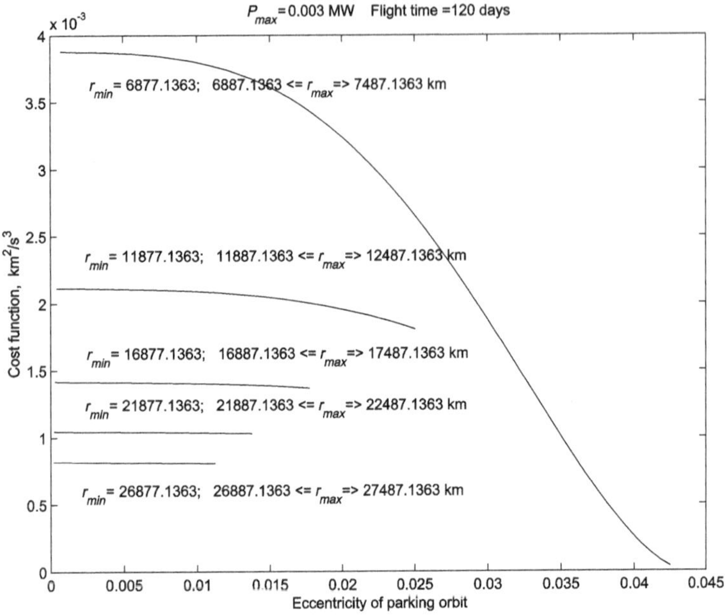

Figure 9.12 Dependence of the functional on eccentricity, perigee and apogee of the final orbit.

illustrated in Fig. 9.13. It is also assumed that the spacecraft initial and final positions are not specified. The analytical solutions, obtained for radius vector and velocity vector on IT (or low-thrust) arcs with power and variable specific impulse, have two constants: d, θ_0. Here Eq. (8.10) is not applicable as the total number of unknowns is less than the number of continuity equations, and this imposes a certain limitation on applicability of the analytical solutions. Based on the main results obtained in the previous section, one can conclude that only one thrust arc can be used in the maneuver. This means that the maneuver trajectory consists of two ZT arcs and one thrust arc, and therefore includes two junction points. According Eq. (2.102), at the junction points $\mathbf{r}$, $\mathbf{v}$, and $\boldsymbol{\lambda}$ must be continuous [3]. Using Eqs. (3.69)–(3.72) and the formulas for elliptical motion, one can obtain the following continuity conditions at these points:

$$\frac{\mu}{p_1}(1+e_1\cos f_1)=\frac{x_{11}}{k^{\frac{1}{4}}},$$
$$\frac{\mu}{p_1}e_1\sin f_1=\frac{x_{21}}{k^{\frac{1}{4}}}, \tag{9.42}$$

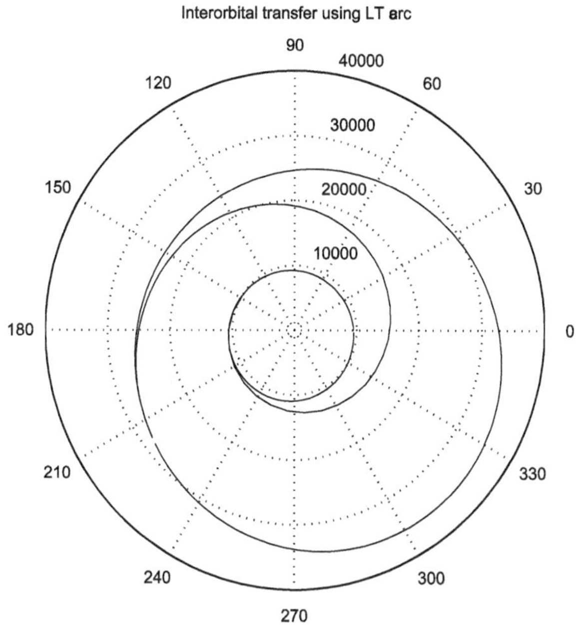

Figure 9.13 Transfer between elliptical orbits.

$$\frac{p_1}{1+e_1\cos f_1}=k^{\frac{1}{2}}x_{31},$$
$$\theta_1=f_1+\omega_1$$

and

$$\begin{aligned}\frac{\mu}{p_2}(1+e_2\cos f_2)&=\frac{x_{12}}{k^{\frac{1}{4}}},\\ \frac{\mu}{p_2}e_2\sin f_2&=\frac{x_{22}}{k^{\frac{1}{4}}},\\ \frac{p_2}{1+e_2\cos f_2}&=k^{\frac{1}{2}}x_{32},\\ \frac{1}{4}\left(\frac{3}{\tan\varphi_1}+\varphi_1\right)-\frac{1}{4}\left(\frac{3}{\tan\varphi_2}+\varphi_2\right)+\theta_0&=f_2+\omega_2,\end{aligned}\tag{9.43}$$

where k is constant, and

$$x_{1i}=\frac{3-z_i}{5z_i-3}d_i,\quad x_{2i}=\frac{6\sqrt{z_i-z_i^2}}{5z_i-3}d_i,\quad x_{3i}=\sqrt{\frac{6\mu}{\bar{P}}z_i\sqrt{z_i}},$$

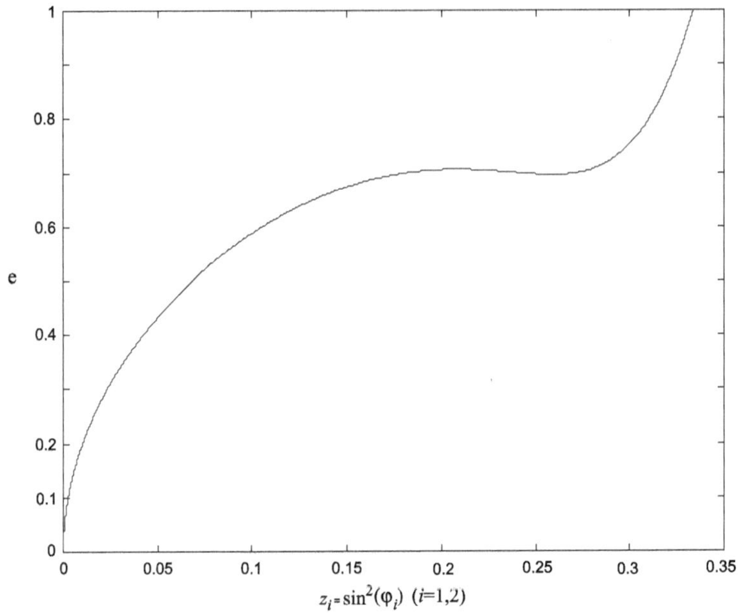

Figure 9.14 Dependence of φ from eccentricity of the final orbit.

and

$$d_i = -\sqrt{\sqrt{\frac{\bar{P}\mu}{6}\frac{1-3z_i}{z_i\sqrt{z_i}}}}, \qquad z_i = \sin^2\varphi_i, \qquad (i = 1, 2).$$

8 equations of Eqs. (9.42) and (9.43) allows us to find 7 unknowns $\theta_1, \varphi_1, f_1, \varphi_2, f_2, k, \theta_0$ and the corresponding solutions must satisfy one additional equation. From Eq. (9.42) and (9.43) it follows that

$$\frac{x_{1i}^2 x_{3i}^2 (x_{1i}^2 + x_{2i}^2)}{\mu^2} - \frac{2x_{1i}^2 x_{3i}}{\mu} + 1 - e_i^2 = 0, \qquad i = 1, 2,$$

from which we can find φ_1 and φ_2. Fig. 9.14 shows the functional relationship between φ and e. It can be seen that the transfer can be accomplished without violating the condition $\sin^2\varphi < 1/3$. From Eqs. (9.42) and (9.43) it follows that

$$\tan f_i = \frac{x_{1i}x_{2i}x_{3i}}{x_{1i}^2 x_{3i} - \mu}, \qquad i = 1, 2,$$

which can serve to find f_1 and f_2. Then

$$\sqrt{k} = \frac{p_1 \mu}{x_{11}^2 x_{31}^2}, \quad \frac{p_1}{p_2} = \frac{x_{11}^2 x_{31}^2}{x_{12}^2 x_{32}^2}. \tag{9.44}$$

This means that the initial elliptical orbit can be given with e_1 and p_1, and the final orbit is given with e_2, but p_2 must satisfy Eq. (9.44). As $k = b/\lambda$, then the initial specific impulse is computed according to Eq. (3.74)

$$I_{sp_0} = \frac{2k}{g m_0}.$$

The constants of integration are

$$\frac{C_1}{\lambda} = \frac{3d_1(1-5z_1)(1-z_1)^{\frac{1}{2}}}{k^{\frac{1}{4}}(3-5z_1)}, \quad \frac{C_3}{\lambda} = -\frac{\bar{P}}{2k}, \quad t_c = \frac{4\bar{P}}{k}.$$

Consequently, the final time and final mass can be computed using the formulas:

$$t_2 = \frac{1}{t_c}\left[\frac{3d_2(1-5z_2)(1-z_2)^{\frac{1}{2}}}{k^{\frac{1}{4}}(3-5z_2)} - \frac{C_1}{\lambda}\right],$$

and

$$\frac{1}{m_2} = \frac{1}{m_0} + \mathcal{J} = \frac{1}{m_0} + \frac{\bar{P}}{2k^2} t_2.$$

The transversality condition in Eq. (2.101) yields

$$\lambda_{72} = -\frac{\partial \mathcal{J}}{\partial m_2} = \frac{1}{m_2^2}.$$

Then from Eq. (3.75) we have $b = \lambda_{72} m_2^2 = 1$. Therefore, from Eq. (3.75) it follows that $\lambda = \frac{2}{I_{sp_0} g m_0} = \frac{1}{k}$. The components of λ at the junction points are [26]:

$$\lambda_{1i} = \lambda \sin \varphi_i, \qquad \lambda_{2i} = \lambda \cos \varphi_i, \quad i = 1, 2.$$

The continuity conditions allow us to find Lawden's constants: [26]:

$$B_i = \frac{\lambda_{1i}}{e_i \sin f_i}, \qquad D_i = (\lambda_{2i} - B_i(1 + e_i \cos f_i))(1 + e_i \cos f_i).$$

Consequently, the solution of the variational problem has been reduced to the solution of the continuity equations at the junction points.

9.3.1 Numerical Example

Consider the transfer between elliptical orbits given by the parameters:

Orbits	p (km)	e	ω (rad)
Initial	10,000	0.1	1.0
Final	28,526.2	0.2	2.25

Also, $m_0 = 52{,}500$ kg, $I_{sp_{min}} = 1000$ s, $I_{sp_{max}} = 35,000$ s, $\bar{P} = 10$ kw, $\epsilon = 0.6$ [21], [38]. Assuming that the transfer is started at $t_1 = 0$, one can find that this maneuver ends at $t_2 = 223{,}627.68$ s (or $t_{2d} = 2.588$ days). The characteristics of the maneuver are given in the table:

Junction points	r (km)	θ (rad)	φ (rad)	v_1 (km/s)	v_2 (km/s)	I_{sp} (s)	m (kg)
First	9991.54	2.58	0.05	0.632	6.313	7752.04	52,500
Second	8425.16	10.09	0.101	0.75	3.738	7820.56	52,040

The total velocity is decreased from $v_{1T} = 6.3450$ km/s to $v_{2T} = 3.8126$ km/s, whereas I_{sp} is increased to 68.52 s. Also, $a = 3.005 \times 10^{-6}$ km^2/s^2, $\mathcal{J} = 1.003 \times 10^{-6}$ km^2/c^3. The fuel expenditure is 0.88% of the initial mass, and the thrust angle is increased from 2.87 to 5.76 degrees with respect to horizon. Using Eqs. (4.81) one can find that the angle between the velocity and horizon is increased from 5.72 to 11.28 degrees (see Fig. 9.15). The constants are very sensitive to the initial conditions and can be computed after the computation of the primer vector. Consequently, one can conclude that the transfer via one thrust arc is possible when the parameter of the final orbit is specified.

9.4 TURNING ELLIPTICAL ORBIT'S PLANE VIA AN INTERMEDIATE THRUST ARC WITH CONSTANT SPECIFIC IMPULSE

Consider the problem of turning elliptical orbit's plane in the central Newtonian field. It is known that this problem can be solved by using one, two or three impulses [139], [88], [167]. Theory of particular solutions has also been used to solve this problem [62], [139]. It was shown that energetically, the use of an IT arc coincides a one-impulsive maneuver, and gives fuel savings compared to three-impulsive maneuver. However, in the studies shown above, it has been possible to obtain solutions only for a limited range of eccentricities. Transfers using arbitrary elliptical orbits are of

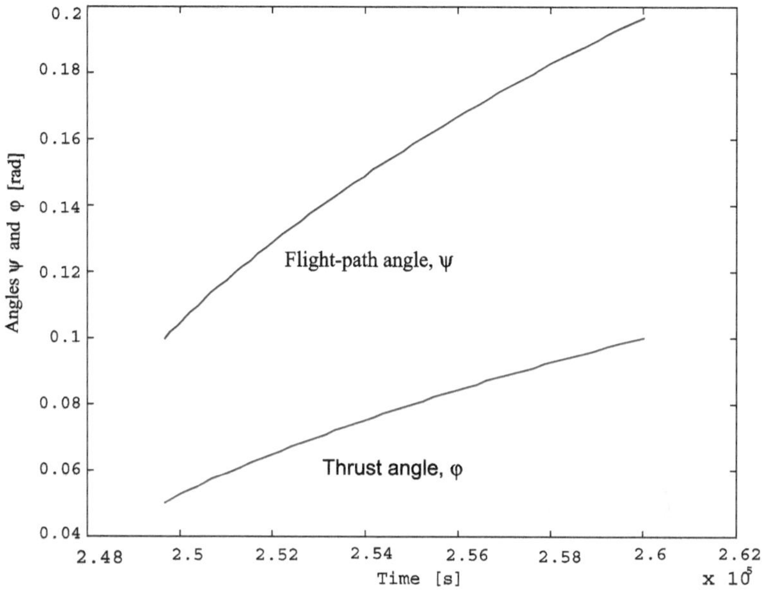

Figure 9.15 Behavior of the thrust angle and angle between the velocity and horizon.

interest. Here it will be shown that the spherical IT arcs can be used to turn the plane of an elliptical orbit with arbitrary eccentricity [65], [148]. The components of the primer vector are determined based on the initial conditions. Consider the problem of turning the plane of an elliptical orbit with eccentricity e by angle α_H with respect to line LH, perpendicular to line of nodes Q_1Q_2 (Fig. 9.16). Let the line of apsides, A_1P_1 of the initial orbit has angle ψ with respect to LH, and the initial and final conditions are given in the form:
At $t=0$:

$$\nu_1=0, \quad \nu_2=\nu_{21}, \quad \nu_3=\nu_{31}, \quad r=r_0, \quad \varphi=\varphi_1,$$
$$\theta=\theta_1, \quad M=M_0, \quad i=i_1, \quad \Omega=\Omega_1. \tag{9.45}$$

At $t=T$:

$$\nu_1=0, \quad \nu_2=\nu_{22}, \quad \nu_3=\nu_{32}, \quad r=r_0, \quad \varphi=\varphi_2,$$
$$\theta=\theta_2, \quad i=i_2, \quad \Omega=\Omega_2. \tag{9.46}$$

According to Fig. 9.16, consider the spherical coordinate system $Or\varphi\theta$. The plane, in which the angle φ is measured, is accepted as a main plane.

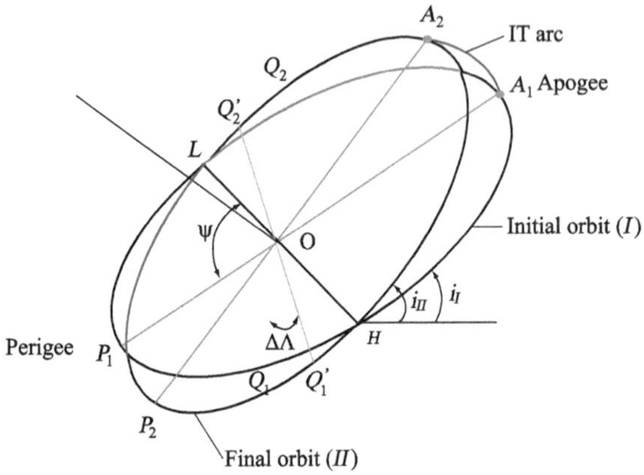

Figure 9.16 Scheme of turning of the orbital plane via spherical IT arc.

Then the position of any orbit with respect to this plane is determined by the angle i and the angular distance, Ω until $Q_1 Q_2$ from some constant orientation. As mentioned earlier, the plane turning can be done, in particular, by one-impulsive maneuver. If the initial and final orbits are denoted by I and II respectively, then this impulse is applied at the point H of orbit I in the orthogonal direction. The magnitude of this impulse is determined by the formulae [167]:

$$\Delta v_H = 2 v_H \sin\frac{\alpha_H}{2}, \tag{9.47}$$

where v_H is the transversal component of the velocity at this point of the orbital plane I, and α_H is the turning angle. As the true anomaly f_H of the point H is equal $f_H = \pi - \psi$, then substituting

$$v_H = \sqrt{\frac{\mu}{r_0(1-e)}}(1 - e\cos\psi)$$

into Eq. (9.47), we obtain

$$\Delta v_H = 2\sqrt{\frac{\mu}{r_0(1-e)}}(1 - e\cos\psi)\sin\frac{\alpha_H}{2}. \tag{9.48}$$

As a result of the impulse, the orbital plane turns to angle α_H about the line LH, which is perpendicular to $Q_1 Q_2$. In this case, the line $Q_1 Q_2$ turns to angle $\Delta\Lambda$, and the angle i_1 is changed to i_2.

Turning of the elliptical orbit's plane can also be conducted by two-impulsive maneuver [167]. The first impulse is applied at the point Λ in the direction orthogonal to the angle i_1. As a result of this impulse, the spacecraft is transferred into some intermediate orbit. Then the second orthogonal impulse is applied at the point H of this orbit. This impulse turns the intermediate orbit to angle i_2 thereby transferring the spacecraft into the orbit II. The magnitudes of the impulses at Λ and Q are determined by the formulas:

$$\Delta v_{12} = 2v_\Lambda \sin\frac{i_1}{2}, \qquad \Delta v_{22} = 2v_Q \sin\frac{i_2}{2},$$

where $\Delta v_{12}, \Delta v_{12}$ are the magnitudes of the impulses, and the second index shows the order of the impulses, v_Λ and v_Q are the transversal components of the velocity at Λ and Q respectively. It can be shown that v_Λ and v_Q are computed according to the formulas:

$$v_\Lambda = 2\sqrt{\frac{\mu}{p}}(1 + e\cos\omega), \qquad v_Q = \sqrt{\frac{\mu}{p}}(1 + e\cos(\omega + \Delta\omega)),$$

where

$$ctg\Delta\omega = \frac{\sin i_2}{tg\alpha_H}.$$

So, the total sum of the impulses is

$$\Delta v_2 = \Delta v_{12} + \Delta v_{22}.$$

The turning maneuver can also be accomplished by applying three impulses [167]. The first impulse is applied at the perigee of the orbit I and it transfers the spacecraft into an intermediate transfer orbit. Second orthogonal impulse is applied at the apogee of this orbit, E, and this impulse serves to turn the orbital plane by α_H with respect to the line perpendicular to the line of nodes. The breaking impulse is applied at the perigee of the resulting orbit, and this impulse transfers the spacecraft into the orbit II coplanar to this orbit. The magnitudes of the impulses are determined according to the formulas: [167]:

$$\begin{aligned}
\Delta v_{13} &= \sqrt{\frac{\mu}{r_1}}\left(\sqrt{\frac{2}{1+\frac{r_1}{r_0}}} - \sqrt{1+e}\right),\\
\Delta v_{23} &= 2v_E \sin\frac{\alpha_H}{2},\\
\Delta v_{33} &= \sqrt{\frac{\mu}{r_1}}\left(\sqrt{\frac{1-e}{\frac{r_1}{r_0}}} - \frac{r_1}{r_0}\sqrt{\frac{2}{1+\frac{r_1}{r_0}}}\right),
\end{aligned}$$

where the second index means the number of impulses, r_0 is the radius of the sphere, $\frac{r_1}{r_0}$ is the ratio of the perigee to apogee, e is the eccentricity of the initial elliptical orbit,

$$v_E = \sqrt{\frac{\mu}{r_1(1+e')}}(1-e')\cos i_2,$$

where e' is the eccentricity of the transfer orbit, and $\frac{r_1}{r_0} = \frac{1-e'}{1+e'}$. The total impulse Δv_3 is determined as

$$\Delta v_3 = \Delta v_{11} + \Delta v_{23} + \Delta v_{33}.$$

The turning of the orbital plane via one impulse in the orthogonal direction represents a practical interest and well known compared to the two- and three-impulsive maneuvers [2]. Optimality of these maneuvers in terms of fuel expenditure depends on several factors, such as the angle between the lines that connect the point at which the impulse is applied and the apogee with the center of ellipse (one-impulsive maneuver); ratios of the orbital inclinations in all three cases, and the velocity at the point at which an impulse is applied. It was shown that a one-impulsive maneuver which starts at a point which is not located on the line of nodes, is fuel optimal compared to the two-impulsive and the three-impulsive maneuvers, but not optimal when the impulsive is applied along the line of nodes [2]. Optimality of the two-impulsive maneuver compared to the three-impulsive maneuver also depends on the turning angle and ellipticity of the orbit. If the orbit is highly elliptical, then the orthogonal impulses at the apogee are more effective and practical [2], [3]. Note that the effectiveness of each of these maneuvers is determined mainly by the characteristic velocity. Comparison of the characteristic velocities given below can determine the effectiveness of using an IT arc compared to one-, two-, and three-impulsive maneuvers [88], [62].

The transfer from the orbit I to the orbit II can also be conducted using an IT arc which lies on the spherical surface. It will be shown below that the IT arc solutions presented in Eqs. (4.31), (4.35), (4.37), (4.40), (4.43), (4.45), and (4.46) can be used to describe the maneuver. The velocity on IT arc is constant and less than a local circular velocity, where $v_1 = 0$. Consequently, the motion on IT arc starts at point P of the orbit I and ends at the point P' of the orbit II. Therefore, at the point A_1 one can start the engine with the thrust magnitude given by the formulae:

$$\Phi = \frac{3\mu}{r_0^2\lambda} m_0 e^{-\frac{3\mu\lambda_1}{r_0^2\lambda} t}.$$

Determine the spherical coordinates φ and θ of the initial and final points, that is A_1 and A_2 of the IT arc depending on the boundary conditions, given by Eqs. (9.45)–(9.46) and the turning angle, α_H. If the angular distance of this point from the node in the orbital plane is denoted by u, then from Fig. 9.16 it can be seen that

$$u_1 = \frac{\pi}{2} + \psi, \qquad u_2 = u_H + \psi. \tag{9.49}$$

For the spherical triangle one can have

$$\sin u_H = \frac{\sin i_1}{\sin i_2}. \tag{9.50}$$

If the relationship between the angles $\alpha_H, i_1, i_2, \Delta\Lambda$ is given as

$$\sin i_1 = tg\alpha_H ctg(\Delta\Lambda), \qquad \sin\alpha_H = \sin i_2 \sin(\Delta\Lambda), \tag{9.51}$$

then using the formulae

$$\sin i_2 = \frac{\sin\alpha_H}{\sin(\arctan(\frac{tg\alpha_H}{\sin i_1}))} \tag{9.52}$$

one can rewrite the second relationship of Eqs. (9.49) using Eq. (9.50) in the form:

$$u_2 = \arcsin\frac{\sin i_1 \sin(\arctan(\frac{tg\alpha_H}{\sin i_1}))}{\sin\alpha_H} + \psi. \tag{9.53}$$

Then on the basis of Eqs. (9.49), (9.51)–(9.53) and

$$\sin\theta = \sin i \sin u, \qquad tg(\varphi - \Omega) = \cos i tgu \tag{9.54}$$

we can obtain the formulas to determine the spherical coordinates of the initial and final points of the IT arc:

$$\begin{aligned}
\varphi_1 &= \arctan(\cos i_1 tgu_1) + \Omega_1, \\
\varphi_2 &= \arctan(\cos i_2 tgu_2) + \Omega_2, \\
\theta_1 &= \arcsin(\sin i_1 \cos\psi),
\end{aligned} \tag{9.55}$$

$$\theta_2 = \arcsin(\sin i_2 \sin u_2),$$

where $\Omega_2 = \Omega_1 + \Delta\Lambda$, and the angle $\Delta\Lambda$, as it follows from Eq. (9.51), is determined as

$$tg(\Delta\Lambda) = \frac{tg\alpha_H}{\sin i_1}. \tag{9.56}$$

The velocity components at the points A_1 and A_2 in the spherical system can be determined by formulas:

$$\begin{aligned}
v_{21} &= \sqrt{\frac{\mu}{r_0}(1-e)}\frac{\cos i_1}{\cos\theta_1},\\
v_{22} &= \sqrt{\frac{\mu}{r_0}(1-e)}\frac{\cos i_2}{\cos\theta_2},\\
v_{21} &= \sqrt{\frac{\mu}{r_0}(1-e)}\frac{\sin i_1 \cos u_1}{\cos\theta_1},\\
v_{22} &= \sqrt{\frac{\mu}{r_0}(1-e)}\frac{\sin i_2 \cos u_2}{\cos\theta_2},
\end{aligned} \tag{9.57}$$

where the angles u_1, θ_1, u_2, θ_2, and i_2 are connected with the boundary conditions and the angle α_H by the relationships given in Eqs. (9.49), (9.51), (9.52), (9.54). Now one can determine the constants of the IT arc solutions. By equating the right hand side of the formulae of determination of the velocity at the IT arc's initial point, computed according to the first two formulas of Eqs. (9.57), one will have

$$\lambda_{10} = \frac{\lambda_0^2}{3}(1-(1-e))[\frac{\cos^2 i_1 + \sin^2 i_1 \cos^2\psi}{\cos^2\theta_1}]. \tag{9.58}$$

Furthermore, by equating v_3 at the initial and final points, and using Eq. (9.57) one can have

$$\begin{aligned}
N_1 a \sin(\varphi_1 - \alpha) &= \sqrt{\frac{\mu}{r_0}(1-e)}\frac{\sin i_1 \cos u_1}{\cos\theta_1},\\
N_1 a \sin(\varphi_2 - \alpha) &= \sqrt{\frac{\mu}{r_0}(1-e)}\frac{\sin i_2 \cos u_2}{\cos\theta_2},
\end{aligned} \tag{9.59}$$

from which it follows that

$$tg\alpha = \frac{\sin\varphi_1 - A\sin\varphi_2}{\cos\varphi_1 - A\cos\varphi_2}, \tag{9.60}$$

$$a = \sqrt{\frac{\mu}{r_0}(1-e)}\frac{\sin i_1 \cos u_1}{N_1 \sin(\varphi_1 - \alpha)\cos\theta_1}, \qquad (9.61)$$

where

$$A = \frac{\sin i_1 \cos u_1 \cos\theta_2}{\sin i_2 \cos u_2 \cos\theta_1}.$$

Besides that, from Eqs. (9.59) it follows that

$$\lambda_5 = \frac{\mu}{r_0}\frac{\sqrt{s_1 \sin^2 2\theta_1 - s_7 \sin^2(\varphi_2 - \alpha)\sin^2 2\theta_2} - (s_2 + s_3\cos^2\theta_2)}{s_6 \sin\theta_2}, \qquad (9.62)$$

where s_1, s_2, s_3 are determined from Eqs. (4.45),

$$s_6 = \frac{\lambda_{10}^2(-3\lambda_{10}^2 + \lambda_0^2)s_5}{9\lambda_0^2(\lambda_0^2 - \lambda_{10}^2)}, \qquad s_7 = (-3\lambda_{10}^2 + \lambda_0^2)^{\frac{3}{2}}.$$

Consequently,

$$T = \sqrt{r_0^3 \mu} \times$$

$$\times \int \sqrt{\frac{s_5 \sin^2 2\theta}{s_1 \sin^2 2\theta - [s_2 + s_3\cos^2\theta + \lambda\lambda_5\sqrt{r_0}\mu(-3\lambda_{10}^2 + \lambda_0^2)^{\frac{3}{2}}\sin\theta]^2}}\,d\theta.$$

The characteristic velocity required for turning of the orbital plane is computed according to the formulae:

$$\Delta V = c\ln\frac{M_0}{M_T} = \frac{3\lambda_{10}v_H}{\lambda_0\sqrt{1-e}} \times$$

$$\times \int \frac{s_1 \sin^2 2\theta - [s_2 + s_3\cos^2\theta + \lambda\lambda_5\sqrt{\frac{r_0}{\mu}}(-3\lambda_{10}^2 + \lambda_0^2)^{\frac{3}{2}}\sin\theta]^2}{s_5 \sin^2 2\theta}\,d\theta. \qquad (9.63)$$

If it is required to minimize

$$J = c\ln\frac{m_0}{m_T},$$

then from the transversality condition yields

$$\lambda_{71} = \frac{c}{m_T}, \qquad \lambda_5 = \frac{\partial J}{\partial\theta_1} = 0, \quad \text{and} \quad \lambda_0 = 1.$$

Then from Eq. (9.63) gives

$$\Delta V = v_H\left[d_1(\theta_2 - \theta_1) + d_2(\cot 2\theta_1 - \cot 2\theta_2) + d_3(\cot\theta_1 - \cot\theta_2)\right],$$

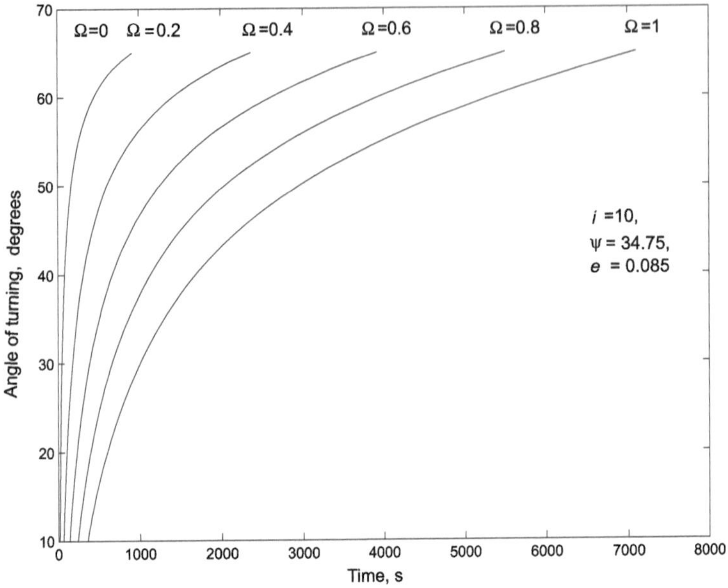

Figure 9.17 Turning angle vs. time for various distances of the perigee (longitude).

where

$$d_1 = \frac{3\lambda_{10}v_H}{4s_5\lambda_0\sqrt{1-e}}(s_1 + s_3^2),$$
$$d_2 = -\frac{3\lambda_{10}v_H}{2s_5\lambda_0\sqrt{1-e}}s_2^2,$$
$$d_3 = \frac{3\lambda_{10}v_H}{4s_5\lambda_0\sqrt{1-e}}(2s_2s_3 + s_3^2).$$

Then from Eq. (4.51) it follows that

$$t_2 = \frac{r_0^2\lambda_0\Delta V}{3\mu\lambda_{10}}.$$

Fig. 9.17 shows that the dependence between the maneuver time and α_H for various $\Omega(=\Omega_1)$ at fixed $i=(i_1)$, ψ, and e. The corresponding spherical IT arc satisfies the necessary conditions of optimality and is an extremal.

9.4.1 Numerical Example

Consider the example with following given values for the problem parameters: The ratio of the characteristic velocity to the circular velocity is com-

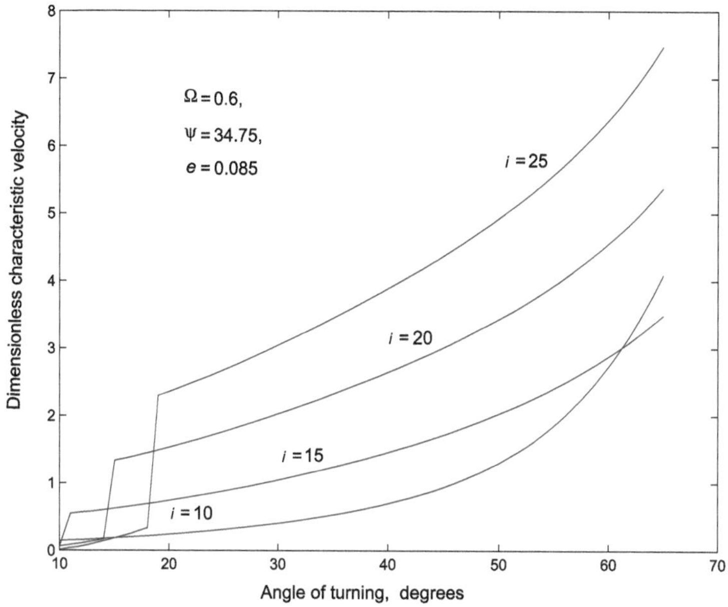

Figure 9.18 Dependence of $\Delta v_{IT}/v_0$ on α_H at various inclinations.

puted for $r_0 = 10{,}000$ km, $e_1 = 0.22$, $0.08 \leq e \leq 0.1$; $0 \leq \Omega \leq 1.0°$, $5° \leq i \leq 25°$, $34.6° \leq \psi \leq 34.8°$. The computations for $e > 0.1$, $\alpha_H < 10°$, $\alpha_H > 65°$, $i > 25°$, $\psi < 34.6°$, $\psi > 35°$ show high values for the characteristic velocity.

In the formulas for impulses the following parameters have been used: $i = 10°$, $\Omega = 0.6°$, $\psi = 34.75°$, $e = 0.085$. It is assumed that if one of the parameters is changed, then the other parameters are considered as constants. The results of the analysis are presented below:

1. The IT arc is more effective than an impulsive maneuver if $10° \leq \alpha_H \leq 40°$, $0 \leq i \leq 15°$ at fixed Ω, ψ, e (see Fig. 9.18).
2. The IT arc is more effective than an impulsive maneuver if $0 \leq e \leq 0.1$ and $10° \leq \alpha_H \leq 52°$.
3. The angle ψ is changed insignificantly. The IT arc is more effective than an impulsive maneuver if $62° < \alpha_H < 65°$.
4. High values of $\Delta V_{IT}/v_0$ correspond to the cases when $0 \leq \Omega \leq 1°$, and α_H are changed independently (Fig. 9.19). In this case, the IT arc is more effective than all three impulsive maneuvers if (Fig. 9.20):

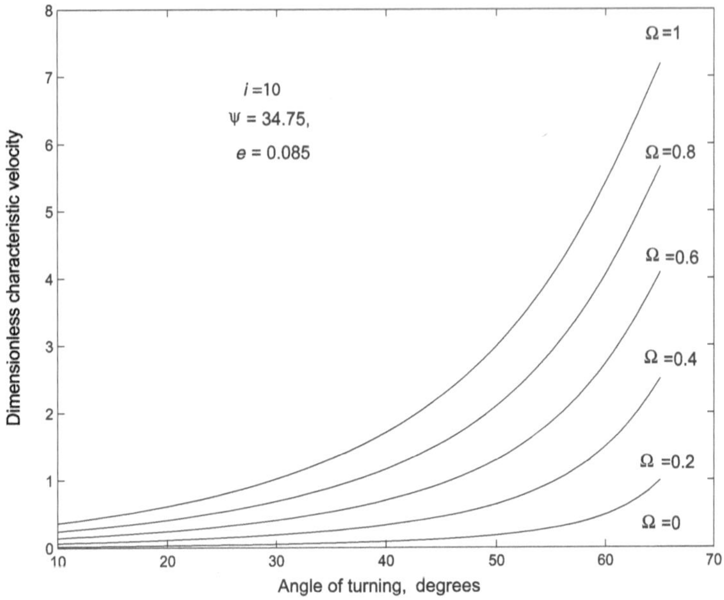

Figure 9.19 Dependence of $\triangle v_{IT}/v_0$ on α_H at various distances of the perigee.

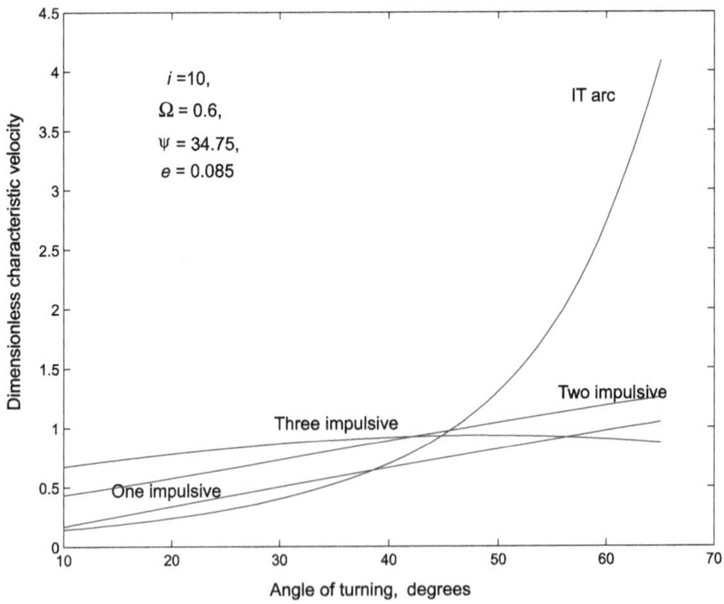

Figure 9.20 Dependence of $\triangle v_{IT}/v_0$ on α_H at various parameters.

$$0 \le \Omega \le 0.2°, \quad 10° \le \alpha_H \le 65°, \quad 0 \le \Omega \le 0.4°, \quad 10° \le \alpha_H \le 55°$$
$$0 \le \Omega \le 0.6°, \quad 10° \le \alpha_H \le 45°, \quad 0 \le \Omega \le 0.8°, \quad 10° \le \alpha_H \le 35°$$
$$0 \le \Omega \le 1.0°, \quad 10° \le \alpha_H \le 25°$$

5. The spherical IT arc is comparable with an impulsive solution if $22° \le \alpha_H \le 23°$, $0.1 \le e \le 0.9$, $20° \le \psi \le 70°$, and $\Omega = 0°$, $10° \le i \le 80°$.
6. The spherical IT arc is more effective than a two-impulsive solution if $\alpha_H \le 25°$, $0.1 \le e_1 \le 0.9$, $\Omega_1 = k\frac{\pi}{2}$, $k = 0, 1, 2, ...$, $10 \le \psi \le 90$, and $10 \le i \le 90$ and is comparable if $25 \le \alpha_H \le 30$.
7. The spherical IT arc is more effective than a three-impulsive solution if $5 \le \alpha_H < 60$, $0.1 < e < 0.3$, $10° \le i < 90°$, $10° \le \psi \le 40°$, and $-360° \le \Omega \le 360°$.
8. If $0 \le e \le 0.1$, then Δv_1 is decreased, Δv_2 and Δv_3 are increased.
9. There exists almost a linear relationship between α_H and $\Delta V_{IT}/v_0$ for $\Omega = 30°,\ 120°$.
10. For any e, i, and ψ, there exist discontinuities in t_2, $\Delta V_{IT}/v_0$, m_2, λ_7 and $\theta_1 = k\frac{\pi}{2}$, $k = 0, 1, 2, ...$ at $10 \le \psi \le 60°$ and $10° \le \imath \le 90°$ (Fig. 9.18).

It can be concluded that at certain values of the parameters of the elliptical orbit, the turning maneuver via an IT arc can be effective or fuel-optimal compared to one-, two-, or three-impulsive maneuvers.

9.5 TRANSFER BETWEEN CIRCULAR ORBITS VIA TWO MAXIMUM THRUST ARCS WITH CONSTANT SPECIFIC IMPULSE

Consider the problem of transfer between coplanar concentric circular orbits with free final time in the central Newtonian field using the analytical solutions obtained in the previous sections for maximum thrust (MT) arcs. These solutions contain four constants of integration, α, a, C_2, ψ_0. Then based on Eq. (8.10) one cam conclude that this transfer can be conducted using two MT arcs. It is assumed that the gravitational acceleration on MT arcs is represented as a linear function of radius vector originated from the center of attraction [91], [142], [158], [85]. The functional to be minimized is the difference between in initial and final mass. The following initial conditions are give:

$$t = 0, \quad r = r_0, \quad v_2 = kr_0, \quad v_1 = 0, \quad M = M_0.$$

The final conditions are:

$$t = t_2, \quad r = r_1 (r_1 \neq r_0), \quad v_2 = kr_1, \quad v_1 = 0,$$

where $k_1 = \sqrt{\frac{\mu}{r_1}}$. In the case of $r_0 = r_1$, the problem considered can be reduced to a rendezvous problem [91]. The phases of motion in the boundary orbits may by arbitrary. Each of these orbits has its own thin spherical (circular) layer. If these is an optimal trajectory in any of these layers, then there will exist at least one MT arc. This is a valid statement if the time of motion on these trajectories takes the values from the interval $(0, \frac{\pi}{k})$, $k = 0, 1, 2, 3, \ldots$. Then if we include both end points of a ZT arc into each layer, then in each layer it will be possible to obtain a trajectory that contain a part of the boundary orbit, MT arc and one end point of the ZT arc mentioned above. This ZT arc can be considered as an elliptical transfer trajectory. Consequently, it is possible to construct a trajectory which consists of three ZT arcs and two MT arcs. The first MT arc transfers the spacecraft from the initial orbit into the elliptical transfer orbit with parameters p, e, ω, and the second MT arc transfers the spacecraft from the transfer orbit into the final orbit. This means that the trajectory contains four junction points. The first junction point belongs to the initial orbit. The second and third points are in the transfer orbit, and the fourth junction point lies in the final orbit. The following conditions must be satisfied at the first junction:

$$0 = \varphi_1 + \psi_1 - \frac{\pi}{2}, \tag{9.64}$$

$$r_0 = a_1 C_{31} \frac{\sin \alpha_1}{\cos \varphi_1}, \tag{9.65}$$

$$0 = a_1 C_{31} \frac{\sin(\alpha_1)}{\cos \varphi_1} [kctg(\alpha_1) + \dot{\varphi}_1 tg\varphi_1], \tag{9.66}$$

$$kr_0 = a_1 C_{31} \frac{\sin \alpha_1}{\cos \varphi_1} \dot{\varphi}_1, \tag{9.67}$$

$$P_1 \sin f_1 = a_1 \sin \alpha_1 \sin \varphi_1, \tag{9.68}$$

$$2P_1 \cos f_1 + R = a_1 \sin \alpha_1 \cos \varphi_1, \tag{9.69}$$

where $\theta_1 = 0$ is the polar angle of the first junction point, and it is accepted zero as the phase of motion in initial orbit is assumed arbitrary, and

$$\dot{\varphi}_1 = -\frac{k \sin 2\varphi_1}{\sin(2\alpha_1)}. \tag{9.70}$$

For the second junction point we have:

$$\chi(t_2) = 0, \tag{9.71}$$

$$\theta_2 = \varphi_1 + \psi_1 - \frac{\pi}{2} = f_2 + \omega, \tag{9.72}$$

$$\frac{p}{1 + e\cos f_2} = a_1 C_{31} \frac{\sin kt_2 + \alpha_1}{\cos\varphi_2}, \tag{9.73}$$

$$\sqrt{\frac{\mu}{p}} e \sin f_2 = a_1 C_{31} \frac{\sin(kt_2 + \alpha_1)}{\cos\varphi_2}[kctg(kt_2 + \alpha_1) + \dot{\varphi}_2 tg\varphi_2], \tag{9.74}$$

$$\sqrt{\frac{\mu}{p}}(1 + e\cos f_2) = a_1 C_{31} \frac{\sin(kt_2 + \alpha_1)}{\cos\varphi_2} \dot{\varphi}_2, \tag{9.75}$$

$$B' e \sin f_2 = a_1 \sin(kt_2 + \alpha_1) \sin\varphi_2, \tag{9.76}$$

$$B'(1 + e\cos f_2) + \frac{D'}{1 + e\cos f_2} = a_1 \sin(kt_2 + \alpha_1) \cos\varphi_2. \tag{9.77}$$

For Eqs. (9.71)–(9.81) the following expressions take place:

$$tg(\varphi_2(t_2)) = \frac{tg\varphi_1 tg\alpha_1}{tg(kt_2 + \alpha_1)} + \frac{cs_1}{a_1 C_{31} ktg(kt_2 + \alpha_1)},$$

$$\dot{\varphi}_2 = -\frac{\cos^2\varphi_1}{\sin^2(kt_2 + \alpha_1)}[ktg\varphi_1 tg\alpha_1 + \frac{cs_1}{a_1 C_{31}}]. \tag{9.78}$$

Here

$$s_1 = \sin\alpha_3 F_2(t_2) - \cos\alpha_3 F_1(t_2),$$

$$\alpha_3 = \alpha_1 + k\frac{M_0}{\bar{m}}, \qquad x_1 = k\frac{M_0}{\bar{m}}, \quad x_2 = k\frac{M_0}{\bar{m}} - t_2,$$

$$F_1(t_2) = Si(x_2) - Si(x_1), \qquad F_2(t_2) = Ci(x_2) - Ci(x_1),$$

where t_2 is the final instant of the first MT arc. Now let us introduce a new variable, τ, which will represent the time of motion on second MT arc, and satisfies the condition

$$\tau_1 < \tau < \tau_2, \tag{9.79}$$

where τ_1, τ_2 are the initial and final times of motion on second MT arc. It can be shown that τ_1 and the current flight time, t are related by

$$\tau = t - t_3, \tag{9.80}$$

here t_3 is the final time of motion in the transfer orbit and the initial time of motion on second MT arc, and measured from the initial point of the first MT arc. Also,

$$\tau_2 = t_4 - t_3, \tag{9.81}$$

where t_4 is the final time of motion on second MT arc and measured from the beginning of the maneuver. Then at the third junction point the following conditions are satisfied:

$$\theta_3 = \varphi_3 + \psi_3 - \frac{\pi}{2} = f_3 + \omega, \tag{9.82}$$

$$\frac{p}{1+e\cos f_3} = a_2 C_{32}\frac{\sin\alpha_2}{\cos\varphi_2}, \tag{9.83}$$

$$\sqrt{\frac{\mu}{p}}e\sin f_3 = a_2 C_{32}\frac{\sin\alpha_2}{\cos\varphi_3}[kctg\alpha_2 + \dot{\varphi}_3 tg\varphi_3], \tag{9.84}$$

$$\sqrt{\frac{\mu}{p}}(1+e\cos f_3) = a_2 C_{32}\frac{\sin\alpha_2}{\cos\varphi_3}\dot{\varphi}_3, \tag{9.85}$$

$$B' e\sin f_3 = a_2 \sin\alpha_2 \sin\varphi_3, \tag{9.86}$$

$$B'(1+e\cos f_3) + \frac{D'}{1+e\cos f_3} = a_2\sin\alpha_2\cos\varphi_3, \tag{9.87}$$

$$a_1\sin(k_1 t_2 + \alpha_1) = a_2\sin\alpha_2. \tag{9.88}$$

Here $\tau_1 = 0$ and

$$\dot{\varphi}_3 = -\frac{\cos^2\varphi_3}{\sin^2\alpha_2}[k_1 tg\varphi_3 tg\alpha_1]. \tag{9.89}$$

The following conditions are satisfied at the fourth junction point:

$$\chi(\tau_2) = 0, \tag{9.90}$$

$$r_1 = a_2 C_{32}\frac{\sin(k_1\tau_2+\alpha_1)}{\cos\varphi_4}, \tag{9.91}$$

$$\theta_4 = \varphi_4 + \psi_4 - \frac{pi}{2}, \tag{9.92}$$

$$0 = a_2 C_{32}\frac{\sin(k_1\tau_2+\alpha_2)}{\cos\varphi_4}[k_1 ctg(k_1\tau_2+\alpha_2) + \dot{\varphi}_4 tg\varphi_4], \tag{9.93}$$

$$\sqrt{\frac{\mu}{r_1}} = a_2 C_{32}\frac{\sin(k_1\tau_2+\alpha_2)}{\cos\varphi_4}\dot{\varphi}_4, \tag{9.94}$$

$$P_2\sin f_4 = a_2\sin(k_1\tau_2+\alpha_2)\sin\varphi_4, \tag{9.95}$$

$$2P_2\cos f_4 + R_2 = a_2\sin(k_1\tau_2+\alpha_2)\cos\varphi_4. \tag{9.96}$$

This system also includes the equalities

$$\begin{aligned} tg\varphi_4 &= \frac{tg\varphi_3 tg\alpha_2}{tg(k_1\tau_2+\alpha_2)} + \frac{cs_2}{a_2 C_{32} k_1 tg(k_1\tau_2+\alpha_2)}, \\ \dot{\varphi}_4 &= -\frac{\cos^2\varphi_4}{\sin^2(k_1\tau_2+\alpha_2)}[k_1 tg\varphi_3 tg\alpha_2 + \frac{cs_2}{a_2 C_{32}}], \end{aligned} \tag{9.97}$$

where

$$s_2 = \sin\alpha_4 F_2 - \cos\alpha_4 F_1,$$
$$\alpha_4 = \alpha_2 + k_1\frac{M_0}{\bar{m}}, \qquad x_3 = k_1\frac{M_0}{\bar{m}} - k_1 t_2, \quad x_4 = k_2\frac{M_0}{\bar{m}} - k_1(t_2+\tau_2),$$
$$F_3(\tau_2) = Si(x_4) - Si(x_3), \qquad F_4(\tau_2) = Ci(x_4) - Ci(x_3).$$

The transversality condition can be written as $\lambda_7(\tau_2) = 1$. Consequently, for the fourth junction point one can have

$$ca_2\sin(k_1\tau_2+\alpha_2) = \frac{M_0}{\bar{m}} - (t_2+\tau_2). \tag{9.98}$$

Here and below, as in the previous sections, we accept that the mass-flow rate in included in the constants $a_1, a_2, C_{31}, C_{32}, P_1, P_2, R_1, R_2, D_1, D_2$. Time between the second and third junctions is determined as

$$t_3 - t_2 = \frac{p^{\frac{3}{2}}(1-e^2)^{\frac{3}{2}}}{\sqrt{\mu}}[E_3 - E_2 - e(\sin E_3 - \sin E_2)], \tag{9.99}$$

where

$$tg\frac{E_2}{2} = \sqrt{\frac{1-e}{1+e}}tg\frac{f_2}{2}, \qquad tg\frac{E_3}{2} = \sqrt{\frac{1-e}{1+e}}tg\frac{f_3}{2}.$$

Consequently, we have Eqs. (9.64)–(9.69), (9.71)–(9.77), (9.81)–(9.87), (9.90)–(9.96), (9.98), which are enough to determine $\varphi_i, f_i, \theta_i, \psi_i, a_j, C_{3j}, P_j, R_j, B, D, t_2, \tau, p, e,$ $(i = 1, ..., 4; j = 1, 2,)$ where $t_1 = 0, \theta_1 = 0$, and (see Fig. 9.21):

- Junction point 1: $\varphi_1, \psi_1, a_1, C_{21}, P_1, f_1, R_1$
- Junction point 2: $\alpha_1, \varphi_2, \psi_2, \theta_2, f_2, \omega, p, e, t_2, B, D$
- Junction point 3: $\varphi_3, \psi_3, \theta_3, f_3, a_2, C_{22}$
- Junction point 4: $\alpha_2, \varphi_4, \psi_4, \theta_4, \tau_2, P_2, f_4, R_2$

Indeed, Eqs. (9.65) and (9.67) with Eq. (9.70) yield

$$\varphi_1 = \frac{\pi}{2} + \alpha_1, \tag{9.100}$$

$$r_0 = -C_{31}a_1. \tag{9.101}$$

Then from the first equation of Eqs. (9.78) it follows that

$$tg\varphi_2 = \frac{1 + \frac{cs_1}{kr_0}}{tg(kt_2 + \alpha_1)}.$$

Substitution of this expression into Eq. (9.73) yields

$$\frac{\rho^2}{(1 + e\cos^2 f_2)^2} = \sin^2(kt_2 + \alpha_1) + \cos^2(kt_2 + \alpha_1)(1 + \frac{cs_1}{kr_0})^2, \tag{9.102}$$

where $\rho = \frac{p}{r_0}$. On the basis of Eqs. (9.78), (9.74), and (9.75) are rewritten as

$$\sqrt{\frac{\mu}{p}} e\sin f_2 = -r_0 \frac{k_1 \sin(k_1 t_2 + \alpha_1)}{\cos\varphi_1} ctg(k_1 t_2 + \alpha_1) \times$$

$$\times [1 - \frac{(1 + e\cos^2 f_2)^2}{\rho^2}(1 + \frac{cs_1}{kr_0})], \tag{9.103}$$

$$\rho^{\frac{1}{2}} = 1 + \frac{cs_1}{kr_0}. \tag{9.104}$$

Then the unknowns α_1, ρ, e, f_2 can be found from Eqs. (9.71), (9.102)–(9.104) in terms of time. From (9.71) one can find the relationship for α_1:

$$\alpha_1 = arctg(-\frac{F_2(t_2)}{F_1(t_2)} - k\frac{M_0}{\bar{m}}. \tag{9.105}$$

Substituting Eq. (9.105) into (9.104), we obtain ρ in terms of t_2:

$$\rho = \rho(t_2) = 1 + \frac{cs_1(\alpha_1(t_2), t_2)}{kr_0}. \tag{9.106}$$

In this case, f_2 and e are found from Eq. (9.102) and (9.103):

$$tgf_2 = \frac{-\rho^{\frac{1}{2}} \sin 2(k_1 t_2 + \alpha_1)(1 - \rho)}{\rho - \Phi_1}, \tag{9.107}$$

$$e^2 = (\frac{\rho}{\Phi_1} - 1)^2 + (\frac{\rho^{\frac{1}{2}} \sin 2(k_1 t_2 + \alpha_1)(1 - \rho)}{\Phi_1})^2. \tag{9.108}$$

Here

$$\Phi_1 = [\sin^2(k_1 t_2 + \alpha_1)(1 - \rho) + \rho]^{\frac{1}{2}}. \tag{9.109}$$

Furthermore, from Eq. (9.90) it follows that

$$\alpha_2 = arctg(-\frac{F_4(\tau_2)}{F_3(\tau_2)} - k\frac{M_0}{\bar{m}} + k_1 t_2. \tag{9.110}$$

Then

$$k_1\tau_2 + \alpha_2 = k_1 T - k_1\frac{M_0}{\bar{m}} + arctg(-\frac{F_4(\tau_2)}{F_3(\tau_2)}, \tag{9.111}$$

where

$$T = t_2 + \tau_2. \tag{9.112}$$

Besides that, from Eq. (9.83) and (9.85) we obtain

$$\sqrt{\frac{\mu}{p^3}}(1 + e\cos f_3)^2 = -k_1\frac{\sin 2\varphi_3}{\sin 2\alpha_2}.$$

From this it follows that

$$tg\alpha_2 tg\varphi_3 = -\frac{\sqrt{\mu p}}{k_1 a_2^2 {C_{32}}^2}. \tag{9.113}$$

From Eqs. (9.92), (9.94), and the second equality of Eqs. (9.97) one can derive that

$$\varphi_4 = k_1\tau_2 + \alpha_2 + \frac{\pi}{2}. \tag{9.114}$$

Then Eq. (9.92) yields

$$a_2 C_{32} = -r_1. \tag{9.115}$$

Consequently, from Eqs. (9.114) and (9.115) with Eqs. (9.97) we obtain

$$\frac{r_1}{p} = (1 + \frac{cs_2(t_2, \tau_2)}{k_1\sqrt{r_1 p}})^2. \tag{9.116}$$

Equating p found from Eqs. (9.104) and (9.116), one can obtain th equation with respect to t_2 and τ_2:

$$\frac{r_1}{r_0} = \frac{(1 + \frac{cs_1(t_2)}{kr_0})^2}{(1 - \frac{cs_2(t_2,\tau_2)}{k_1 r_1})^2}. \tag{9.117}$$

Now we can find the second equation containing t_2 and τ_2. From Eq. (9.84) and (9.113) it follows that

$$e\sin f_3 = -\frac{(x^2-1)\sin\alpha_2\cos\alpha_2}{x^2\sqrt{\cos^2\alpha_2 + x^2\sin^2\alpha_2}}, \tag{9.118}$$

where $x = -\frac{r_1}{p}$. From Eqs. (9.83), (9.89), and (9.113) we obtain

$$e\cos f_3 = -(\frac{1}{x\sqrt{\cos^2\alpha_2 + x^2\sin^2\alpha_2}}. \tag{9.119}$$

From Eqs. (9.118) and (9.119) it can be seen that

$$tgf_3 = \frac{(x^2-1)\sin\alpha_2\cos\alpha_2}{x(1+x\sqrt{\cos^2\alpha_2 + x^2\sin^2\alpha_2}},$$

$$e^2 = \frac{(x^2-1)^2\sin^2\alpha_2\cos^2\alpha_2}{x^4(\cos^2\alpha_2 + x^2\sin^2\alpha_2)} + \frac{1-x\sqrt{\cos^2\alpha_2 + x^2\sin^2\alpha_2}}{x^2(\cos^2\alpha_2 + x^2\sin^2\alpha_2)}. \tag{9.120}$$

By equating the values of e, determined from Eqs. (9.108), (9.120) one can find the equation that contain t_2 and τ_2:

$$\frac{(\rho^2-\Phi)^2}{\Phi^2} - \frac{\rho(1-\rho)\sin^2 2(kt_2+\alpha_1)}{\Phi^2} =$$

$$= \frac{(x^2-1)^2\sin^2\alpha_2\cos^2\alpha_2}{x^4(\cos^2\alpha_2 + x^4\sin^2\alpha_2)} + \frac{1-x\sqrt{\cos^2\alpha_2 + x^2\sin^2\alpha_2}}{x^2(\cos^2\alpha_2 + x^2\sin^2\alpha_2)}, \tag{9.121}$$

where α_1, α_2 are determined from Eqs. (9.105) and (9.110). Eqs. (9.117) and (9.121) in principle, are enough to determine t_2 and τ_2. However, these equations are not solvable analytically in a closed-form as they contain special functions, and therefore they can be solved using numerical methods. Let

$$t_2 = t_2^*(r_0, r_1, \frac{M_0}{\bar{m}}, c), \quad \tau = \tau^*(r_0, r_1, \frac{M_0}{\bar{m}}, c).$$

Then from Eqs. (9.105) and (9.110) it follows that

$$\alpha_1^* = arctg(-\frac{F_2(t_2^*)}{F_1(t_2^*)}) - k\frac{M_0}{\bar{m}}, \tag{9.122}$$

$$\alpha_2^* = arctg(-\frac{F_4(\tau_2^*)}{F_3(\tau_2^*)}) - k_1\frac{M_0}{\bar{m}} + k_1 t_2^*. \tag{9.123}$$

Then let $\alpha_1^*, \alpha_2^*\ t_2^*\ \tau_2^*$ satisfy the inequalities:

$$
\begin{aligned}
&n\pi < /\alpha_1^*/ < \pi(\frac{1}{2}+n),\\
&\pi(\frac{1}{2}+n) < /kt_2^*+\alpha_1^*/ < \pi(n+1),\\
&n\pi < /\alpha_2^*/ < \pi(\frac{1}{2}+n),\\
&\pi(\frac{1}{2}+n) < /k_1\tau_2^*+\alpha_2^*/ < \pi(n+1),\\
&\frac{M_0}{\bar{m}} > t_2^*,\\
&\frac{M_0}{\bar{m}} - t_2^* > \tau_2^*.
\end{aligned}
$$

In this case, it is easy to obtain other solutions of the problem. From Eqs. (9.89), (9.104), (9.107), and (9.108) it follows that

$$
\begin{aligned}
p^* &= r_0(1+\frac{cs_1^*}{kr_0},\\
\varphi_1^* &= \frac{\pi}{2}+\alpha_1^*,\\
tgf_2^* &= \frac{-(\rho^{\frac{1}{2}})^*\sin 2(k_1t_2^*+\alpha_1^*)(1-\rho^*)}{\rho^*-\Phi_1^*},\\
e^* &= [(\frac{\rho^*}{\Phi^*}-1)^2+(\frac{(\rho^{\frac{1}{2}})^*\sin 2(kt_2^*+\alpha_1^*)(1-\rho^*)}{\Phi^*})^2]^{\frac{1}{2}}.
\end{aligned}
\tag{9.124}
$$

Furthermore, Eqs. (9.81), (9.81), (9.89), (9.96), (9.98), (9.99), (9.101)–(9.103), and (9.115) yield

$$
\begin{aligned}
tg\varphi_2^* &= -\frac{1+\frac{cs_1^*}{kr_0}}{tg(kt_2^*+\alpha_1^*)},\\
x &= \frac{\sqrt{\mu}(\sqrt{p^*}-\sqrt{r_1})}{cs_2^*},\\
a_2^* &= \frac{\frac{M_0}{\bar{m}}-T^*}{tg(k_1\tau_2^*+\alpha_2^*)},\\
tg\varphi_3^* &= -\frac{\sqrt{\mu p^*}}{k_1x^{*2}tg\alpha_2^*},\\
tgf_3^* &= \frac{p^*\cos\varphi_3^*}{e^*x^*\sin\alpha_2^*}-1,
\end{aligned}
$$

$$
\begin{aligned}
a_1^* &= \frac{a_2^* \sin\alpha_2^*}{\sin(kt_2^* + \alpha_1^*)}, \\
C_{31}^* &= -\frac{r_0}{a_1^*}, \\
C_{32}^* &= -\frac{r_1}{a_2^*}, \\
tg\varphi_4^* &= -\frac{tg\varphi_3^* tg\alpha_2^*}{tg(k_1\tau_2^* + \alpha_2^*)} + \frac{cs_2^*}{k_1 a_2^* C_{32}^* tg(k_1\tau_2^* + \alpha_2^*)}, \\
t_3^* &= t_2^* + \frac{p^{*\frac{3}{2}}(1 - e^{*2})^{\frac{3}{2}}}{\sqrt{\mu}}[E_3^* - E_2^* - e^*(\sin E_3^* - \sin E_2^*)], \\
T^* &= t_2^* + \tau_2^*, \\
t_4^* &= t_3^* + T^* - t_2^*.
\end{aligned} \tag{9.125}
$$

Consequently, Eqs. (9.68), (9.69), (9.76), (9.77), (9.86), (9.87), (9.95), (9.96) allow us to determine the angles f_1, f_4, and the constants of the primer vector on ZT arcs. After some simplifications we obtain:

$$
\begin{aligned}
ctgf_1^* &= \frac{1}{2} ctg2\varphi_1^*, \\
ctgf_4^* &= \frac{1}{2} ctg2\varphi_4^*, \\
P_1^* &= a_1^* \frac{\sin\alpha_1^* \sin\varphi_1^*}{\sin f_1^*}, \\
P_2^* &= a_2^* \frac{\sin\alpha_2^* \sin\varphi_4^*}{\sin f_4^*}, \\
R_1^* &= P_2^*(tg\varphi_1^* \sin f_1^* - \cos f_1^*), \\
R_2^* &= P_2^*(tg\varphi_4^* \sin f_4^* - \cos f_4^*), \\
B^{*2} &= \frac{a_1^{*2} \sin^2(kt_2^* + \alpha_1^*)}{e^{*2}\sin^2 f_2^* + [\Phi_2^* + \frac{[e^{*2}(\sin^2 f_3^* - \sin^2 f_2^*) + (\Phi_3^2 - \Phi_2^2)]^{\frac{1}{2}}}{\Phi_2^*(\frac{1}{\Phi_2^{*2}} - \frac{1}{\Phi_3^{*2}})}]}, \\
D^{*2} &= \frac{B^{*2}[e^{*2}(\sin^2 f_3^* - \sin^2 f_2^*) + (\Phi_3^2 - \Phi_2^2)]}{\frac{1}{\Phi_2^{*2}} - \frac{1}{\Phi_3^{*2}}},
\end{aligned} \tag{9.126}
$$

Now it is possible to determine the polar coordinates of the junction points, and the angles between the thrust vector and horizontal axis for MT arcs. These unknowns can be determined from Eqs. (9.64), (9.72), (9.82), (9.91):

$$\begin{aligned}
\psi_1^* &= \frac{\pi}{2} - \varphi_1^*, \\
\theta_2^* &= \varphi_2^* - \varphi_1^*, \\
\omega_2^* &= \theta_2^* - f_2^*, \\
\theta_3^* &= \theta_2^* + f_3^* - f_2^*, \\
\psi_2^* &= \theta_2^* + f_3^* - f_2^* - \varphi_3^* + \frac{\pi}{2}, \\
\theta_4^* &= f_4^* + \omega_2^*.
\end{aligned} \tag{9.127}$$

Consequently, the solution of the variational problem is reduced to the solution of certain number of continuity equations that take place at the junction points.

Now consider the condition that allows to accomplish this maneuver with two MT arcs. Consider Eq. (9.117). At given $r_0, r_1, \frac{M_0}{\bar{m}}, c$, from this equality it follows that

$$\frac{kr_0 + c\min s_1(t_2)}{k_1 r_1 - c\min s_2(t_2, \tau_2)} \le \left(\frac{r_1}{r_0}\right)^{\frac{1}{2}} \le \frac{kr_0 + c\max s_1(t_2)}{k_1 r_1 - c\max s_2(t_2, \tau_2)}. \tag{9.128}$$

Also it can be shown that

$$\begin{aligned}
\min s_1(t_2) &= \int_{x_0}^{\alpha_1+\pi} \frac{\sin(\alpha_1 - x)}{x} dx, \\
\max s_2(t_2, \tau_2) &= \int_{x_0}^{\alpha_1+\pi} \frac{\sin(\alpha_1 - x)}{x} dx.
\end{aligned} \tag{9.129}$$

Analyzing $s_2(t_2, \tau_2)$, one can have

$$\begin{aligned}
\min s_2(t_2, \tau_2) &= \int_{x_3}^{\alpha_4+\pi} \frac{\sin(\alpha_4 - x)}{x} dx, \\
\max s_2(t_2, \tau_2) &= \int_{x_3}^{\alpha_4} \frac{\sin(\alpha_4 - x)}{x} dx,
\end{aligned} \tag{9.130}$$

where

$$\begin{aligned}
x_0 &= k_1 \frac{M_0}{\bar{m}}, \\
x_3 &= k_1 \frac{M_0}{\bar{m}} - k_1 t_2, \\
x_4 &= k_2 \frac{M_0}{\bar{m}} - k_1 (t_2 + \tau_2), \\
\alpha_4 &= \alpha_2 + k_1 \frac{M_0}{\bar{m}} - k_1 t_2.
\end{aligned}$$

Using the trapezoid method, Eqs. (9.129) and (9.130) can be rewritten in the form:

$$\min s_1(t_2) \approx \frac{\bar{m}(\alpha+\pi)\sin\alpha}{2kM_0} > -\frac{\pi}{4}\frac{\bar{m}}{kM_0},$$
$$\max s_1(t_2) \approx \frac{\bar{m}\alpha\sin\alpha}{2kM_0} < -\frac{\pi}{4}\frac{\bar{m}}{kM_0},$$
$$\min s_2(t_2,\tau_2) \approx \frac{\bar{m}(\alpha_2+\pi)\sin\alpha_2}{2k_1(M_0-\bar{m}t_2)} > -\frac{\pi}{4}\frac{1}{k_1(\frac{M_0}{\bar{m}}-\frac{\pi}{2k_1})}, \tag{9.131}$$
$$\max s_2(t_2,\tau_2) \approx \frac{\bar{m}\alpha_2\sin\alpha_2}{2k_1(M_0-\bar{m}t_2)} < \frac{\pi}{4}\frac{1}{k_1(\frac{M_0}{\bar{m}}-\frac{\pi}{2k_1})}. \tag{9.132}$$

Substitution of these expressions into Eq. (9.128) yields

$$\frac{kr_0-\frac{\pi}{4}\frac{\bar{m}}{kM_0}}{k_1r_1+\frac{\pi}{4}\frac{1}{k_1(\frac{M_0}{\bar{m}}-\frac{\pi}{2k_1})}} < \left(\frac{r_1}{r_0}\right)^{\frac{1}{2}} < \frac{kr_0+\frac{\pi}{4}\frac{\bar{m}}{kM_0}}{k_1r_1-\frac{\pi}{4}\frac{1}{k_1(\frac{M_0}{\bar{m}}-\frac{\pi}{2k_1})}}. \tag{9.133}$$

Furthermore, from Eqs. (9.104) and (9.116) it follows that

$$p_{1\,\min}+p_{2\,\min} \le 2p < p_{1\,\max}+p_{2\,\max}, \tag{9.134}$$

where

$$p_{1\,\min} = r_0\left(1+\frac{c\min s_1}{kr_0}\right)^2,$$
$$p_{2\,\min} = r_1\left(1-\frac{c\min s_2}{k_1r_1}\right)^2,$$
$$p_{1\,\max} = r_0\left(1+\frac{c\max s_1}{kr_0}\right)^2,$$
$$p_{2\,\max} = r_1\left(1+\frac{c\max s_2}{k_1r_1}\right)^2.$$

Here $\min s_1(t_2)$, $\max s_1(t_2)$, $\min s_2(t_2,\tau_2)$, $\max s_2(t_2,\tau_2)$ are evaluated with Eqs. (9.131). So, the transfer between the circular orbits with two MT arcs is possible if the ratio of the radii of the boundary orbits satisfies the condition in Eq. (9.133). The semi-latus rectum of the transfer orbit satisfies Eq. (9.134). As there are two gravitational fields in this problem, the losses of the gravitational acceleration at the second and third junction points are determined according to the formulas:

$$/\Delta g_1/ = /\frac{1}{r^3(f_2)}-\frac{1}{r_0^3}/, \qquad /\Delta g_2/ = /\frac{1}{r_1^3}-\frac{1}{r^3(f_3)}/, \tag{9.135}$$

where

$$r(f_2) = \frac{p}{1 + e\cos f_2}, \qquad r(f_3) = \frac{p}{1 + e\cos f_3}.$$

As in this problem the thrust generated by the chemical propulsion system is considered, then for evaluation of the relative error in the functional and in the dynamical model, $O(\Delta r/r_0)$, one can use the following formulas for the first and second junction points:

$$O(\frac{\Delta r(f_2)}{r_0}) \sim \frac{r(f_2) - r_0}{r_0},$$
$$O(\frac{\Delta r(f_3)}{r_1}) \sim \frac{r_1 - r(f_3)}{r_1}.$$

9.5.1 Numerical Example

It is required to transfer the spacecraft from the initial orbit with $r_0 =$ 6600 km to the final orbit with $r_1 = 26{,}400$ km with the use of a propulsion system with $c = 3$ km/s. In this case, $k = 0.0011775$ rad/s, and $k_1 = 0.00014718$ rad/s. Let $m_0 = 10{,}000$ kg and $\beta = m_0/100$ kg/s. The transfer trajectory is illustrated in Fig. 9.21. Based on the analytical solutions obtained above, at the junction points we have:

1. $\varphi_1 = -0.03256$, $\psi_1 = 1.60336$, $\alpha_1 = -1.60336$, $a_1 = 1042.6035$, $C_{21} = -6.3303$,
   $f_1 = -0.03253$, $P_1 = -1042.6023$, $R_1 = 1042.6023$
2. $\varphi_2 = 0.03279$, $\psi_2 = 1.60336$, $\theta_2 = 0.06535$, $f_2 = 0.00085$, $\omega = 0.0645$, $p = 10{,}561.7006$, $e = 0.59993$, $t_2 = 49.67158$, $B = -66{,}988.3772$, $D = -169{,}809.6909$
3. $\varphi_3 = -0.00096$, $\psi_3 = 4.77782$, $\theta_3 = 3.20607$, $f_3 = 3.14157$, $a_2 = -1042.2544$, $C_{22} = 1.0003$, $\alpha_2 = -1.57231$
4. $\varphi_4 = 0.0013$, $\psi_4 = 4.77782$, $\theta_4 = 3.20832$, $\tau_2 = 19.06081$, $f_4 = 0.7854$ $P_2 = -1042.2553$, $R_2 = -1042.2562$.

The maneuver time is $T_{NT} = 10{,}546.26$ s. The sizes of the thin spherical layers are $\Delta r_2 = |r(f_2) - r_0| = 1.3307$ km and $\Delta r_3 = |r_1 - r(f_3)| = 0.01814$ km. The relative errors in the functional and in the dynamical model are of the following orders:

$$O(\frac{\Delta r(f_2)}{r_0}) = \frac{r(f_2) - r_0}{r_0} = 2.02e - 04,$$
$$O(\frac{\Delta r(f_3)}{r_0}) = \frac{r_1 - r(f_3)}{r_1} = 6.87e - 07.$$

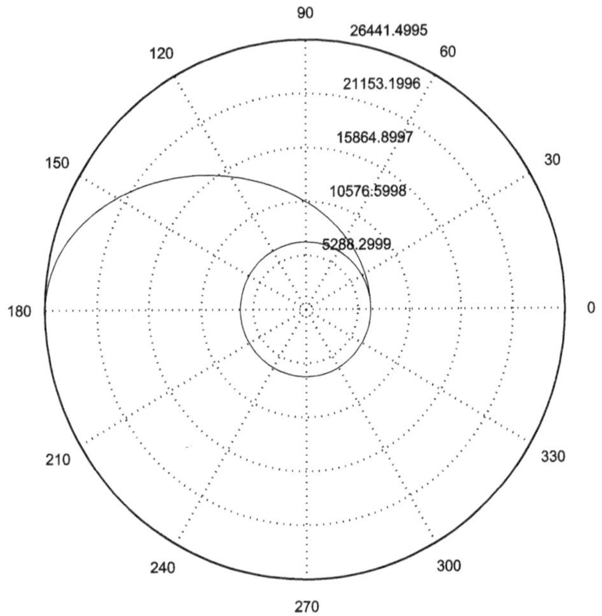

Figure 9.21 Transfer trajectory with two MT arcs.

Note that these values mean that in this problem the assumption on linearity of the gravitational field is valid. The losses of the gravitational acceleration at the second and third junction points are (in $1/\text{km}^3$):

$$|\epsilon_2| = 2.10e - 15,$$
$$|\epsilon_3| = 1.120e - 19.$$

In the case of constant exhaust velocity, the quantity $\Delta V = c\ln(m_0/m_f)$ can be accepted for computation of the fuel expenditure and referred to as the characteristic velocity [3]. For the corresponding Hohmann impulsive transfer we have

$$\frac{\Delta V_{IMP}}{v_0} = (1 - \frac{1}{\rho})\sqrt{\frac{2\rho}{1+\rho}} + \sqrt{\frac{1}{\rho}} - 1 = 0.44868.$$

The analytical solutions allow us to obtain that

$$\frac{\Delta V_{MT}}{v_0} = 0.448797.$$

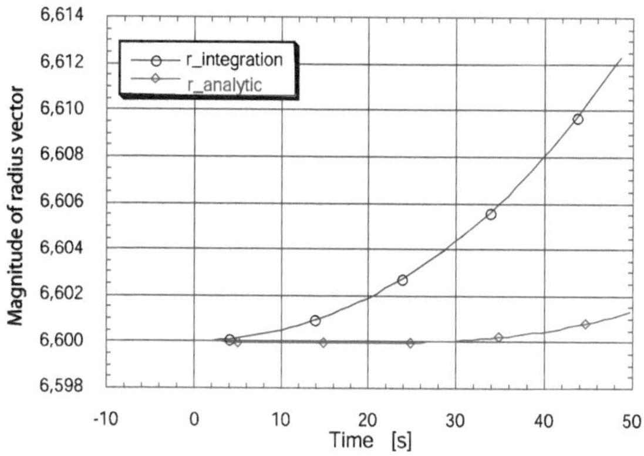

Figure 9.22 First MT arc: magnitude of radius vector vs. time.

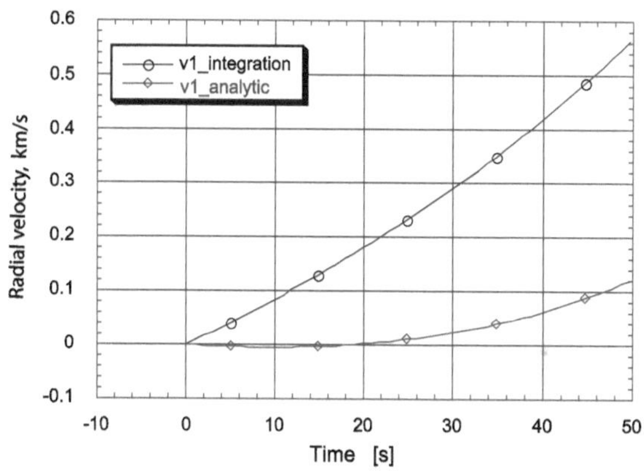

Figure 9.23 First MT arc: radial velocity vs. time.

The results of the computations, presented in Figs. 9.22–9.31, are based on the use of the thrust angle's time history in the numerical integrations of the equations of motion. The numerical integrations are carried out using the mathematical library IMSL and Runge-Kutta algorithm of a fifth order with fixed step size. Figs. 9.22–9.26 show the behavior of radius and velocity vectors, the thrust angle and the switching function during the first MT arc. Fig. 9.22 show that the difference between the analytical and numerical computations is 11.0 km, whereas the difference in velocity components in both cases is less than 0.004. It can be seen from Figs. 9.25 and 9.30 that the thrust angle is less than 1.87°. The analytical results for the second MT arc are presented in Figs. 9.27–9.31. The analysis show

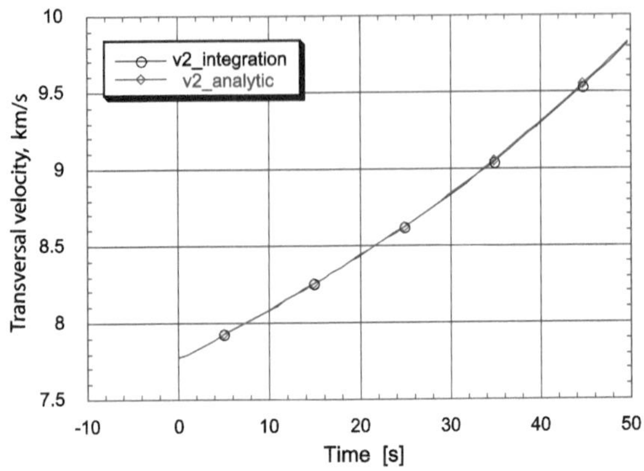

Figure 9.24 First MT arc: transversal velocity vs. time.

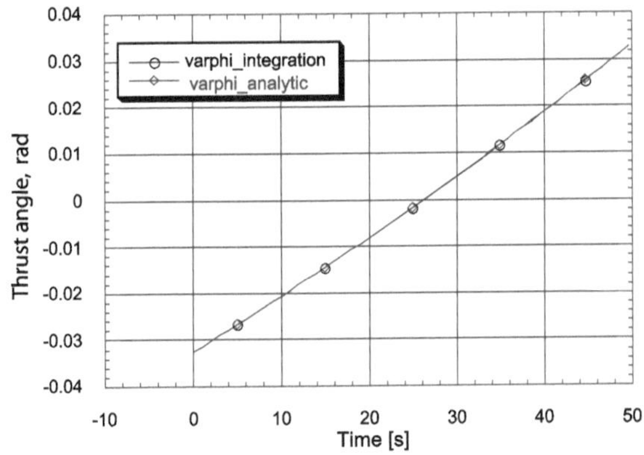

Figure 9.25 First MT arc: thrust angle vs. time.

that the characteristic velocity of the transfer is sensitive to $\rho = \frac{r_1}{r_0}$ and less sensitive to $\frac{m_0}{beta}$ and the exhaust velocity. The results for $2 \le \rho \le 10$ and $m_0/150 \le \beta \le m_0/60$ (which corresponds to $2g \le T/m_0 \le 5g$) are presented in Figs. 9.32 and 9.33. It was shown that $\Delta V_{MT}/v_0$ differs from $\Delta V_{IMP}/v_0$ no more than 0.0005 for the given interval $2 \le \rho \le 10$ (see Fig. 9.34). It can be concluded that the fuel expenditure in the case of MT with $\beta =$ *const* is comparable with the fuel expenditure in the case of the Hohmann impulsive transfer [167]. Note that the eccentricity of the transfer orbit is increased when the ratio of the radii of the orbits is increased.

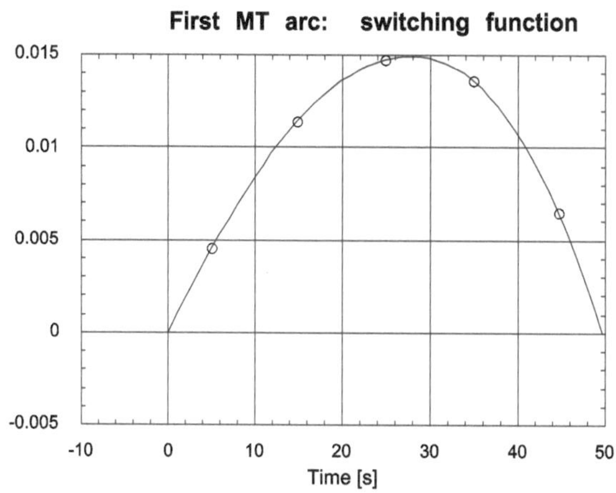

Figure 9.26 First MT arc: switching function vs. time.

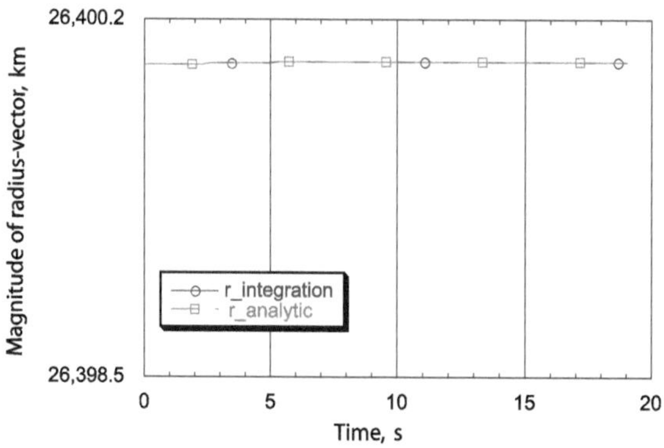

Figure 9.27 Second MT arc: magnitude of radius vector vs. time.

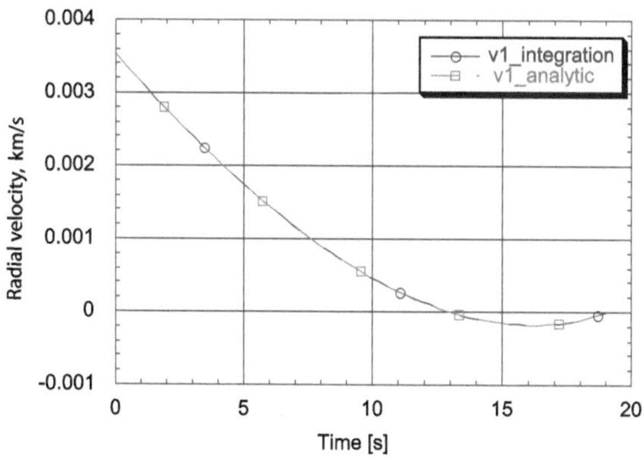

Figure 9.28 Second MT arc: radial velocity vs. time.

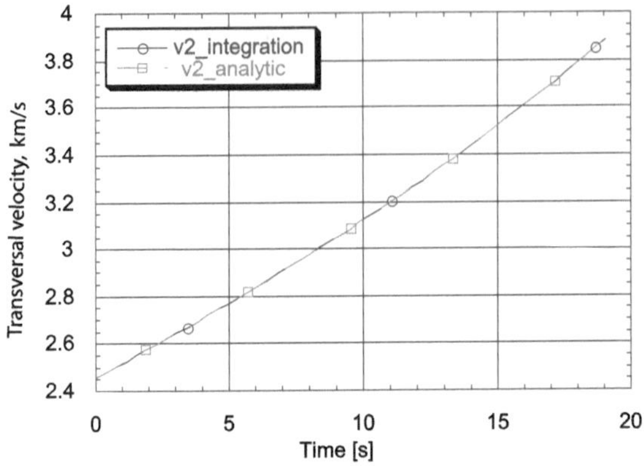

Figure 9.29 Second MT arc: transversal velocity vs. time.

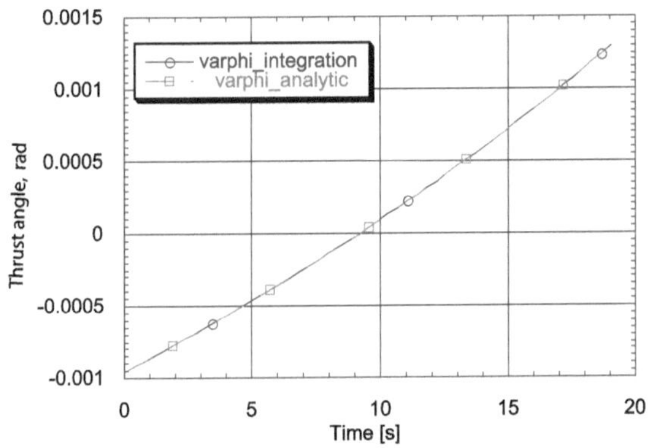

Figure 9.30 Second MT arc: thrust angle vs. time.

9.6 TRANSFER BETWEEN CIRCULAR AND HYPERBOLIC ORBITS VIA A MAXIMUM THRUST ARC WITH CONSTANT SPECIFIC IMPULSE

In this section, we consider the free final time minimum fuel transfer between circular and hyperbolic orbits in the Newtonian field using a MT arc. It is assumed that gravitational acceleration on the MT arc is a linear function of the radius vector and the hyperbolic orbit is tangent to the initial circular orbit at its perigee. The schematic picture of the terminal orbits is given in Fig. 9.35. It should be noted that the first assumption does not necessarily require the final orbit be tangential to the initial orbit. This problem can be analyzed by applying the closed-form analytical

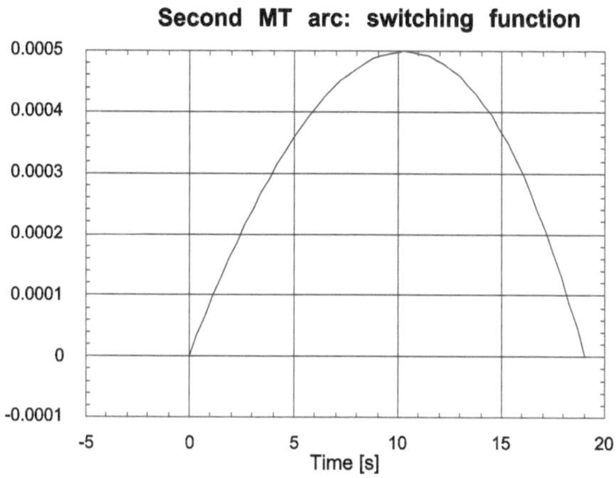

Figure 9.31 Second MT arc: switching function vs. time.

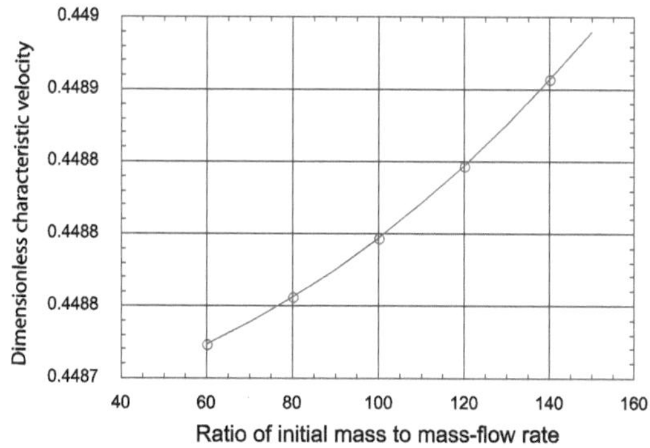

Figure 9.32 Functional dependence between ΔV_{MT} and $\frac{m_0}{\beta}$.

solution given in the previous section. The trajectory consists of three arcs, namely, NT-MT-NT, where the NT arcs represent the terminal orbits. The time of motion on these arcs is not fixed *a priori*. The optimal trajectory contains two switching points, the first is on the initial circular orbit, the second is on the final hyperbolic orbit. The assumption of a linear central gravity field applies only in the thin layer around the initial orbit in which the propulsion system is operating. The minimizing functional is the difference between initial and final masses, that is $\mathcal{J} = m_0 - m_1$. At $t = t_1 = 0$, the initial conditions are: $r = r_0$, $v_{11} = 0$, $v_{21} = kr_0$, $m = m_0$ and the final time, $t = t_2$, the conditions are given as

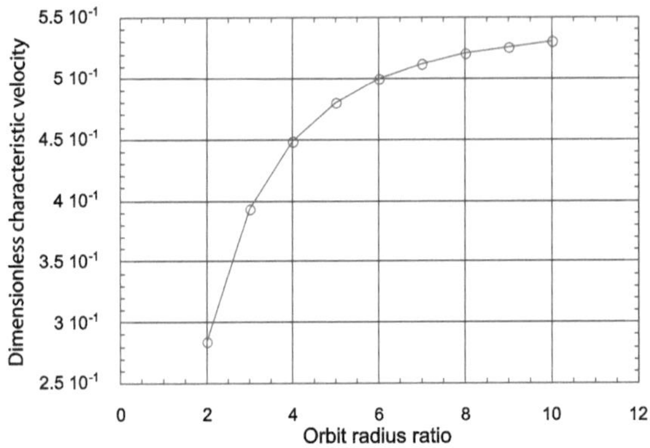

Figure 9.33 Functional dependence between ΔV_{MT} and the ratio of radii of the boundary orbits.

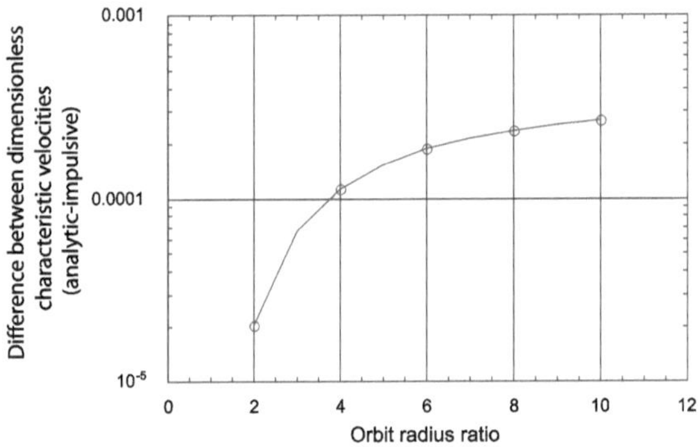

Figure 9.34 Functional dependence between the difference, ΔV_{MT}-ΔV_{IMP} and the ratio of radii of the boundary orbits.

$$
\begin{aligned}
t &= t_2, \\
r &= \frac{p}{1+e\cos f}, \\
v_1 &= \sqrt{\frac{\mu}{p}}e\sin f, \\
v_2 &= \sqrt{\frac{\mu}{p}}(1+e\cos f),
\end{aligned}
$$

where $k=\sqrt{\mu/r_0^3}$. On the optimal trajectory, at the switching points, the radius, velocity, and primer vectors must all be continuous and the switching function $\chi=0$ [3]. Using the analytical solutions given in the previous

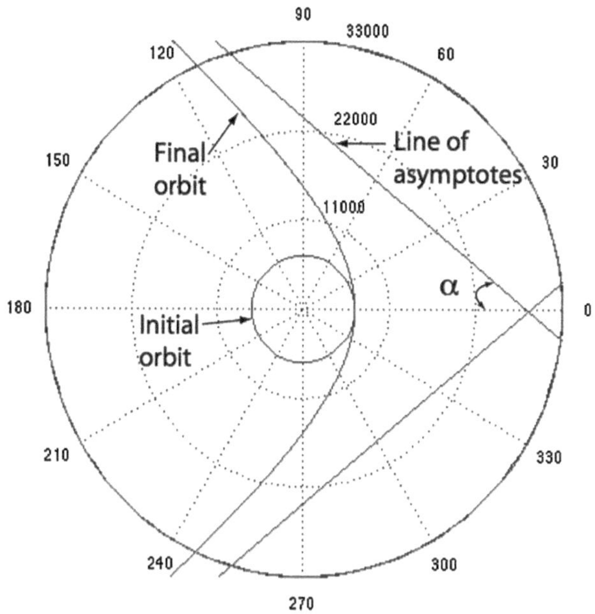

Figure 9.35 Maneuver geometry.

section, we can obtain following continuity conditions at the switching points:
At the first switching point:

$$\begin{aligned} 0 &= \varphi_1 + \psi_1 - \frac{\pi}{2}, \\ r_0 &= aC_2 \frac{\sin\alpha}{\cos\varphi_1}, \\ kr_0 &= aC_2 \frac{\sin\alpha}{\cos\varphi_1} \dot{\varphi}_1, \\ P_1 \sin f_1 &= a \sin\alpha \sin\varphi_1, \\ 2P_1 \cos f_1 + R_1 &= a \sin\alpha \cos\varphi_1, \end{aligned}$$

where f_1 is the true anomaly, $\theta_1 = 0$ is the polar angle of the initial point of the MT arc, and

$$\dot{\varphi}_1 = -\frac{k \sin 2\varphi_1}{\sin 2\alpha} \tag{9.136}$$

At the second switching point:

$$
\begin{aligned}
F_1 \sin\alpha_1 + F_2 \cos\alpha_1 &= 0, \\
\theta_2 &= \varphi_2 + \psi_2 - \frac{\pi}{2} = f_2 + \omega, && (9.137)\\
\frac{p}{1+e\cos f_2} &= aC_2 \frac{\sin(kt_2+\alpha)}{\cos\varphi_2}, && (9.138)\\
\sqrt{\frac{\mu}{p}} e \sin f_2 &= aC_2 \frac{\sin(kt+\alpha)}{\cos\varphi_2}[k\cot(kt+\alpha) + \dot{\varphi}_2 \tan\varphi_2], \\
& && (9.139)\\
\sqrt{\frac{\mu}{p}}(1+e\cos f_2) &= aC_2 \frac{\sin(kt_2+\alpha)}{\cos\varphi_2}\dot{\varphi}_2, && (9.140)\\
P_2 e \sin f_2 &= a\sin(kt_2+\alpha)\sin\varphi_2, \\
P_2(1+e\cos f_2) + \frac{R_2}{1+e\cos f_2} &= a\sin(kt_2+\alpha)\cos\varphi_2,
\end{aligned}
$$

where

$$
\begin{aligned}
\tan\varphi_2 &= \frac{\tan\varphi_1 \tan\alpha}{\tan(kt_2+\alpha)} + \frac{cs}{aC_2 k \tan(kt_2+\alpha)}, \\
\dot{\varphi}_2 &= -\frac{k\sin 2\varphi_2}{\sin 2(kt_2+\alpha)}, \\
s &= F_2 \sin\alpha_1 - F_1 \cos\alpha_1, \\
\alpha_1 &= \alpha + \frac{km_0}{\beta}, \\
x_1 &= \frac{km_0}{\beta}, \\
x_2 &= \frac{km_0}{\beta} - kt_2, \\
F_1 &= Si(x_2) - Si(x_1), \\
F_2 &= Ci(x_2) - Ci(x_1)
\end{aligned}
$$

and (p, e, ω) are parameters of the elliptic transfer orbit, t_1 and t_2 are the initial and final times on the MT arc, respectively, and $\psi_2 = \psi_1$ since ψ is a constant. P_1, P_2, R_1, R_2 are known as Lawden's integration constants. Without loss of generality we can choose $t_1 = 0$.

From the transversality condition (taking into account the arbitrary phasing on the boundary orbits) it follows that

$$
\lambda_{71} = -\frac{\partial \mathcal{J}}{\partial m_1} = 1 \qquad (9.141)
$$

at $t = t_2$. That means that at the second switching point we have

$$ca\sin(kt_2 + \alpha) = m_0 - \beta t_2 \tag{9.142}$$

From the conditions at the first switching point it can be shown that

$$\begin{aligned}
\alpha_1 &= \arctan(-\frac{F_2}{F_1}) - \frac{km_0}{\beta}, \\
\alpha &= \alpha_1 - \frac{km_0}{\beta}, \\
f_1 &= \arctan(-\frac{\sin 2\alpha_1}{2}), \\
\varphi_1 &= \alpha + \frac{\pi}{2}, \\
\dot{\varphi}_1 &= k, \\
r_0 &= -C_2 a_1
\end{aligned} \tag{9.143}$$

and from the conditions at the second switching point one can obtain

$$\tan\varphi_2 \tan(kt_2 + \alpha) = -(1 + \frac{cs}{kr_0}). \tag{9.144}$$

Substituting the integration constant from Eq. (9.143) into Eq. (9.138) for the radius vector magnitude yields

$$1 + e\cos f_2 = -\frac{p}{r_0}\frac{\cos\varphi_2}{\sin\gamma} \tag{9.145}$$

where $\gamma = kt_2 + \alpha$. Substitution of Eq. (9.145) into Eq. (9.140) yields

$$\tan^2\varphi_2 \tan^2\gamma = \frac{p}{r_0} \tag{9.146}$$

Then, from Eq. (9.144) and Eq. (9.146) it follows that

$$\frac{p}{r_0} = \left(1 + \frac{cs}{kr_0}\right)^2 \tag{9.147}$$

At given p, r_0, c, m_0, and β, Eq. (9.146) can be used to find the final time t_2. From Eqs. (9.139)–(9.140), it can be found that

$$e\sin f_2 = -\sqrt{\frac{p}{r_0}}(\frac{\cos^2\gamma - \sin^2\varphi_2}{\cos\gamma\cos\varphi_2}),$$

$$e\cos f_2 = -\sqrt{\frac{p}{r_0}\frac{\sin\varphi_2}{\cos\gamma}} - 1,$$

from which, by denoting the first equation by e_1, and the second equation by e_2, one can obtain expression for f_2 and relationship for p, r_0, c, m_0, β:

$$\tan f_2 = \frac{e_1}{e_2}, \qquad e^2 = e_1^2 + e_2^2$$

Consequently, the unknowns at the second switching point, $r_2, v_1, v_2, \theta_2, \varphi_2$ can be computed based on Eqs. (9.137)–(9.138), (9.140), and (9.144). The thrust angle at the first switching point, φ_1, is found from Eq. (9.143), and $\psi_1 = \psi_2 = \frac{\pi}{2} - \varphi_1$. The integration constants may be computed using the expressions:

$$a = \frac{m_0 - \beta t_2}{c\sin(kt_2 + \alpha)}, \qquad C_2 = -\frac{r_0}{a}$$

Then, we can find the initial primer vector magnitude as $\lambda_0 = a\sin\alpha$. Lawden's integration constants can be found from following equations:

$$\begin{aligned}
P_1 &= a\sin\alpha,\\
P_2 &= \frac{a\sin\gamma\sin\varphi_2}{e\sin f_2},\\
R_1 &= -a\sin\alpha,\\
R_2 &= (1 + e\cos f_2)[P_1(1 + e\cos f_2) + a\sin\gamma\cos\varphi_2].
\end{aligned}$$

The relative errors associated with the linear approximation can be computed using Eqs. (6.15)–(6.16).

9.6.1 Numerical Example

The optimal transfer problem discussed in the previous section is analyzed in this section using numerical values of the transfer parameters. It should be noted that the continuity conditions at the switching points allow us to calculate the integration constants and final time in order to construct time histories and the transfer trajectory. For the simple case following values have been considered: $r_0 = 6600$ km, $\mu = 398{,}600.1445$ km^3/s^2, $k = 0.0011775$ rad/s, $m_0 = 10{,}000$ kg, $\beta = m_0/100$ kg/s, $c = 3$ km/s, $e = 1.2$, $p = 2.3$. The variable ω can be chosen arbitrarily.

Using the formulas derived in the previous section, the following numerical results have been obtained. Integration constants are:

$$a = -2625.9, \quad \alpha = -1.6236, \quad C_2 = 2.5135, \quad \lambda_0 = 2622.2.$$

At the first switching point:

$$t_0 = 0, \quad r_1 = r_0 = 6600, \quad \theta_1 = 0, \quad v_1 = 0,$$
$$v_2 = 7.7714, \quad m_0 = 10{,}000, \quad \varphi_1 = -0.05285, \quad \psi_1 = 1.6236.$$

At the second switching point:

$$t_2 = 73.778, \quad r_2 = 6605.0, \quad \theta_2 = 0.10442, \quad v_1 = 0.3432,$$
$$v_2 = 11.777, \quad m_2 = 2622.2, \quad \varphi_2 = 0.05157,$$
$$\psi_2 = 1.6236, \quad f2 = 0.051541$$

In order to obtain a more complete qualitative picture of the transfer, the following quantities have been used: angle of asymptotes, α_h, energy constant, K, semi-major axes, a_h, hyperbolic mean motion, n_h, and the dimensionless characteristic velocity of the analytical and impulsive transfer between terminal orbits, $\Delta v/v_0$. These values can easily be calculated using the formulas:

$$\begin{aligned}
\alpha_h &= \arctan\sqrt{e^2-1} = 39.715, \\
K &= v_p^2 - \frac{2\mu}{r_p} = 18.118, \\
r_p &= \frac{p}{1+e} = 6600.0, \\
v_p &= \sqrt{\frac{\mu}{p}}(1+e) = 11.7858, \\
a_h &= \frac{p}{e^2-1} = \frac{\mu}{K} = 22{,}000.0, \\
n_h &= \sqrt{\mu a_h^3} = 0.00019350, \\
\frac{\Delta v}{v_0} &= 0.51673, \\
\frac{\Delta v_{imp}}{v_0} &= \sqrt{1+e} - 1 = 0.51657,
\end{aligned}$$

where r_p and v_p are the perigee distance of the hyperbolic orbit and total velocity at this point of the orbit, and where $v_0 = \sqrt{\mu/r_0}$. For comparison, the difference in the dimensionless analytical and impulsive characteristic velocities is $1.6e-4$. The relative errors associated with the linear approximation can be computed based on the formulas given in the previous section. The errors are as follows: in the gravitational acceleration $\epsilon_g = 7.8933e-15$; in the dynamic model $O\left(\Delta r(f_2)/r_0^4\right) = 2.6351e-15$; in

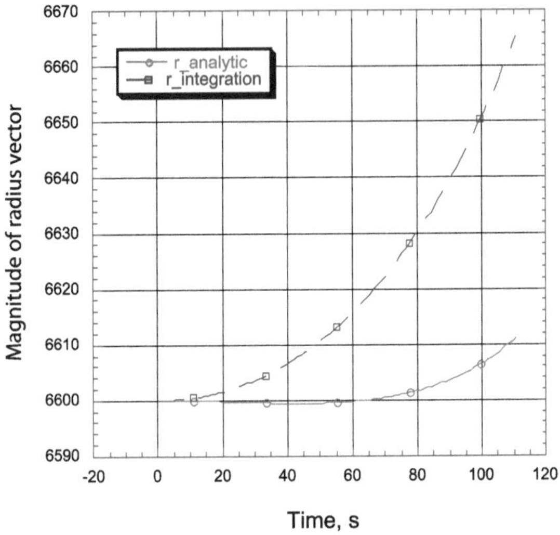

Figure 9.36 Time history of *r*.

the performance functional $O\left(\Delta r(f_2)/r_0\right)^2 = 1.1478e - 7$. For comparison purposes, the equations of motion in the Newtonian field can be integrated numerically if final time and thrust direction history are available. These values were taken from the analytical results. A fifth-order fixed-step Runge-Kutta method was used as the direct method of integration. The absolute error in all computations was less than 10^{-11}. Figs. 9.36–9.39 show the time histories of the magnitude of the radius vector, components of the velocity vector, and the thrust angle. Figs. 9.36–9.38 also show the integrated time histories of the position and velocity vectors for comparison with the analytical results. The results show that the relative differences in the magnitude of the radius vector is approximately 0.825%, and the differences in the radial and transversal velocities are 12.47% and 0.011%, respectively.

In order to analyze the possible relationships between the transfer variables and understand their behavior during the maneuver, following sets of initial conditions have been used:

$$\frac{m_0}{60} \leq \beta \leq \frac{m_0}{150} \qquad 2.1 \leq \frac{p}{r_0} \leq 3.8, \qquad 1 \leq e \leq 2.8$$

where m_0 and r_0 were fixed at the values given above. The computations and analysis of the results indicate an almost linear relationship of the flight

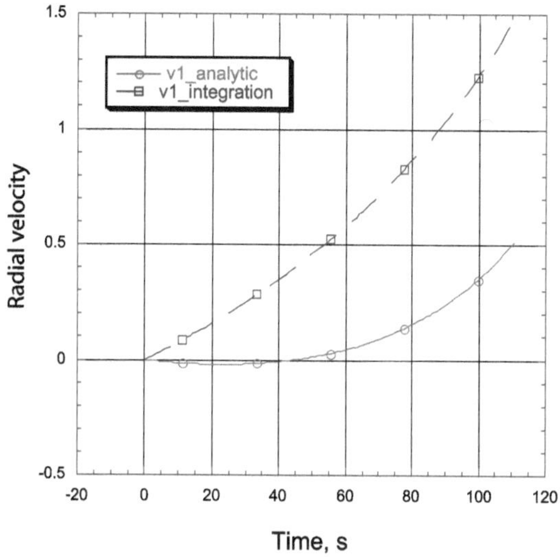

Figure 9.37 Time history of v_1.

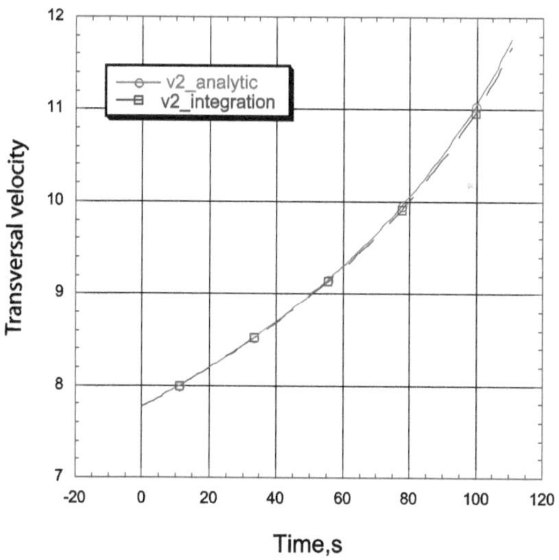

Figure 9.38 Time history of v_2.

duration with the ratio of the initial mass to mass-flow rate, m_0/β, and with the flight duration and the polar angle (under the conditions that the eccentricity, ratio p/r_0, and mass are fixed). The same almost linear relationship (see for example, Fig. 9.40) applies to the eccentricity with the ratio p/r_0

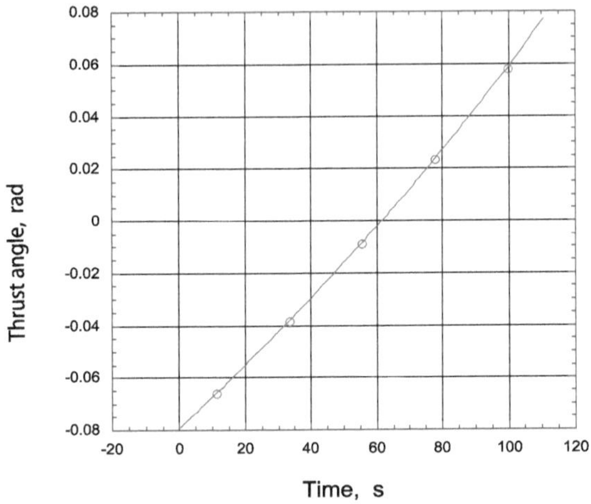

Figure 9.39 Time history of φ.

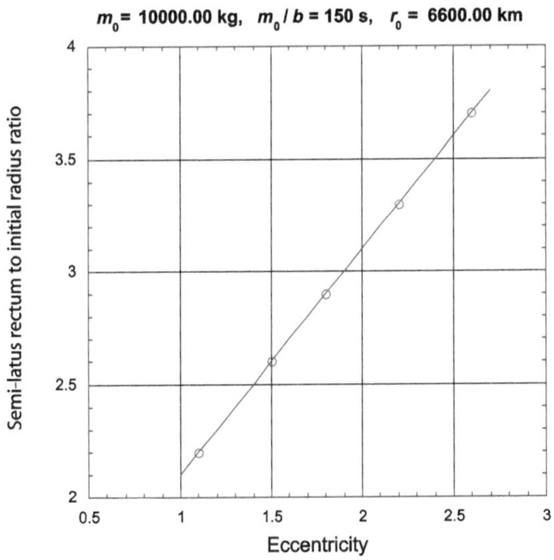

Figure 9.40 Relationship between p/r_0 and e.

and the eccentricity with the energy constant (under the conditions that the initial mass and the initial radius are fixed). The eccentricity, energy constant and dimensionless characteristic velocity, $\Delta v/v_0$, are less sensitive to changes in the ratio m_0/β and, except for the $\Delta v/v_0$, these parameters behave in a similar manner with respect to the ratio m_0/β at fixed eccen-

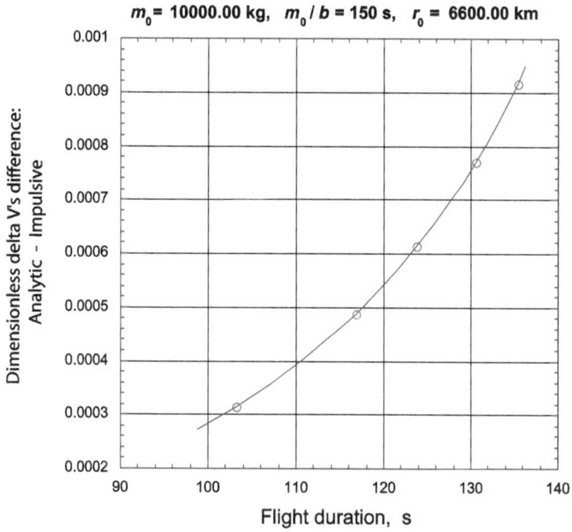

Figure 9.41 $\Delta v - \Delta v_{imp}/v_0$ vs. flight duration.

tricity, ratio p/r_0, initial mass and initial radius. It was also observed that the thrust angle is almost linear with respect to ratio m_0/β and the behavior of the initial and final thrust angles are nearly symmetric with respect to axes of the ratio m_0/β. Larger values of the ratio result in larger values of the thrust angle. The eccentricity increases monotonically with the final total velocity and flight duration, in which the radial component increases more slowly than the transversal velocity. There exists only one set of eccentricity, ratio p/r_0, and flight duration at fixed m_0/β, initial mass and initial radius. The dimensionless characteristic velocity of the impulsive transfer, denoted by $\Delta v_{imp}/v_0$, was also compared with the analytical performance. The results indicate that the difference between $\Delta v/v_0$ and $\Delta v_{imp}/v_0$ slowly increases with the flight duration, eccentricity, and ratio p/r_0, and this difference has a nearly linear relationship with the latter parameters and is less than 0.001 for all ranges of parameters considered in these computations (see Fig. 9.41).

The analysis confirms that the Newtonian field can be approximated by a linear field to describe analytic MT arcs if the spacecraft remains in the thin spherical layer around the initial orbit and can be used to analyze optimal transfer maneuver performed within the boundaries of the layer. Although the transfer parameters satisfy the necessary conditions of optimality, the tests for sufficiency conditions require further investigations.

The variational problem on the determination of optimal spacecraft trajectories has been considered. A problem statement that involves both low and high thrust propulsion systems has been discussed. It is shown that a linear field approximation can be applied to the problem. The analytical solutions for maximum thrust arcs have been presented. The optimal transfer between circular and hyperbolic orbits in a Newtonian field is analyzed. The corresponding analytical solutions can be used to describe all parameters of the maneuver and impulsive transfer can be replaced by the maximum thrust arc that lies in the thin layer around the initial orbit.

9.7 PLANETARY DESCENT AND LANDING TRAJECTORIES WITH CONSTANT THRUST ACCELERATION

In this section, the analytical solutions trajectories presented in Chapter 7 will be used to simulate a powered Mars descent and landing trajectory with an optimal constant thrust acceleration vector. The results for the position, velocity and mass obtained here are comparable to those of Ref. [123], especially the results of case 1 considered in this reference. As is shown in the solutions, the main procedure is to search for the constants a_1, b_1, a_2, b_2, a_3, and b_3 for given simulation time, t_1 and control parameter, u. Assuming that the control parameter is the thrust acceleration, the following inequality is used in the simulations:

$$\frac{\rho_1}{m} \leq \sigma \leq \frac{\rho_2}{m}$$

with

$$m_{dry} \leq m_1 \leq m < m_0 = m_{wet}, \qquad m_1 = m(t_1),$$

where ρ_1 and ρ_2 are the boundary values of thrust. So practically,

$$u \leq u_{max} = \sigma_{max} = \frac{\rho_2}{m_{wet}}.$$

As the mass changes exponentially, the simulation time was computed according to the condition:

$$[t_1]_{min} = \frac{1}{\alpha u_{max}} \ln \frac{m_0}{m_1}.$$

9.7.1 Numerical Example

This example simulates a powered Mars descent and landing trajectory using the initial conditions provided in Ref. [123]. The following values of the parameters and terminal conditions are used:

$$g0 = 3.7114, \quad m_{dry} = 1505, \quad m_{wet} = 1905, \quad \alpha = 4.5 \times 10^{-4},$$
$$\rho_1 = 4972, \quad \rho_2 = 13{,}260, \quad \sigma_{max} = \frac{\rho_2}{m_{dry}}, \quad \sigma_{min} = \frac{\rho_1}{m_{wet}},$$
$$u_{max} = \sigma_{max}, \quad u_{min} = \sigma_{min}.$$
$$x_{10} = 1500.0, \quad x_{20} = 500.0, \quad x_{30} = -2000.0,$$
$$v_{10} = -75.0, \quad v_{20} = 0, \quad v_{30} = 100.0$$
$$x_{11} = 0, \quad x_{21} = 0, \quad x_{31} = 0, \quad v_{11} = 0, \quad v_{21} = 0, \quad v_{31} = 0,$$
$$15 \le a_1 \le 35, \quad 1 \le b_1 \le 10,$$
$$a_2 = 0, \quad b_2 = 0,$$
$$-13 \le a_3 \le -11, \quad -10 \le b_3 \le -1,$$
$$u_{min} \le u \le u_{max},$$
$$0 \le t \le 55.$$

In particular, the following values have been used:

$$a_1 = 35, \quad b_1 = 4, \quad a_2 = 0, \quad b_2 = 0,$$
$$a_3 = -13, \quad b_3 = -5, \quad u = 5.69, \quad 0 \le t \le 50$$

Here, the position, velocity, accelerations and mass are measured in *meters*, *meters/s*, *meters/s*^{2} and *kg* respectively, and these units will be omitted for simplicity. Only elementary algebraic functions and a very simple search algorithm were used to simulate the descent trajectories. The algorithms used in the simulations do not use any iterations, function calls or approximations except for the uniform gravity field, which is a reasonable approximation for maneuvers in the vicinity of the Mars surface. The preliminary results of the simulations are presented in Figs. 9.42–9.45. Figs. 9.42 and 9.43 demonstrate that the analytical solutions can generate feasible and extremal descent and landing trajectory envelops and the parameter domains. This means that any of the values of the coefficients from the given domains can provide an extremal or optimal trajectory. The extremality is understood in the sense of satisfying the necessary conditions of optimality with performance index given as $J = u(t_f - t_i)$, where $u = const$. In this sense, these descent trajectories considered can be optimal if the final conditions are

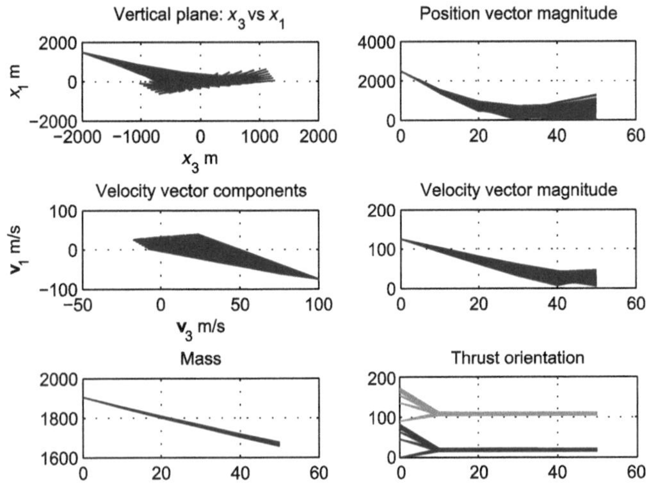

Figure 9.42 Descent trajectory envelops and the parameter domains: $15 \le a_1 \le 35$, $5 \le b_1 \le 10$, $a_2 = 0$, $b_2 = 0$, $-12 \le a_3 \le -8$, $-7 \le b_3 \le 0$, $5.7 \le u \le 6.2$.

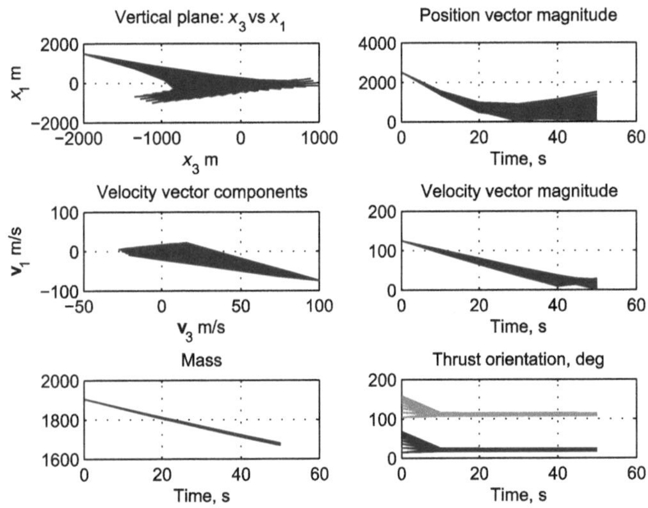

Figure 9.43 Descent trajectory envelops and the parameter domains: $15 \le a_1 \le 35$, $5 \le b_1 \le 10$, $a_2 = 0$, $b_2 = 0$, $-13 \le a_3 \le -11$, $-10 \le b_3 \le -1$, $5.6 \le u \le 5.9$.

met exactly with appropriate selection of the control, u. Feasibility of the trajectories is understood in the sense of connecting the initial and final conditions with some nonzero landing errors in the final position and velocity vectors, and satisfying the mass, control and time constraints as well

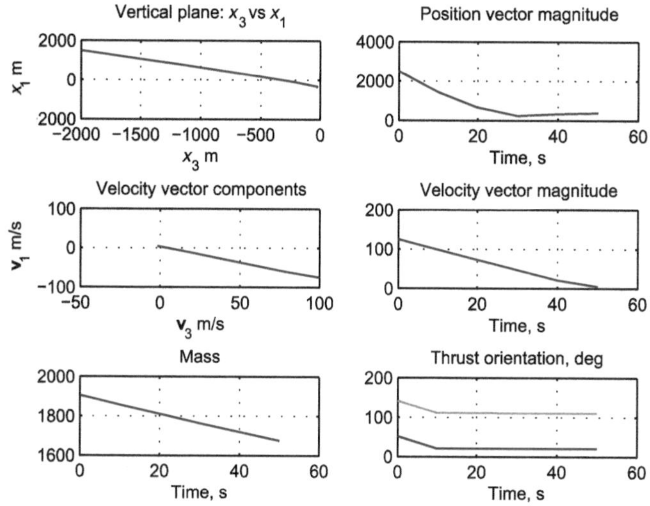

Figure 9.44 Descent trajectory parameters: $a_1 = 35, b_1 = 4, a_2 = 0, b_2 = 0, a_3 = -13, b_3 = -5, u = 5.69$.

as an engineering intuition. Although the purpose of the optimal control problem was to minimize J, the algorithms were also designed to minimize the landing errors in position and velocity. This is done by searching for not only the constants a_1, b_1, a_2, b_2, a_3, and b_3, but also for the appropriate simulation time, t_1 and the control parameter, u to achieve the minimum landing errors. The visual analysis of the boundaries of the trajectory envelops expose the largest landing errors of 1200 m in position and as high as 20 m/s in velocity. But it can be easily seen that by appropriate selection of the coefficients and the control parameter and the final time, these errors can be minimized to a some degree of satisfaction of the final conditions. Indeed, Figs. 9.44 and 9.45 show that both position and velocity errors at landing can be decreased or in some cases, eliminated using a local search of some parameters. For Fig. 9.44, the final values of the parameters are:

$$\mathbf{r}_1(-384.3, 500.0, -34.5), \quad \mathbf{v}_1(5.0, 0, -1.7), \quad m_1 = 1676.1, \quad t_1 = 50.0$$

For Fig. 9.45, the final values of the parameters are:

$$\mathbf{r}_1(0.0,\ 500.0,\ 661.2), \quad \mathbf{v}_1(-7.6,\ 0,\ 0.9), \quad m_1 = 1699.8, \quad t_1 = 44.4$$

More detailed search resulted in more accurate landing parameters which are given in the Table 9.1. The last lane of the table shows that the safe

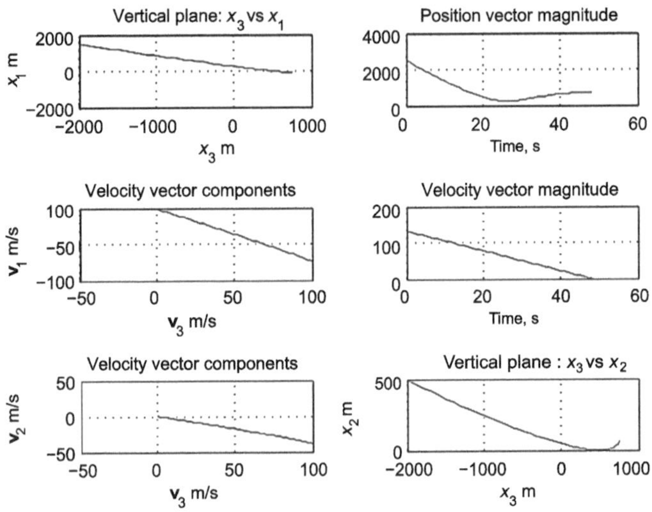

Figure 9.45 Descent trajectory parameters: $a_1 = 35$, $b_1 = 21$, $a_2 = 5$, $b_2 = 10$, $a_3 = -14.5$, $b_3 = -2.0$, $u = 5.723$.

Table 9.1 Set of selected coefficients and parameters

a_1	a_3	b_1	b_3	u	t_1	$\|\mathbf{r}_1\|$	$\|\mathbf{v}_1\|$
37.70	4.90	−14.40	−4.10	5.600	49.60	464.8702	0.6984
37.79	4.90	−14.40	−4.10	5.600	49.60	464.8361	0.4789
37.00	4.70	−14.00	−4.20	5.600	49.59	463.9562	0.0290
38.00	4.80	−14.40	−4.10	5.600	49.60	465.3398	0.0024

landing with almost zero velocity, $|\mathbf{v}_1| = 0.0024$ m/s can be provided with the positional error of $|\mathbf{r}_1| = 465.34$ m at landing. The final mass is computed to be less than 1676 kg, which means that less than 229 kg of fuel is used to perform the descent and landing maneuver. The total maneuver time was found to be 49.60 s. The studies show that more narrow and a wider local (or even global) search for the coefficients, the maneuver time and the control parameter will yield more accurate and safe landings.

9.8 COMPARISON OF ANALYTICAL AND NUMERICAL SOLUTIONS

9.8.1 One Class of Thrust Arcs With Constant Specific Impulse and Variable Power

One of such thrust arcs can be described by the following formulas [40]:

$$
\begin{aligned}
r &= \frac{\mu}{n^2} L_1(s), && (9.148)\\
v_1 &= nL_2(s), && (9.149)\\
v_2 &= nL_3(s), && (9.150)\\
\theta &= \theta_0 - 3\cot\varphi - 4\varphi, && (9.151)\\
m &= m_0 \exp h(\varphi), && (9.152)\\
t &= t_0 + 27\frac{\mu}{n^3}\int_{\varphi_0}^{\varphi} \frac{s^7(3-5s^2)}{(1-3s^2)^2}\, d\varphi, && (9.153)\\
\beta &= -\frac{n^4}{27\mu}\frac{m}{c}\frac{(1-3s^2)^3(9-25s^2+20s^4)}{s^{11}(3-5s^2)^3}, && (9.154)\\
a_t &= \frac{n^4}{27\mu}\frac{(1-3s^2)^3(9-25s^2+20s^4)}{s^{11}(3-5s^2)^3}, && (9.155)
\end{aligned}
$$

where

$$
L_1(s) = 9\frac{s^6}{1-3s^2}, \quad L_2(s) = 2\frac{k(1-2s^2)}{s^2(3-5s^2)}, \quad L_3(s) = \frac{3-13s^2+12s^4}{3s^3(3-5s^2)},
$$

$$
s = \sin\varphi, \qquad h(\varphi) = \frac{1}{c}\left[n\frac{k(10s^4-7s^2+1)}{s^3(3-5s^2)} + d\right], \qquad d = \frac{C_1}{\lambda} = const,
$$

n, θ_0, m_0, d, λ, t_0, t_c, C_1 are constants, c is constant exhaust velocity. Note that these solutions are valid in the case when $\sin^2\varphi < \frac{1}{3}$, or when $|\varphi| < 0.6155$. The corresponding Lagrange multipliers can be computed using the formulas:

$$
\begin{aligned}
\lambda_1 &= \lambda s,\\
\lambda_2 &= \lambda k,\\
\lambda_4 &= \lambda\frac{n^3}{27\mu}\frac{k(1-3s^2)}{s^9},\\
\lambda_5 &= \lambda n,\\
\lambda_7 &= \lambda\frac{c}{m}.
\end{aligned}
\qquad (9.156)
$$

9.8.2 One Class of Thrust Arcs With Variable Specific Impulse and Maximum Power

According to the analysis of the variation problem, there exist at least two classes of the thrust arcs described analytically. These classes represent spiral trajectories around the center of attraction. One of these classes can be described in the form:

$$r = \frac{\mu}{d^2}F_1, \tag{9.157}$$

$$v_1 = dF_2, \tag{9.158}$$

$$v_2 = dF_3, \tag{9.159}$$

$$\theta = \frac{1}{4}\left(\frac{3}{\tan\varphi_0} + \varphi_0\right) - \frac{1}{4}\left(\frac{3}{\tan\varphi} + \varphi\right) + \theta_0, \tag{9.160}$$

$$\frac{1}{m} = \frac{1}{m_0} + \frac{P_{max}\lambda^2}{2b^2}t, \tag{9.161}$$

$$I_{sp} = \frac{2b}{g\lambda m}, \tag{9.162}$$

$$\frac{m_0}{m_f} = \frac{I_{sp,f}}{I_{sp,0}}, \tag{9.163}$$

$$t = \frac{1}{t_c}\left[\frac{3zk(1-5s)}{3-5s} - \frac{C_1}{\lambda}\right] + t_0, \tag{9.164}$$

where

$$s = \sin^2\varphi, \qquad k = \cos\varphi, \qquad z = \sqrt{\frac{\mu}{r}(1-3s)},$$

$$F_1(s) = s^{\frac{3}{4}}, \qquad F_2(s) = \frac{6sk}{3-5s}\sqrt{\frac{1-3s}{s^{\frac{3}{4}}}}, \qquad F_3(s) = \frac{3-s}{3-5s}\sqrt{\frac{1-3s}{s^{\frac{3}{4}}}},$$

d, θ_0, m_0, t_0, b, λ are constants. The Lagrange multipliers are described by the formulas:

$$\lambda_1 = \lambda\sqrt{s}, \tag{9.165}$$

$$\lambda_2 = \lambda k, \tag{9.166}$$

$$\lambda_4 = -\frac{k}{r}\sqrt{\frac{\mu}{r}(1-3s)}\lambda, \tag{9.167}$$

$$\lambda_5 = 0, \tag{9.168}$$

$$\lambda_7 = \frac{b}{m^2}. \tag{9.169}$$

Constants of integration are determined in the form:

$$d = a^2,$$

$$a = \left(\frac{\mu\alpha}{3}\right)^{\frac{1}{4}},$$

$$\begin{aligned}
\alpha &= \frac{P_{max}}{2}\frac{\lambda}{b} - \frac{C}{\lambda}, \\
\frac{C_1}{\lambda} &= \frac{3k_0 - z_0(1 - 5s_0)}{3 - 5z_0}, \\
\frac{\bar{C}}{\lambda} &= \frac{C}{\lambda} - \frac{P_{max}}{2}\frac{\lambda}{b}, \\
t_c &= \frac{5}{2}\frac{P_{max}\lambda}{b} - 3\frac{\bar{C}}{\lambda}.
\end{aligned} \tag{9.170}$$

9.8.3 Comparison for Low Thrust Arcs

Consider the maneuver of transfer from the Earth orbit to a Mars orbit using a propulsion with limited power and limited specific impulse. It is assumed that the spacecraft is launched into the Earth orbit using a propulsion of high thrust. The spacecraft starts the maneuver with low thrust on a spiral trajectory. The specific impulse is considered constant. The last time instant of motion on the spiral trajectory is the initial time of the Earth-Mars heliocentric trajectory. The specific impulse is assumed constant on the heliocentric trajectory. On both spiral and heliocentric trajectories the power is equal to its maximum value.

The spacecraft parameters are selected according to the references [38] and [168]. Consider the maneuver of escape from the Earth gravity field in which only the Earth gravitational attraction and thrust are taken into account. The maneuver starts from the earth orbit and lasts until the spacecraft reaches the boundaries of the Earth sphere of influence. Then the Earth-Mars transfer is analyzed. The total sum of masses, M_W, of all components of the propulsion system and payload mass, M_{PL}, and also initial mass, M_0 are known. The final mass, M_1 and the tank mass, M_P are subject to determination. The other spacecraft parameters, including the thrust angle and the coordinates of the junction points are also subject to determination. The measurement units are those given in the SI system, but the angles are measured in radians, the distance is measured in kilometers, and the power is measured in kilowatts.

9.8.4 Escape From the Earth Gravity Field

The results of computations and their comparison with existing results.

1. The solution of the maneuver problem is based on the use of the analytical solutions for thrust arcs with constant specific impulse and variable power given above. The control parameters are the thrust angle (between the thrust vector and the perpendicular to the radius vector) and the

mass-flow rate. In Ref. [38], the results of numerical solution of the problem, including both geocentric and heliocentric trajectories using indirect method of optimization. The following values of propulsion parameters and coefficients of effectiveness have been used in this problem [38]:

$M_0 = 525{,}000$

$M_{PL} = 35{,}000$

$M_W = 60{,}000$

$P_{max} = 10{,}000$

$\eta = 0.6$

$\alpha = 0.6$

$I_{sp} \in 1000 \leq I_{sp} \leq 35{,}000$

The parameters of the initial orbit:

$h_0 = 320$

$r_e = 6378.0$

$r_0 = r_{min} = r_e + h_0$

$e \in 0.01 \leq e \leq 0.12$

The computations have been conducted according to Eqs. (9.148)–(9.155). The change in payload mass is not considered in these computations. It has been shown that when the perigee of the initial orbit is given, the duration of the maneuver depends on the eccentricity of the orbit. For given perigee, r_{min} of the orbit, this dependence is given in the table below:

Eccentricity	Flight time
0.01	334.090
0.01	167.308
0.03	111.7091
0.05	67.222
0.07	48.1482
0.1	33.8325
0.11	30.793
0.12	28.259

Besides that, it has been shown that there exists a monotonic dependency of fuel expenditure from the specific impulse, I_{sp} (see Fig. 9.46). As it can be seen from Fig. 9.46, the value $I_{sp} = 1000\ s$, used in Ref. [38], corresponds to maximum fuel expenditure. If this value of I_{sp} is accepted, then for the final fuel mass one can obtain $M_P = 450{,}000$, that means that $M_1 = 75{,}000 < M_W + M_{PL}$ which is not acceptable. That's why it is accepted that $I_{sp} > 1000$. The other results are given in the table below, where

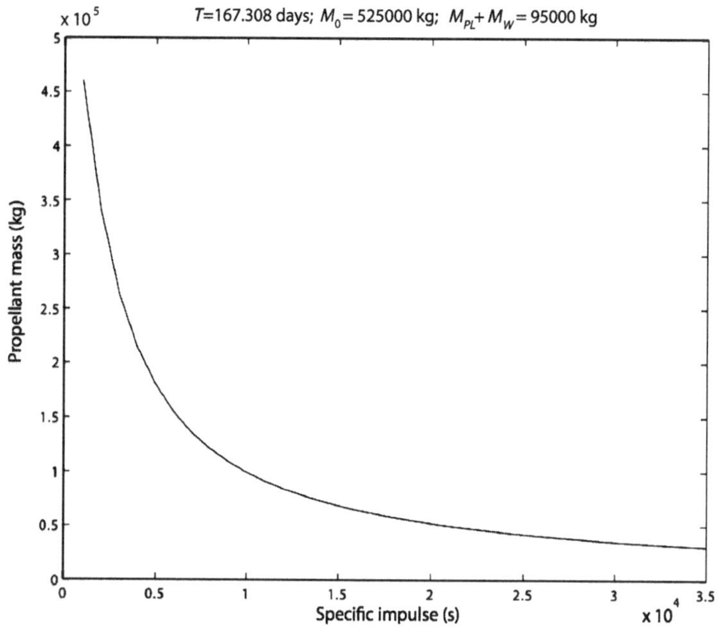

Figure 9.46 Fuel expenditure vs. specific impulse. Horizontal axis: specific impulse; vertical axis: mass.

the results of the analytical solutions are presented on the left side, and the optimal solutions are given on the right side [38]. The thrust arcs described by the analytical solutions can not be connected to a circular orbit, and therefore, in the computations it is accepted that $e > 0$. The analysis show that the thrust arc is rapidly increases in the beginning of the maneuver, and then slowly increases until the end of the maneuver. The optimal solution shows that the thrust angle oscillates about zero, and then rapidly increases. The mass is changed nonlinearly and linearly according to analytical and numerical optimization respectively. But the final masses are almost the same. The qualitative difference in the mass behavior is that in the analytical solutions, the thrust is variable, and in the numerical integration, the thrust is constant. The escape trajectories in both solutions have a spiral form. The number of revolutions about the center of attraction is less than the number of revolutions according to the numerically integrated solutions. This is associated with the differences in the initial and final thrust angles and ellipticity of the maneuver. When the eccentricity of the final orbit is decreased, the number of revolutions is increased.

Parameters	Vadali S.R. and etc.	Azimov, D.M.
Initial orbit	$r_0 = 6698,\ e = 0$	$r_{min} = 6698,\ e = 0.11$
Flight duration	30	30.793
Specific impulse	1000	2345.8
Final mass	225,000	224,998.2
Necessary fuel amount	300,000	300,001.8

2. Tim Dawn's work is devoted to optimization of the Earth-Mars-Earth transfer trajectories using indirect methods [168]. The following parameters used in the trajectory computations [168]:

$M_0 = 283{,}292$

$M_{PL} = 22{,}154$

$M_W = 43{,}527$

$M_S = 5000$

$P_{max} = 10{,}882$

$\eta = 0.6$

$\alpha = 0.6$

$I_{sp} \in 1000 \leq I_{sp} \leq 35{,}000$

Parameters of the initial orbit:

$h_0 = 407$

$r_e = 6378.0$

$r_0 = r_{min} = r_e + h_0$

$e \in 0.01 \leq e \leq 0.12$

The computations have been conducted according to Eqs. (9.148)–(9.155). The results of the computations are given below in the table. Analysis show that the results of the analytical solutions and numerical integrations are very close [168]. Note that the lesser eccentricity of the initial orbit yields longer escape from the gravity field.

Parameters	Dawn, T.	Azimov, D.M.
Initial orbit	$h_0 = 407,\ r_e = 6378.0,$ $e = 0$	$h_0 = 407,\ r_e = 6378.0,$ $e = 0.07$
Flight time	76	76.6
Specific impulse	6000	1832
Fuel tank mass	10,077	9562.5
Necessary fuel amount	189,480	191,249.2

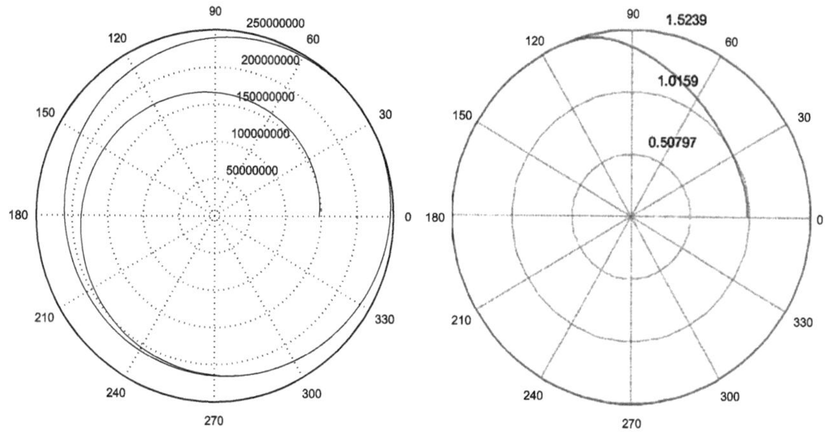

Figure 9.47 Earth-Mars heliocentric trajectory. Left: proposed by Azimov, D.M. Right: reported by Vadali, S. in Ref. [38].

9.8.5 Earth-Mars Heliocentric Transfer

Results of the computations and comparison with existing results.

1. The solution of this maneuver problem is based on the use of analytical solutions for thrust arcs with variable specific impulse and maximum power given in Eqs. (9.157)–(9.164). The control parameters are the thrust angle and specific impulse. The following parameters are used [38]:

$M_0 = 224{,}998.2$
$M_{PL} = 35{,}000$
$M_W = 60{,}000$
$P_{max} = 10{,}000$
$\alpha = 0.6$
$\eta = 0.68$

The computation results are given in the table below and illustrated in Figs. 9.47. Analysis show that the analytical and numerical solutions are close or comparable. Particular interest is shown to the behavior of the specific impulse. In the case of analytical solutions, this parameter is monotonically increased all the time of the maneuver. In the case of numerical optimization, the specific impulse is increased in the beginning of the maneuver, then decreases in the second half of the maneuver. This difference in specific impulse can be explained by the fact that the analytical solutions represent a motion with a constant thrust acceleration. This means that the specific impulse has to be monotonically increasing, and the mass is a monotonically decreasing function of time. It was shown that the parameters of motion with constant thrust acceleration are close to the parameters of optimal motion with variable specific impulse. The plots of mass and specific impulse show a monotonic character in their behavior, although their values differ. Unlike in the numerical optimization, the number of revolutions about the Sun in the analytical solutions depends on the difference of the thrust angles.

Parameters	Vadali, S.R. and etc.	Azimov, D.M.
Flight time	114	110
Initial specific impulse	1000	2192.15
Final specific impulse	1750	4917.07
Initial mass	225,000	224,998.2
Final mass	127,331	126,607
Necessary fuel amount	97,669	98,391
Number of revolutions	0.33	0.847

2. Analysis show that the results of the analytical and numerical solutions are very close on main parameters. The input parameters to the computations are those obtained from Ref. [169]. Whiffen and Sims' work is devoted to the application of the new algorithm titled "Static and Dynamic Control" designed for escape and capture maneuvers using low thrust. This algorithm utilizes the direct method of optimization. The comparison of the results is given in the table below.

Parameters	Whiffen, G.J. and Sims, J.A.	Azimov, D.M.
Maximum power	2.3	2.3
Flight time	223.309	223
Initial specific impulse	not known	1068.5
Final specific impulse	not known	1242.4
Initial mass	475.205	472.093
Final mass	406.156	406
Necessary fuel amount	69.04	69.093

CHAPTER 10

Conclusions

The following main results have been obtained:

- The variational problem of determining optimal trajectories of a point mass, which is an alternative to the Lawden's problem, and unlike it, allows us to construct trajectories not only with constant specific impulse, but also with variable specific impulse and maximum power. Only the thrust arcs with constant specific impulse and maximum power can be optimal. The thrust arcs with variable specific impulse and variable power can not be extremals of the problem. The number of thrust arcs and the structure of the optimal trajectory depend on the number of constants of motion;
- The accuracy of Battin's guidance law (the thrust arcs are replaced by impulses, and the gravitational acceleration is represented by a constant vector) can be significantly improved by using maximum thrust arcs of a non-zero duration instead of impulses and representation of the gravitational acceleration by a linear function of radius vector of the point mass;
- Synthesis of a trajectory can be conducted using only particular solutions of the Hamiltonian system of equations of the variational problem, and the general solution of this system has only a theoretical importance;
- In the design of a transfer trajectory from the boundaries of the sphere of influence to a given low Earth elliptical parking orbit, to minimize the radiation dose it is recommended to divide the trajectory into two parts: descent trajectory until the altitudes below 800 km, and the transfer trajectory from this altitude to the parking orbit. A minimum-time flight through the radiation zone yields a significant reduction of the accumulated dose and a significant fuel expenditure. The longer transfer times lead to more economically effective maneuvers;
- There exist explicit functional dependence of the radiation dose accumulated at given altitude from the instant change of the dose at this altitude, area bounded by the axis of altitude and curve of the dose change. This dependence is a critical relationship in the strategy of minimization of the radiation dose and fuel expenditure;

Analytical Solutions for Extremal Space Trajectories
DOI: http://dx.doi.org/10.1016/B978-0-12-814058-1.00010-2

- The methodology of analytical determination of extremal trajectories, including determination of the number and sequence of the thrust arcs has been developed;
- The method of application of the analytical solutions for thrust arcs in the synthesis of optimal orbital transfer trajectories has been proposed;
- It has been shown that, in essence, the solution of the variational problem can be reduced to the solution of a certain number of nonlinear continuity equations;
- The detailed solutions of the following problems have been presented:
  - transfer between elliptical orbits using IT arcs with constant specific impulse;
  - transfer between elliptical orbits using IT arcs with variable specific impulse. The duration of the transfer and the number of revolutions depend on the thrust angle;
  - transfer from a given position into a given elliptical orbit with variable specific impulse;
  - turning of the plane of given elliptical orbit via an IT arc with constant specific impulse. For some turning angles and sizes of the elliptical orbit, this turning maneuver is energetically better for 0.1% compared to one-, two-, and three-impulsive turning maneuvers;
  - transfer between circular orbits using two MT arcs with constant specific impulse. This transfer can be considered optimal as the absolute difference between the characteristic velocities on MT arcs and impulsive transfer is 0.0001. The similar quantity in the case of transfer between circular and hyperbolic orbits is 0.0006.

APPENDIX

DETERMINATION OF THE RADIAL COMPONENT OF PRIMER VECTOR

$$\lambda_1 = \frac{W_1 + W_2}{W_3}, \qquad \lambda_2 = \sqrt{(\lambda^2 - \lambda_1^2)}, \qquad \lambda = const$$

where

$$\begin{gathered}
W_1 = -(C^2\mu^2 r^4 + 2C^2\mu r^5 v_1^2 - 2C^2\mu r^5 v_2^2 + C^2 r^6 v_1^4 - 10C^2 r^6 v_1^2 v_2^2 \\
+C^2 r^6 v_2^4 - 2C\lambda_5\mu^2 r^3 v_2 + 8CC^2\lambda_5\mu r^4 v_1^2 v_2 + 4C\lambda_5\mu r^4 v_2^3 + 10C\lambda_5 r^5 v_1^4 v_2 \\
+8C\lambda_5 r^5 v_1^2 v_2^3 - 2C\lambda_5 r^5 v_2^5 - 2\lambda_5^2\mu^2 r^2 v_1^2 + \lambda_5^2\mu^2 r^2 v_2^2 - 4\lambda_5^2\mu r^3 v_1^4 - \\
6\lambda_5^2\mu r^3 v_1^2 v_2^2 - 2\lambda_5^2\mu r^3 v_2^4 - 2\lambda_5^2 r^4 v_1^6 - 3\lambda_5^2 r^4 v_1^4 v_2^2 + \lambda_5^2 r^4 v_2^6 + \\
4\lambda_2\mu^3 r v_1^2 + 12\lambda_2\mu^2 r^2 v_1^4 - 4\lambda_2\mu^2 r^2 v_1^2 v_2^2 + 12\lambda_2\mu r^3 v_1^6 + 8\lambda_2\mu r^3 v_1^4 v_2^2 \\
-4\lambda_2\mu r^3 v_1^2 v_2^4 + 4\lambda_2 r^4 v_1^8 + 12\lambda_2 r^4 v_1^6 v_2^2 + 12\lambda_2 r^4 v_1^4 v_2^4 + 4\lambda_2 r^4 v_1^2 v_2^6 - \\
6\mu^4\gamma_1 - 6\mu^4\gamma_2 - 24\mu^3 r v_1^2\gamma_1 - 24\mu^3 r v_1^2\gamma_2 + 24\mu^3 r v_2^2\gamma_1 + 24\mu^3 r v_2^2\gamma_2 - \\
36\mu^2 r^2 v_1^4\gamma_1 - 36\mu^2 r^2 v_1^4\gamma_2 + 24\mu^2 r^2 v_1^2 v_2^2\gamma_1 + 24\mu^2 r^2 v_1^2 v_2^2\gamma_2 - 36\mu^2 r^2 v_2^4\gamma_1 - \\
36\mu^2 r^2 v_2^4\gamma_2 - 24\mu r^3 v_1^6\gamma_1 - 24\mu r^3 v_1^6\gamma_2 - 24\mu r^3 v_1^4 v_2^2\gamma_1 - 24\mu r^3 v_1^4 v_2^2\gamma_2 + \\
24\mu r^3 v_1^2 v_2^4\gamma_1 + 24\mu r^3 v_1^2 v_2^4\gamma_2 + 24\mu r^3 v_2^6\gamma_1 + 24\mu r^3 v_2^6\gamma_2 - 6r^4 v_1^8\gamma_1 - \\
6r^4 v_1^8\gamma_2 - 24r^4 v_1^6 v_2^2\gamma_1 - 24r^4 v_1^6 v_2^2\gamma_2 - 36r^4 v_1^4 v_2^4\gamma_1 - 36r^4 v_1^4 v_2^4\gamma_2 - \\
24r^4 v_1^2 v_2^6\gamma_1 - 24r^4 v_1^2 v_2^6\gamma_2 - 6r^4 v_2^8\gamma_1 - 6r^4 v_2^8\gamma_2)^{\frac{1}{2}} \\
W_2 = \sqrt{3}(C\mu r^2 + Cr^3 v_1^2 - Cr^3 v_2^2 - \lambda_5\mu r v_2 + \lambda_5 r^2 v_1^2 v_2 + \lambda_5 r^2 v_2^3) \\
W_3 = 2\sqrt{3}(\mu^2 + 2\mu r v_1^2 - 2\mu r v_2^2 + r^2 v_1^4 + 2r^2 v_1^2 v_2^2 + r^2 v_2^4)
\end{gathered}$$

NOMENCLATURE

A_1, A_2	apogee points
$\mathbf{a}$	thrust acceleration
$a, \alpha, \lambda_0, \lambda_{10}$	constants
$B, D, P_1, R_1, P_2, R_2, \alpha_1, \alpha_2, a_1, a_2$	primer vector's constants
C	constant for Hamiltonian
C_{21}, C_{22}	integration constants
c	exhaust velocity [km/s]
e, e'	eccentricities
$\mathbf{e}(e_1, e_2, e_3),\quad \mathbf{l}(l_1, l_2, l_3)$	unit thrust vectors
F	Lagrange function
F_1	difference of values of integral sine [rad]
F_2	difference of values of integral cosine [rad]
f_H	true anomaly [degree]
$f_i, (i = 1, ..., 7)$	right sides of equations of motion
$\mathbf{x}(x_1, ..., x_7)$	state vector
$\mathbf{v}(x_1, x_2, x_3),\quad \mathbf{v}(v_1, v_2, v_3)$	velocity vector
$\mathbf{r}(x_4, x_5, x_6),\quad \mathbf{r}(r, 0, 0)$	radius vector
P, P_{max}	power [kw]
$m(= x_7)$	mass [kg]
I_{sp}	specific impulse [s]
t	flight time [s], [days]
t_0, t_2	initial and final times of flight
T, T_{min}, T_{max}	thrust values
η, γ	auxiliary variables
$u_i, (i = 1, ..., 7),\; P, I_{sp}, e_1, e_2, e_3$	controls
$\boldsymbol{\lambda}(\lambda_1, \lambda_2, \lambda_3)$	primer vector
$\boldsymbol{\lambda_r}$	vector conjugate to radius vector
$\lambda_i, (i = 4, ..., 7),\quad \mu_j, (j = 1, 2, 3)$	Lagrange multipliers
g	gravitational acceleration [km/s^2]
H	Hamiltonian
$\mathcal{J}$	functional
k, k_1	Shuler's frequency [rad/s]
p	semi-latus rectum, parameter [km]
t_0, τ_1	initial times of first and second maximum thrust arcs [s]
t_2, τ_2	final times of first and second maximum thrust arcs [s]
t_3	initial time of second maximum thrust arc computed from the initial time of motion [s]
β	mass-flow rate [kg/s]
δr	thin of spherical layer [km]
$\Delta g_2, \Delta g_3$	errors in gravitational acceleration at second and third switching points [km/s^2]
Δr	difference in values of radius vector [km]
ΔV	velocity rate [km/s]

θ	polar angle, azimuth [rad]
μ	gravitational parameter [km^3/s^2]
φ	thrust angle (between primer vector and local horizontal) [rad]
χ	switching function
ψ	angle between primer vector and inertial axis, and between A_1P_1 and LH [rad]
δ	ascending node [degree]
i	orbit inclination [degree]
α_H	turn angle [degree]
Ω	longitude [degree]
ω	argument of perigee [degree]
u	angular distance measured from node [degree]
r_0	altitude of perigee [km]
r_1	altitude of apogee [km]
P_1, P_2	perigee points
Q_1, Q_2	nodes
$\Delta\Lambda$	angle between Q_1Q_2 and $Q_1'Q_2'$ [degree]
LH	axis of turn
H, E	points at which impulse is applied

REFERENCES

[1] D.E. Okhotsimiskii, T.M. Eneev, Nekotorye zadachi zapuska isskustvennykh sputnikov zemli, Uspekhi Fizicheskih Nauk 63 (1a) (1957).

[2] J.H. Irving, Space Technology, Wiley and Sons, New York, 1959, pp. 3.05–10.16, Edited by Seifert.

[3] D.F. Lawden, Optimal Trajectories for Space Navigation, Butterworths, London, 1963, pp. 55–99.

[4] W.G. Melbourne, C.G. Sauer Jr., Optimum thrust programs for power-limited propulsion systems, Acta Astronautica VIII (4) (1962) 205–227.

[5] C.A. Larson, J.E. Prussing, Optimal orbital rendezvous using high and low thrust, in: AAS/AIAA Space Flight Mechanics Meeting, San Diego, California, 1989, August 7–10, pp. 513–532, Paper AAS 89-354.

[6] G. Colasurdo, L. Casalino, Trajectories towards near-Earth-objects using solar electric propulsion, in: AAS/AIAA Astrodynamics Specialist Conference, Girdwood, Alaska, Aug. 1999, AAS Paper 99-339, Also in: Advances in the Astronautical Sciences, Part I, vol. 103, Univelt Inc., San Diego, California, 2000, pp. 593–607.

[7] J.P. Marec, Optimal Space Trajectories, Elsevier Scientific Publishing Company, Amsterdam, 1979, pp. 7–101.

[8] V.N. Lebedev, Computation of Spacecraft Motion with Low-Thrust, Computational Center of Academy of Sciences of USSR, Moscow, 1968.

[9] V.V. Salmin, Optimization of Space Transfers with Low-Thrust, Mashinostroenie, Moscow, 1987.

[10] S.D. Grishin, Yu.A. Zakharov, V.K. Odolevskii, Studies of Spacecraft with Low-Thrust Propulsion, Mashinostroenie, Moscow, 1990.

[11] F.R. Chang-Diaz, M.M. Hsu, E. Braden, I. Johnson, T.F. Yang, Rapid Mars Transits with Exhaust Modulated Plasma Propulsion, Technical report N. TN-3539, NASA, 1995.

[12] W.G. Melbourne, Interplanetary Trajectories and Payload Capabilities of Advanced Propulsion Vehicles, JPL Technical Report, JPL, Pasadina, California, 1961, pp. 32–68.

[13] T.M. Eneev, P.Z. Akhmetshin, V.A. Egorov, G.B. Efimov, Dynamics of Systems with Low-Thrust Propulsion, Keldysh Institute of Applied Mathematics, Moscow, 2002, p. 53.

[14] E.L. Akim, V.A. Stepaniants, A.G. Tuchin, Accuracy of Orbit Determination for Low-Thrust Trajectory to Mars, Advances in the Astronautical Sciences, 1998, pp. 1–13, AAS 98-387.

[15] P.Z. Akhmetshin, Expeditions to Main Asteroid Belt with Low-Thrust Via Gravitational Maneuver at Mars, Keldysh Institute of Applied Mathematics, Moscow, 2002, p. 145.

[16] V.V. Malyshev, V.E. Usachov, Y.D. Tychinski, The guidance strategy for the Russian solar probe within "Fire" mission, in: 48-th International Astronautical Congress, Italy, Turin, 1997.

[17] V.V. Malyshev, V.E. Usachov, Y.D. Tychinski, Analysis and optimization of guidance of a solar probe with allowance for stochastic and uncertain disturbances, Journal of Computer and Systems Sciences International 38 (4) (1999).

[18] Yu.A. Ryjov, V.V. Malyshev, K.M. Pichkhadze, V.E. Usachov, Analysis and synthesis of missions for direct investigations of solar crown, Izvestiya of the Academy of Sciences of Russia, Theory of Systems and Controls (4) (2001) 131–152.
[19] R.Z. Akhmetshin, G.B. Eneev, G.B. Efimov, On the possibility of asteroid rendezvous with sample return to the Earth abstract, in: IAU Colloquium, 1996, p. 1.
[20] V. Coverstone-Carroll, S.N. Williams, Optimal low-thrust trajectories using differential inclusion concepts, The Journal of the Astronautical Sciences 42 (4) (1994 October-December) 379–393.
[21] E.M. Braden, B.F. Cockrell, A.J. Bordano, Power-Limited Human Mission to Mars, JSC, vol. 28436, Lyndon B. Jonson Space Center, Aerospace and Flight Mechanics Division, Houston, September 1998, pp. 1–35.
[22] J. Betts, Survey of numerical methods for trajectory optimization, AIAA Journal of Guidance, Control, and Dynamics 21 (2) (1998) 193–207, http://dx.doi.org/10.2514/2.4231.
[23] G.L. Grodzovskii, Yu.N. Ivanov, V.V. Tokarev, Optimization problems in the mechanics of low-thrust space flight, in: Proceedings of the Second All-Union Conference on Theoretical and Applied Mechanics, Israel Program for Scientific Translation, Jerusalem, 1968, pp. 201–221.
[24] N. Markopoulos, Analytically exact non-Keplerian motion for orbital transfers, in: AIAA/AAS Astrodynamics Conference, AIAA-94-3758-CP, Scottsdale, Arizona, 1994, August 1–3, pp. 383–412.
[25] B.D. Tapley, W.E. Miner, W.F. Power, The Hamilton-Jacobi method applied to the low-thrust trajectory problem, in: Proceedings of Astrodynamics Congress, Belgrad, 1968, pp. 293–305.
[26] R.H. Bishop, D.M. Azimov, Analytical trajectories for extremal motion with low-thrust exhaust-modulated propulsion, Journal of Spacecraft and Rockets 381 (61) (2001) 897–903, See also: New analytic solutions to the fuel-optimal orbital transfer problem using low-thrust exhaust modulated propulsion, in: AAS/AIAA Space Flight Mechanics Meeting, Clearwater, FL, 23–26 January, 2000, Paper AAS 00-131.
[27] V.V. Malyshev, V.E. Usachev, Yu.D. Tychinskii, Optimization of structural dynamical system modeling a flight to solar crown with controllable electro-jet thrust vector and gravimaneuvers at planets, in: Systems Analysis and Control of Space Complexes, 4-th International Conference, Kremea, Evpatoria, June 26–July 2, 1999.
[28] V.V. Ivashkin, A.V. Chernov, Determination of Optimal Space Trajectories to an Asteroid Approaching the Earth Using Low-Thrust, Preprint KIAM N. 19, Keldysh Institute of Applied Mathematics, Moscow, 1997.
[29] V.V. Ivashkin, A.V. Zaytsev, A.V. Chernov, Optimal flights to near-Earth asteroid, Acta Astronautica 44 (5) (1999) 219–229.
[30] H. Seywald, Trajectory optimization based on differential inclusion, Journal of Guidance, Control and Dynamics 17 (3) (May-June 1994) 480–487.
[31] C.A. Kluever, Optimal low-thrust interplanetary trajectories by direct method techniques, The Journal of the Astronautical Sciences 45 (3) (July-September 1997) 247–262.
[32] General Mission Analysis Tool (GMAT), Product Description, Goddard Space Flight Center, http://gmat.gsfc.nasa.gov, 2007.
[33] G.G. Fedotov, On using the possibilities of combination of high-and low-thrust for flights to Mars, Space Research 39 (6) (2001) 613–621.

[34] C. Ocampo, COPERNICUS, A General Spacecraft Trajectory Design and Optimization System, The University of Texas at Austin, Dept. Aerospace Engineering and Engineering Mechanics, 25 October 2001.
[35] W. Sheel, B.A. Conway, Optimization of very-low thrust many-revolution spacecraft trajectories, AIAA Journal of Guidance, Control and Dynamics 17 (6) (1994) 1275–1282.
[36] J.D. Thorne, C.D. Hall, Minimum-time continuous thrust orbit transfers, The Journal of the Astronautical Sciences 45 (4) (1997) 411–432.
[37] S. Tang, B.A. Conway, Optimization of low-thrust interplanetary trajectories using collocation and nonlinear programming, AIAA Journal of Guidance, Control and Dynamics 18 (3) (1995) 599–604.
[38] S.R. Vadali, R. Nah, E. Braden, I.L. Jpohnson Jr., Fuel-optimal planar interplanetary trajectories using low-thrust exhaust-modulated propulsion, in: AAS/AIAA Space Flight Mechanics Meeting, Breckenridge, Colorado, 1999, February 7–10, Paper AAS 99-132.
[39] T. Crain, R.H. Bishop, W. Fowler, R. Kenneth, Optimal Interplanetary Trajectory Design Via Hybrid Genetic Algorithm: Recursive Quadratic Program Search, Advances in the Astronautical Sciences, 1999, pp. 449–465, AAS 99-133.
[40] D.M. Azimov, Extremal analytical solutions for intermediate-thrust arcs in a Newtonian field, AIAA Journal of Guidance, Control and Dynamics 33 (5) (September-October 2010) 1550–1565.
[41] H.G. Kelly, R.E. Kopp, H.G. Moyer, Singular extremals, in: G. Leitmann (Ed.), Topics in Optimization, Academic Press, 1967, pp. 63–102.
[42] X.M. Robbins, Optimality of intermediate thrust arcs of rocket trajectories, AIAA Journal 3 (8) (1965) 139–145.
[43] D.F. Lawden, Necessary conditions for optimal rocket trajectory, Quarterly Journal of Mechanics and Applied Mathematics 12 (1959) 476.
[44] D.F. Lawden, Interplanetary rocket trajectories, in: Space Trajectories, Mir, Moscow, 1963, pp. 177–242.
[45] H.J. Kelly, Method of Gradients. 6. Optimization Techniques, Academic Press, New York, 1962, pp. 206–252.
[46] B.M.F. de Vinbeke, J. Geerts, Optimization of Multiple Impulse Orbital Transfers by Maximum Principle, Rept. OA-4, 1964.
[47] C.D. Johnson, Singular solutions in problem of optimal control, in: C.G. Leondes (Ed.), Advances in Control Systems. Theory in Applications, vol. 2, Academic Press, New-York–London, 1965, pp. 209–269.
[48] R. Gabasov, F.M. Kirillova, Osobye Optimalnye Upravleniya, Nauka, Moskva, 1965, p. 256.
[49] R. Gabasov, F.M. Kirillova, To the theory of high-order necessary conditions of optimality, Differential equations VI (4) (1970) 665–676.
[50] B.S. Goh, The second variation for the singular Bolza problem, SIAM Journal on Control 4 (2) (1966) 309–325.
[51] B.S. Goh, Necessary conditions for singular extremals involving multiple control variables, SIAM Journal on Control 4 (4) (1966) 716–731.
[52] S.T. Zavalishin, Supplements to the lawden's theory, Applied Mathematics and Mechanics (5) (1989) 731–738.
[53] H. Kelley, Necessary conditions for singular extremals based on the second variation, AIAA Journal 8 (6) (1964) 26–29.

[54] H.J. Shelley, A transformation approach to singular subarcs in optimal trajectory and control problems, SIAM Journal on Control 2 (2) (1964) 234–240.
[55] D.J. Bell, D.H. Jacobson, Singular Optimal Control Problems, Academic Press, New York, 1975, pp. 61–151.
[56] V.F. Krotov, Methods of solving of variational problems based on the sufficient conditions of absolute minimum. I, Remote Control III (12) (1962) 1571–1583.
[57] V.F. Krotov, Methods of solving of variational problems based on the sufficient conditions of absolute minimum. II, Remote Control IV (5) (1962) 581–598.
[58] V.I. Gurman, To the question of optimality of singular regimes of rockets' motion in the central field, Kosmicheskie Issledovaniya IV (4) (1965) 499–509.
[59] A.G. Azizov, N.A. Korshunova, On an analytical solution of the optimum trajectory problem in a gravitation field, Celestial Mechanics 38 (4) (1986) 297–306.
[60] N.A. Korshunova, Questions of integration of equations of the variational problem in the central field, in: Abstract of Candidate Dissertation in Physics and Mathematics, Moscow State University, Moscow, 1977, pp. 1–24.
[61] D.M. Azimov, Intermediate thrust arcs in Mayer's variational problem, Journal of Applied Mathematics and Mechanics 64 (1) (2000) 87–95.
[62] A.G. Azizov, N.A. Korshunova, Application of Levi-Civita's method in analysis of optimal trajectories, Space Research 17 (3) (1979) 378–386.
[63] N.A. Korshunova, On optimal trajectories of rockets in a central field, Mechanics and Applied Mathematics (476) (1975) 85–94.
[64] D.M. Azimov, To the Question on Optimality of Intermediate Thrust Arcs. Algorithms and Numerical Solutions to the Problems of Computational and Applied Mathematics, Tashkent State University, Tashkent, 1988, pp. 6–9.
[65] D.M. Azimov, Investigation of optimal trajectories in a central Newtonian field, in: Abstract of Kandidate Dissertation, University of Peoples' Friendship Named After P. Lumumba, Moscow, 1991, 13 pp.
[66] A.G. Azizov, N.A. Korshunova, Variational Problems of Space Flight Mechanics, Tashkent State University, 1991, pp. 1–84.
[67] D.M. Azimov, Analytical solutions for intermediate thrust arcs of rocket trajectories in a Newtonian field, Journal of Applied Mathematics and Mechanics 60 (3) (1996) 421–427.
[68] D.M. Azimov, New classes for intermediate-thrust arcs of flight trajectories in a Newtonian field, AIAA Journal of Guidance, Control and Dynamics 23 (1) (2000) 142–145.
[69] B. Bonnard, M. Chyba, Singular Trajectories and Their Role in Control Theory, Math-Vematiques and Applications, vol. 40, Springer-Verlag, Berlin, 2003.
[70] D.G. Hull, Necessary conditions for a strong relative minimum for singular optimal control, Journal of Optimization Theory and Applications 137 (1) (2008) 157–169.
[71] H.J. Oberle, Numerical computation of singular control functions in trajectory optimization, AIAA Journal of Guidance, Control and Dynamics 13 (1990) 153–159.
[72] D.F. Lawden, Calculation of singular extremal rocket trajectories, AIAA Journal of Guidance, Control and Dynamics 15 (6) (1992) 1361–1365.
[73] F. Bonnans, P. Martinon, E. Trelat, Singular arcs in the generalized Goddard's problem, Institut National De Recherche En Informatique Et En Automatique, INRIA, 30 March, 2007, 27 pp.
[74] P. Tsiotras, H.J. Kelley, Goddard problem with constrained time of flight, AIAA Journal of Guidance, Control and Dynamics 15 (2) (1992) 289–295.

[75] I.C. Park, H. Yan, Q. Gong, M. Ross, Numerical verification of singular arcs on optimal moon-to-Earth transfer, in: AAS/AIAA Space Flight Mechanics Meeting, San Diego, CA, 2010, AAS 10-104.
[76] I.C. Park, H. Yan, Q. Gong, M. Ross, Necessary conditions for singular arcs for general restricted multi-body problem, in: 49th IEEE Conference on Decision and Control, Atlanta, GA, December 15–17, 2010, pp. 5816–5821.
[77] D.M. Azimov, F. Aragon-Sanabria, Extremal control and guidance solutions for orbital transfer with intermediate thrust, Journal of the Astronautical Sciences 62 (4) (December 2015) 351–383.
[78] N.A. Korshunova, D.M. Azimov, Analytical solutions for thrust arcs in a field of two fixed centers, AIAA Journal of Guidance, Control, and Dynamics 37 (5) (2014) 1716–1718.
[79] D.M. Azimov, Analytical synthesis of optimal trajectory control and guidance solutions with applications to autonomous systems, in: The 46th ISCIE International Symposium on Stochastic Systems Theory and Its Applications, SSS 2014, Kyoto, Japan, Oct. 31–Nov. 2, 2014, 8 pp.
[80] D.M. Azimov, R.G. Mukharlyamov, Analytical approach to analysis of extremal trajectories and stability of programmed motion, Vestnik of Russian University of Peoples-Friendship 4 (2012).
[81] D.M. Azimov, On one case of integrability of atmospheric flight equations, Journal of Aircraft 48 (5) (2011) 1722–1732.
[82] D.M. Azimov, Extremal trajectories for space navigation and guidance, in: Proceedings of the International Conference on the Analysis and Mathematical Applications in Engineering and Science, Curtin University Sarawak, Miri, Malaysia, 19–22 Jan. 2014, pp. 92–101.
[83] A.M. Letov, Flight Dynamics and Control, Nauka, Moscow, 1969, p. 259.
[84] E. Rowland Burns, A Study on the Optimal Rocket Trajectory, AIAA Paper, N. 20, 1971, pp. 1–8.
[85] V.A. Ilin, G.E. Kuzmak, Optimal Transfers of Spacecraft, Mir, Moscow, 1976, p. 744.
[86] O.F. Makarov, On structure of control regimes with energetically minimum expenditures in a gravitational field, in: Problems of Mechanics of Controlable Motion, Perm State University, Perm, 1989, pp. 87–89.
[87] A.I. Lurie, Thrust programming in a central gravitational field, in: Topics in Optimization, Academic Press, 1967, pp. 103–146.
[88] N.A. Korshunova, Analytical theory of a point's trajectory optimization in gravitational fields, in: Abstract of Dissertation for D.Sc. Degree, Department of Mechanics-Mathematics, Moscow State University Named After Lomonosov, M.V., Moscow, 1992, pp. 1–24.
[89] V.M. Alekseev, V.M. Tikhomirov, S.V. Fomin, Optimal Control, Nauka, Moscow, 1979, pp. 370–407.
[90] A.A. Milyutin, N.P. Osmolovski, Calculus of Variations and Optimal Control, Translations in Mathematical Monographs, vol. 180, Amer. Math. Soc., Providence, Rhode Island, 1998, pp. 7–148.
[91] V.S. Novoselov, Analytical Theory of Optimization in the Gravitational Fields, Leningrad State University, Leningrad, 1972, p. 317.
[92] Variation problems with constraints on control, in: G. Leitmann (Ed.), Methods of Optimization with Applications to Aerospace Systems, Mir, Moscow, 1965, pp. 209–243.

[93] G. Leitmann, The Calculus of Variations and Optimal Control: An Introduction, Springer Science + Business Media, LLC, 1981, 311 pp.

[94] E. Bryson Jr., Yu-Chi Ho, Applied Optimal Control. Optimization, Estimation and Control, Ginn and Company, a Xerox Company, Waltham, Massachussets, 1969, pp. 177–211.

[95] V.F. Krotov, Global Methods in Optimal Control Theory New York, Marcel Dekker, Inc., New York, 1996, pp. 133–179.

[96] J.M. Longuski, J.J. Guzmán, J.E. Prussing, Optimal Control with Aerospace Applications, Springer, 2014.

[97] G.A. Bliss, Calculus of Variations, The Open Court Publishing Company, The Mathematical Association of America, Chicago, Illinois, 1925, pp. 128–136.

[98] R.H. Battin, An Introduction to the Mathematics and Methods of Astrodynamics, AIAA Education Series, New York, 1987, pp. 550–566.

[99] G.A. McCue, Quazilinearization determination of optimum finite-thrust orbital transfers, AIAA Journal 5 (4) (1967) 755–763.

[100] K.R. Brown, E.F. Harrold, G.W. Johnson, Rapid Optimization of Multiple-Burn Rocket Flights, N. CR-1430, NASA, 1969.

[101] R.G. Mukralyamov, On construction of systems of differential equations of mechanical systems' motion, Differential equations 3 (23) (2003) 343–353.

[102] R. Fletcher, M.J.D. Powell, A rapidly convergent descent method for minimization, The Computer Journal 6 (2) (1963) 163–168.

[103] K.G. Grigorev, I.S. Grigorev, Investigation of optimal transfer trajectories of spacecraft with high limited thrust between satellite orbits around Earth and Moon, Space Research 35 (1) (1997) 52–75.

[104] A.E. Bryson, W.F. Denham, A steepest ascent method for solving optimum programming problems, Journal of Applied Mechanics (87) (1962) 247–257.

[105] K.G. Grigorev, I.S. Grigorev, Optimal trajectories of transfers of spacecraft with high limited thrust between a Earth satellite orbit and Moon, Space Research 32 (6) (1994) 108–129.

[106] K.G. Grigorev, A.V. Fedina, Optimal transfers of spacecraft with high limited thrust between complanar circular orbits, Space Research 33 (4) (1995) 403–416.

[107] A.L. Kornhauser, P.M. Lion, G.A. Hazelrigg, An analytic solution for constant-thrust, optimal cost, minimum-propellant space trajectories, AIAA Journal N.7 (1971) 1234–1239.

[108] H.M. Robbins, An analytical study of the impulsive approximation, AIAA Journal 4 (8) (1966) 1417–1423.

[109] J.F. Andrus, I.F. Burns, J.Z. Woo, Nonlinear Optimal Guidance Algorithms, N. CR-86160, NASA, 1969.

[110] C. Marchal, J.P. Marec, C.B. Winn, Survey paper-synthesis of the analytical results on optimal transfers between Keplerian orbits. ONERA tire a part, N. 515, 1967.

[111] C. Marchal, Transferts Optimax Entre Orbites Elliptiques (duree indifferente), These de doctorat, AO 1609, Faculte des Sciences de Paris, 1967.

[112] D.M. Azimov, Extremal trajectories for space navigation and guidance, in: Proceedings of the International Conference on the Analysis and Mathematical Applications in Engineering and Science, Sarawak, Miri, Malaysia, Curtin University, 19–22 January 2014, pp. 92–101, 10 pp.

[113] D.M. Azimov, Extremal control and guidance solutions for orbital transfer with intermediate thrust, in: 23rd AAS/AIAA Space Flight Mechanics Meeting, Kauai, Hawaii, February 10–14, 2013, 20 pp.
[114] D.M. Azimov, Real-time guidance, control and targeting algorithms for autonomous planning and execution of space operations, in: The 45th ISCIE International Symposium on Stochastic Systems Theory and Its Applications, SSS 2013, University of Ryukyus, Okinawa, Japan, November 1–2, 2013, 6 pp.
[115] A.G. Azizov, N.A. Korshunova, To the Question on Determination of Analytical Solutions on Maximum Thrust Arcs. Controllable Dynamical Systems and Their Applications, Tashkent State University, Tashkent, 1987, pp. 10–13.
[116] V.K. Isaev, Pontryagin maximum principle and optimal thrust programming of rockets, Remote Control 21 (8) (1961) 986–1001.
[117] D.E. Jezewski, J.M. Stoolz, A closed-form solution for minimum-fuel, constant-thrust trajectories, AIAA Journal 8 (7) (1970) 1229–1234.
[118] D.E. Jezewski, Optimal analytic multi-burn trajectories, AIAA Journal 10 (5) (1972) 680–685.
[119] D. Jezewski, B. Rozendal, Effective method of computation of optimal N-impulsive space trajectories, AIAA Journal 6 (11) (1968) 138–145.
[120] J.F. Andrus, Elliptic integral solutions to a class of space flight optimization problems, AIAA Journal 14 (8) (1976) 1026–1030.
[121] G. Leitmann, An Introduction to Optimal Control, McGraw-Hill Book Company, New York, 1968, pp. 13–26.
[122] A. Wolf, J. Tooley, S. Ploen, M. Ivanov, B. Acikmese, K. Gromov, Performance trades for mar pin-point landing, in: IEEE Aerospace Conference, Inst. of Electrical and Electronics Engineers, 2006, Paper 1661.
[123] L. Blackmore, B. Açikmeşe, D.P. Scharf, Minimum-landing-error powered-descent guidance for Mars landing using convex optimization, AIAA Journal of Guidance, Control and Dynamics 33 (4) (July-August 2010) 1161–1171.
[124] D.M. Azimov, Enhanced Apollo-class real-time targeting and guidance for powered descent and precision landing, in: AIAA-2013-5258, AIAA Guidance, Navigation and Control Conference, Boston, MA, August 19, 2013, 24 pp.
[125] T.L. Yang, Optimal Control for a Rocket in a Three-Dimensional Central Force Field, Technical Memorandum, TM-69-2011-2, Belcomm Inc., May 29, 1969, 51 pp.
[126] D.M. Azimov, Extremal analytical and guidance solutions for powered descent and precision landing, in: 23rd International Symposium on Space Flight Dynamics, Pasadena, California, October 29–November 2, 2012, 24 pp.
[127] D.M. Azimov, To the Question on Determination of Junction Points on Optimal Trajectories. Dynamical Systems and Their Applications, Tashkent State University, Tashkent, 1990, pp. 18–20.
[128] I.S. Grigorev, K.G. Grigorev, On conditions of maximum principle in optimal control problems for a set of dynamical systems and their applications to solutions of spacecraft optimal control problems, Space Research 41 (3) (2003) 307–331.
[129] A. Miele, Variational calculus in applied aerodynamics and flight mechanics, in: G. Leitmann (Ed.), Methods of Optimization with Applications to Aerospace Systems, Mir, Moscow, 1965, pp. 99–163.
[130] D.F. Lawden, Analytical Methods of Optimization, Hafner Press, a Division of Macmillan Publishing Co., Inc., New York, 1975, pp. 122–153.

[131] D.G. Hull, Optimal Control Theory for Applications, Springer-Verlag, Inc., New York, 2003, pp. 258–301.

[132] A. Miele, An extension of the theory of the optimum burning program for the level flight of a rocket-powered aircraft, Journal of Aeronautical Sciences (24) (1957) 874–884.

[133] A. Miele, General variational theory of the flight paths of rocket-powered aircraft, missiles and satellite carriers, Astronautica Acta (4) (1958) 264–288.

[134] A.V. Arutyunov, Optimality Conditions. Abnormal and Degenerate Problems, Kluwer Academic Publishers, The Netherlands, Dordrecht, 2000, pp. 89–123.

[135] G.A. Korn, T.M. Korn, Mathematical Handbook for Scientists and Engineers, McGraw-Hill Book Company, Inc., 1961.

[136] A.V. Dmitruk, Jacobi type conditions for the problem of Bolza with inequalities, Mathematical Notes of the Academy of Sciences of USSR 35 (1984).

[137] D.M. Azimov, H.R. Bishop, New classes of optimal analytical space trajectories, in: The World Space Congress 2002, Astrodynamics Symposium, IAC-02-A.6.01, 2002, pp. 1–10.

[138] D.M. Azimov, H.R. Bishop, Optimal Transfer Between Circular and Hyperbolic Orbits Using Analytical Maximum Thrust Arcs, Advances in the Astronautical Sciences, vol. 112, 2002, pp. 671–689, See also: D.M. Azimov, H.R. Bishop, optimal transfer between circular and hyperbolic orbits using analytical maximum thrust arcs, in: AAS/AIAA Space Flight Mechanics Meeting, San Antonio, Texas, 2002, January 27–30, Paper AAS 02-155, 20 pp.

[139] N.A. Korshunova, On some particular solutions of the problem for a point with variable mass, Reports of Academy of Sciences of Uzbekistan (8) (1975) 13–15.

[140] G.N. Duboshin, Celestial Mechanics. Main Problems and Methods, Fizmatgiz, Moscow, 1963.

[141] D.M. Azimov, On determination of optimal trajectories in a linear central field, UzNIINTI. Dep. in UzNIINTI, Tashkent, December 5, 1988, 887–Uz88. 1988, 22 pp.

[142] D.M. Azimov, B.I. Musabekov, A new analytical solutions for maximum thrust arcs and their application to optimal transfer between coplanar orbits, in: Proceedings of Ankara International Aerospace Conference and Symposia, Ankara, Turkey, 1996, September 19–21, pp. 193–200.

[143] D.M. Azimov, R.H. Bishop, Planetary capture using low-thrust propulsion, in: 16th International Symposium on Space Flight Dynamics, Pasadena, California, USA, December 3–6, 2001, pp. 1–31.

[144] R.H. Bishop, D.M. Azimov, T. Dawn, M. DeRidder, E. Braden, S. Vadali, K. Aroonwilairut, G. Condon, T. Crain, Mars sample return, in: The SEP Earth Return NASA JSC, July 13, 2001, Presentation at NASA JSC/UT Meeting, Space Flight Mechanics Division, NASA Lyndon Johnson Space Center, Houston, USA, pp. 1–43.

[145] D.M. Azimov, R.H. Bishop, Earth-mars minimum fuel trajectories via circular and spiral low thrust arcs, in: Texas A and M University, NASA JSC/A and M/UT Meeting, College Station, Texas, USA, 2000, December 1, pp. 1–26.

[146] I.S. Gradshtein, I.M. Ryzhik, Table of Integrals, Series, and Products, Academic Press, New-York, 1965, pp. 5, 928.

[147] R.H. Bishop, D.M. Azimov, C. Ocampo, G. Condon, T. Crain, et al., in: Low Thrust Trajectory Work at JSC, Lunar Institute, December 5, 2001, Presented to the Mars Program Systems Engineering Team, NASA JSC, Houston, USA, pp. 1–48.

[148] D.M. Azimov, R.H. Bishop, Turning elliptical orbital planes via intermediate thrust spherical arcs, Journal of Spacecraft and Rockets 25 (2) (2002) 358–367, See also: Analytic solution to the problem of turning of an elliptic orbital plane via spherical intermediate thrust arc, in: AAS/AIAA Space Flight Mechanics Meeting, Clearwater, Florida, 23–26 January, 2000, Paper AAS 00-208.
[149] D.M. Azimov, H.R. Bishop, Planetary Capture Using Constant Specific Impulse and Variable Power Propulsion, Advances in the Astronautical Sciences, vol. 115, 2003, See also: D.M. Azimov, H.R. Bishop, Planetary capture using constant specific impulse and variable power propulsion, in: Texas A and M University/AAS John L. Junkins Astrodynamics Symposium, College Station, Texas, USA, May 23–24, 2003, Paper AAS 02-155, 18 pp.
[150] D.M. Azimov, Survey of studies on investigation of thrust arcs in gravitational fields, Remote Control (11) (2005) 14–34.
[151] S. Pines, Constants of the motion for optimum thrust trajectories in a central gravitational field, in: G. Leitmann (Ed.), Topics in Optimization, Academic Press, New-York, 1967.
[152] D.F. Lawden, Optimal powered arcs in an inverse square law field, Journal of American Rocket Society 31 (4) (1961) 566–568.
[153] D.F. Lawden, Optimal intermediate thrust arcs in a gravitational field, Astronautica Acta 8 (2) (1962) 106–123.
[154] P.E. Kopp, H. Moyer, Necessary conditions of optimality of singular extremals, AIAA Journal 3 (8) (1965) 81–94.
[155] D.M. Azimov, Six classes of trajectories of motion with intermediate thrust in the Newtonian field, Reports of Academy of Sciences of Uzbekistan (2) (1999) 13–16.
[156] M.I. Zelikin, N-order variant of Lawden's problem, in: 6-th All Union Meeting on Theoretical and Applied Mechanics, Proceedings, 24–30 September, 1986, Tashkent, 1986, 63 pp.
[157] D.M. Azimov, Integration of Hamilton-Jacobi equation for intermediate thrust arcs using methods of analytical mechanics, Problems of Mechanics (1) (1998) 3–7.
[158] D.M. Azimov, R.H. Bishop, Extremal rocket motion with maximum thrust in a linear central field, Journal of Spacecraft and Rockets 38 (5) (2001) 765–776.
[159] K.A. Ehricke, Space Flight. Dynamics, vol. 2, Van Nostrand, New-York, 1962, pp. 498–504.
[160] D.M. Azimov, Intermediate thrust arcs for a point moving in the linear central field, in: Problems of Mechanics, 6, Academy of Sciences of Uzbekistan, Tashkent, 1998, pp. 3–8.
[161] D.M. Azimov, On one class of maximum thrust arcs in the central fields, Problems of Mechanics (2–3) (1999) 3–9.
[162] D.M. Azimov, R.H. Bishop, Transfer between circular and hyperbolic orbits using analytical maximum thrust arcs, Journal of Spacecraft and Rockets 40 (3) (2003) 433–436.
[163] D.M. Azimov, Thrust arcs of extremal trajectories in a linear central field, Remote Control (10) (2005) 3–23.
[164] D.M. Azimov, R.H. Bishop, Optimal trajectories for space guidance, in: Annals of New York Academy of Sciences, vol. 1065, New York, NY, 2005, pp. 189–209.
[165] D.M. Azimov, R.H. Bishop, Optimal trajectories for space guidance, Annals of New York Academy Sciences (1065) (2005) 1–21.

[166] R.H. Bishop, D.M. Azimov, W. Fowler, VASIMR class analytic guidance, in: Preliminary Investigations on Optimal Trajectories, Advanced Space Propulsion Laboratory, 2001, pp. 1–17, Presentation at NASA JSC/UT Meeting, NASA Lyndon Johnson Space Center, Houston, USA, January 18.
[167] K. Ehricke, Space Flight, vol. 2, part 3, 1966, p. 571.
[168] T. Dawn, Low-Thrust Piloted, 1-Year Roundtrip Mars Mission in 2018, Technical report, NASA/JSC/EG5-Advanced Mission Design Branch, Human Exploration and Development of Space (HEDS), February 12, 2001.
[169] G.J. Whiffen, J.A. Sims, Application of the SDC optimal control algorithm to low-thrust escape and capture trajectory optimization, in: AAS/AIAA Space Flight Mechanics Meeting, San-Antonio, Texas, 27–30 January, 2002, 22 pp.

INDEX

E

F

G

H

I

FSC
www.fsc.org
MIX
Papier | Fördert gute Waldnutzung
FSC® C083411

Kolibri 360 GmbH
Ferdinand-Jühlke-Straße 7
99095 Erfurt, Deutschland
produktsicherheit@kolibri360.de